T0290944

An Introduction to Rehabilitation Engineering

Series in Medical Physics and Biomedical Engineering

Series Editor: **J G Webster**, University of Wisconsin-Madison, USA

Other books in the series:

The Physics of Modern Brachytherapy for Oncology
D Baltas, N Zamboglou, and L Sakelliou

Electrical Impedance Tomography
D Holder (Ed)

Contemporary IMRT
S Webb

The Physical Measurement of Bone
C M Langton and C F Njeh (Eds)

Therapeutic Applications of Monte Carlo Calculations in Nuclear Medicine
H Zaidi and G Sgouros (Eds)

Minimally Invasive Medical Technology
J G Webster (Ed)

Intensity-Modulated Radiation Therapy
S Webb

Physics for Diagnostic Radiology
P Dendy and B Heaton

Achieving Quality in Brachytherapy
B R Thomadsen

Medical Physics and Biomedical Engineering
B H Brown, R H Smallwood, D C Barber, P V Lawford and D R Hose

Monte Carlo Calculations in Nuclear Medicine
M Ljungberg, S-E Strand and M A King (Eds)

Introductory Medical Statistics 3rd Edition
R F Mould

Ultrasound in Medicine
F A Duck, A C Barber and H C Starritt (Eds)

Design of Pulse Oximeters
J G Webster (Ed)

The Physics of Medical Imaging
S Webb

Series in
Medical Physics and Biomedical Engineering

An Introduction to Rehabilitation Engineering

Edited by

Rory A. Cooper
University of Pittsburgh and
U. S. Department of Veterans Affairs
Pennsylvania, USA

Hisaichi Ohnabe
University of Pittsburgh
Pennsylvania, USA

Douglas A. Hobson
University of Pittsburgh
Pennsylvania, USA

Taylor & Francis is an imprint of the
Taylor & Francis Group, an informa business

Cover image by Christine Heiner, Joe Olson, Jeremy Puhlman, Corey

Blauch, and Garrett Grindle.

CRC Press
Taylor & Francis Group
6000 Broken Sound Parkway NW, Suite 300
Boca Raton, FL 33487-2742

Printed in the United States of America on acid-free paper
10 9 8 7 6 5 4 3

International Standard Book Number-10: 0-8493-7222-4 (Hardcover)
International Standard Book Number-13: 978-0-8493-7222-3 (Hardcover)

Library of Congress Cataloging-in-Publication Data

An introduction to rehabilitation engineering / [edited by] Rory A. Cooper, Hisaichi Ohnabe, Douglas A. Hobson.
p. ; cm.
Includes bibliographical references and index.
ISBN-13: 978-0-8493-7222-3 (hardcover : alk. paper)
ISBN-10: 0-8493-7222-4 (hardcover : alk. paper)
1. Rehabilitation technology. 2. Biomedical engineering. I. Cooper, Rory A. II. Ohnabe, Hisaichi. III. Hobson, Douglas A. [DNLM: 1. Rehabilitation. 2. Biomedical Engineering–instrumentation. 3. Electronics, Medical–instrumentation. 4. Prostheses and Implants. 5. Self-Help Devices. 6. Sensory Aids. WB 320 I618 2006] I. Title.

RM950.I58 2006
617'.03–dc22 2006021440

Visit the Taylor & Francis Web site at
http://www.taylorandfrancis.com

and the CRC Press Web site at
http://www.crcpress.com

Dedication

This book is dedicated to the memory of Thomas J. O'Connor, Ph.D., a graduate of the University of Pittsburgh, Department of Rehabilitation Science and Technology. Dr. O'Connor was committed to studying technologies to expand community participation and the quality of life of people with disabilities. As a person with a disability, he had additional insight and understanding of the challenges faced. He was a scholar, a friend, and a colleague, who, due to his untimely death, never reached his full potential. The editors and all of the authors have agreed to have their portion of the royalties from the sales of this book go to the Thomas J. O'Connor Fund at the University of Pittsburgh. These funds will support student involvement in rehabilitation engineering and assistive-technology-related education, research, and development to carry on Dr. O'Connor's legacy.

Foreword

Rehabilitation Engineering is a nascent and developing field compared to most other engineering disciplines, yet there is compelling evidence that the practice of rehabilitation engineering has its roots in antiquity. Examples include use of a "stick" as an aid for ambulation, the attachment of wheels to a chair and many other implementations to compensate for functional deficits. The utilization of a pole as a walking aid appears in an Egyptian stele in the Carlsberg Sculpture Museum (Copenhagen) that dates to 1500 BC. This same simple appliance may be observed in use in developing countries today. While prosthetics is often viewed as separate and distinct from rehabilitation engineering — perhaps most prominently by prosthetists — it is an obvious and natural subset of the larger concept of rehabilitation engineering.

Historically, warfare has provided a stimulus for advances in rehabilitation engineering, and it is not surprising to learn that medieval armorers as the first prosthetists were also the first rehabilitation engineers. The term "Rehabilitation Engineering" is of relatively recent vintage. Jim Reswick — a seminal contributor and pioneer for modern rehabilitation engineering — attributes this term to James Garrett, the Chief of Research and Development at SRS (Social and Rehabilitation Service), one of the principal early sources of support for rehabilitation engineering research and development (c. 1970).

The modern era of rehabilitation engineering began with the establishment of "Rehabilitation Engineering Centers" (RECs) with support from the Federal Government through SRS in the early 1970s. A program for "Rehabilitation Engineering Centers of Excellence" was proposed at a conference in Annapolis, Maryland, in 1970. These centers were defined in 1971 by an expert panel appointed by Elliott Richardson, Secretary of the U.S. Department of Health, Education and Welfare. The REC program was subsequently included in the Rehabilitation Act of 1973 — a "watershed" event for rehabilitation — and was mandated to receive 25% of agency (SRS) research funding (later reduced). The REC program was initiated with the establishment of five centers at academic institutions with prominent engineering and medical engagement in rehabilitation. It is interesting to note that these original institutions were selected to submit applications. These initial five RECs were awarded to Rancho Los Amigos Hospital, USC for *FES of Nerves and Muscles*; Moss Rehabilitation Hospital, Temple and Drexel Universities for *Neuromuscular Control Systems*; Texas Institute for Rehabilitation Research, Baylor and Texas A&M Universities for *Effects of Pressure on Tissue*; Harvard and MIT for *Sensory Feedback Systems*; and University of Virginia for *Technology for People with Spinal Cord Injuries*. Substantial programs in rehabilitation engineering were also established by the Veterans Administration in selected VA Centers. By 2000, the National Institute for Disability and Rehabilitation Research (NIDRR) was supporting 17 RERCs (Rehabilitation Engineering Research Centers). NIDRR was created, first as NIHR (National Institute for Handicapped Research) in the newly founded U.S. Department of Education as the successor agency for research programs originally established in SRS.

Research and development was the dominant focus for rehabilitation engineering during the decade of the 1970s and resulted in the formation of *RESNA*, the Rehabilitation Engineering Society of North America in 1980. While R&D has continued as a major part of the rehabilitation field, an increased emphasis on the delivery of services began to emerge in the mid-1980s. By the mid 1990s service delivery had emerged as the major emphasis for both RESNA and for the field. Formal training programs in rehabilitation engineering and assistive technology were being offered at major universities. To date most of these training programs are attached to more traditional academic and professional training departments.

Those of us who have observed and contributed to the field of rehabilitation engineering over the past 20 to 30 years have seen extensive changes in both practices and in research and development. We have witnessed periods of rapid development resulting from federal and state support for research, service delivery and training. Rehabilitation engineering in many respects has been a creation of government, and while it is indisputable that government has been an advocate and benefactor for our field, it has also frequently been an adversary. Government has supported R&D for "orphan" technologies, and it has supported consumer advocacy. But government has also limited access to technologies and services. Agencies (e.g., VA, Medicare, etc.) have frequently refused to support new, advanced technologies even when they have been proven to be more effective and provide for long-term functional and economic benefit. Disconnects between R&D agencies and support service agencies (SSA, CMS, etc.) have created interesting ironies, but they also have resulted in the formation of interesting alliances. We have responded by engaging in educational and informational efforts to inform policy, regulatory and legislative bodies. We have established and participated in the development of standards both to improve and guarantee quality, and also to inform and assist government agencies with formulation of more enlightened policies for acquisition and approval.

The world of research and service for rehabilitation engineering and assistive technology is an ever-changing landscape. We must be both vigilant and diligent if we are to continue to be viable as a profession and field of endeavor in the complex and often convoluted environment of healthcare and its sub-fields of rehabilitation and rehabilitation engineering. This must be a primary consideration in our education and training programs. If we are to do this well we must have reference and teaching materials that accurately reflect the state of our field. *An Introduction to Rehabilitation Engineering* reflects the current state of art, science, engineering and technology for the fields of rehabilitation engineering and assistive technology. This book serves as both a reference and a tutorial for our field. The fact that it is entitled as an *Introduction* suggests that it can be a starting point for those new to rehabilitation engineering. I would call your attention to the systematic presentation of objectives and corresponding presentation of information in each section. This book is also sufficiently comprehensive in coverage to serve as a reference to those with extensive of experience and accomplishments. The authors have taken care in providing explanations, and they make clear distinctions in definitions — for example, in distinguishing between *rehabilitation engineering* and the related field of *assistive technology*.

The maturity and value of a field is proportional to its collected store of knowledge and practices. These are in turn reflected by the literature of the field. I trust that you will find this addition to our published knowledge to be an important and substantial contribution that will serve both those new to rehabilitation engineering and long-serving veteran clinicians and scientist-engineers.

The editors and authors of *An Introduction to Rehabilitation Engineering* all have demonstrated their competence and expertise as clinicians, or engineers and scientists and in many instances in all of these domains. I feel privileged to know these dedicated and accomplished professional personally.

It has been my pleasure and honor to review and comment on this important contribution to the field and discipline of rehabilitation engineering. I commend this reference to you enthusiastically and with abiding satisfaction.

Clifford E. Brubaker, Ph.D.
Professor and Dean
School of Health and Rehabilitation Sciences
University of Pittsburgh

Preface

This book was written as a collaborative effort by many of our friends and colleagues. It is intended to be used by advanced undergraduate students, graduate students, and interested professionals to introduce themselves to the exciting area of rehabilitation engineering and assistive technology. The book includes 18 chapters representing the major areas of research, development, and service delivery for rehabilitation engineers. Each chapter provides a list of learning objectives and study questions to help guide the reader. At the end of the book, there is also a list of key terms and their definitions. We hope that this work will serve as a textbook in a university course and as a reference.

The first chapter provides a brief introduction to rehabilitation engineering and assistive technology. Included in Chapter 1 are key topics in the International Classification of Function, design processes, and issues related to service delivery, research, and development. The participatory action design process is described because it provides a theme throughout the book. A new model, described as the PHAATE, is presented to incorporate the critical aspects of rehabilitation engineering and assistive technology.

Clinical practice and application of rehabilitation engineering and assistive technology are the ultimate goals. Eventually, all of the models, devices, and policies are geared toward maximizing community participation through appropriate, if not optimal, service delivery. Chapter 2 covers the models for assistive technology service delivery, and briefly describes areas of practice and provides some guidance regarding good clinical practice. Credentialing is presented along with an explanation as to its importance.

Assistive technology and rehabilitation engineering of devices for people with disabilities takes three basic approaches: design for use by the broadest possible population, design for subpopulations, and design for the individual. Chapter 3 describes helpful design tools and principles of universal design. An advantage of the universal design approach is that it helps to lower cost and increase access to products. However, some individuals need customized products, and hence orphan technologies are also covered.

For consumers to benefit from assistive devices; they have to be transferred to industry by some means or sufficient information has to be made available for clinical rehabilitation engineers to replicate the work. Chapter 4 describes various technology-transfer mechanisms and models. Basic principles, such as consumer pull, are presented. Technology transfer is an area that has received substantial interest from the U.S. Federal government; hence, some of these mechanisms are described as well.

Quality assurance and the ability to objectively compare the properties of assistive devices are critical in the efforts to continuously improve products and services. National and international standards are important components of assuring and advancing quality of assistive devices. Chapter 5 describes the process for creating assistive device standards, and provides several examples of existing or proposed standards. While standards are important and necessary, they are not sufficient by themselves to ensure adequate quality and appropriate matching of the technology to the individual. It is necessary to remain up to date with current scientific, engineering, and clinical literature and practices.

The concept of the human body in a seated position seems simple, yet it is complex. A reason for this is that the human body is not designed to sit, at least not for prolonged periods of time, in the same position. The human body alters its position frequently from various lying, sitting, and standing positions, depending on the task being performed, be it an active or a resting task. For people with disabilities who are unable to walk and require the use of wheeled mobility and seating devices, sitting is necessary, and identifying the most effective seating system for a specific individual's needs can be challenging. The purpose of Chapter 6 is to review the principles of seating biomechanics and the components of seating systems designed to accommodate specific needs based on the person's preferences, functional capabilities, and environmental demands (both physical and social).

Because body weight is transmitted through the bony prominences such as the scapula, sacrum, greater trochantors, ischial tuberosities, and heels, it can cause significant concentrations of pressure at the skin surface and in the underlying soft tissue. The pressure peaks and gradients surrounding these peaks can put the soft tissue at risk of breakdown. Wheelchair users, especially those having limited ability to reposition them and/or have a loss of sensation in the areas where weight is being supported, are at high risk for developing pressure ulcers. Chapter 7 provides the material necessary to gain a basic understanding of soft tissue biomechanics and for maintaining soft tissue integrity.

Wheelchairs and scooters make up a significant portion of assistive devices in use today. There is anticipated continued growth in the wheelchair market due to aging populations, increasing longevity, increased incidence of recovery from trauma, and more effective therapies for chronic diseases. Chapter 8 describes the basic design and service delivery principles of wheelchairs and scooters. There is an overwhelming need for continued advancements in wheelchair technology and for skilled rehabilitation engineers, suppliers, and practitioners to ensure that individuals receive the most functionally appropriate mobility device that meets their needs.

Electrical stimulation therapy involves the application of external electrical energy to tissue, which results in the stimulation of a physiological response. The degree and type of response depends on the mode, quantity and quality of energy, and the anatomical location of the application site. Physiological responses evoked by electrical stimulation range from relaxation of spasticity, controlled muscle contractions, increased production of endorphins, increased muscle fiber recruitment, circulatory stimulation, and enhancement of reticuloendothelial response. Chapter 9 provides an overview of functional electrical stimulation and its applications.

Transportation is a key component to full integration into the community. Accessible public transportation is necessary to provide persons with disabilities the same opportunities as others in employment, education, religious worship, and recreation. Chapter 10 includes description of accessible transportation legislation and the means of safely transporting people in wheelchairs.

Rehabilitation robotics is a combination of industrial robotics and medical rehabilitation, which encompasses many areas including mechanical and electrical engineering, biomedical engineering, prosthetics, autonomous mobility, artificial intelligence, and sensor technology. The success of robotics in the industrial arena opens up opportunities to significantly improve quality of life of people with disabilities, including integration into employment, therapy augmentation, and so on. These opportunities could be realized to the full provided rehabilitation robotic systems can be developed to meet the needs of people with disabilities in their daily living activities. Chapter 11 provides information about the various applications of robotics in medical rehabilitation.

Level of amputation is the one of the most important characteristics in determining postamputation function. In upper-limb amputations, preservation of the thumb to allow opposition with the remaining fingers will preserve some fine motor skills. Longer residual limbs provide an extended lever arm to power prosthesis and allow for more area of contact to secure the prosthesis. Chapter 12 includes basic information on prosthetic design and usage.

Chapter 13 provides a general overview of the principles, design, fabrication, and function of spinal, lower-extremity, and upper-extremity orthoses. It is acknowledged that entire books and atlases have been dedicated to this subject. Orthotics is considered an art and a skill, requiring creativity and knowledge in anatomy, physiology, biomechanics, pathology, and healing. Utilizing these, practice is required for one to evolve into a competent clinician who is comfortable with designing and fabricating orthoses to meet the unique needs of each individual client.

In engineering terms, sensory loss can be regarded as a reduction of information input, reducing access to the information needed to perform any given task or to interact with the environment. In visual enhancement, the challenge is complicated by the fact that vision loss has to be regarded as multidimensional, with several different (not necessarily independent) vision variables caused by different disease processes. In visual substitution, the challenge is to present the necessary information effectively under the constraints imposed by the much lower effective "bandwidth" of the other senses compared to the vision channel. Chapter 14 provides an overview of visual impairment and technologies to enhance function for people with impaired vision.

The World Health Organization estimates that 250 million people in the world function with a disabling hearing loss. The presence of an uncorrected hearing loss serves to reduce an individual's level of participation in work, home, school, family, and social domains. The consequences may result in poorer academic outcomes, decreased vocational opportunities, social isolation, and depression. Chapter 15 describes the assistive technology for this population, which is referred to as hearing assistance technology (HAT). HAT devices are designed to either enhance the degraded auditory signal caused by the hearing loss or to substitute the degraded auditory signal via alternative modes of information input (mainly visual or tactile).

People with disabilities have engaged rehabilitation engineers in problem solving related to the accessibility and usability of telecommunications, computers, and the World Wide Web. This chapter will focus on problems people with disabilities have when using and accessing the Web. These problems or their solutions may include computers and telecommunication systems. On the client side, rehabilitation engineering has made important contributions to the development of assistive technology. Web-related assistive technology is described in Chapter 16; as well as other important technologies.

Chapter 17 focuses on augmentative and alternative communication (AAC) technology specific to improving communication function and participation. This chapter introduces the basic elements of a comprehensive AAC assessment and the role of rehabilitation engineers in making decisions about AAC technology. The significance of language issues and AAC language representation methods is described, emphasizing the need for AAC technology to support the spontaneous generation of language in order to optimize communication function and participation. Finally, this chapter covers performance and outcomes measurement and the role rehabilitation engineers in collecting and analyzing data to support the design of and evaluation for AAC systems.

Play and leisure activities are much more interest-driven than other human occupations. They are chosen for their own sake for the excitement, challenge, enjoyment, or feelings of competence that they inspire. The more our life is filled with things that interest us, the higher our satisfaction with our quality of life. The human need for recreation is especially important in the presence of physical, sensory, or cognitive impairment that affect the ability to function in everyday activities and environments. Chapter 18 provides an introduction to adaptive sports and recreation.

After the chapters, there is a list of selected terms and their definitions to help guide readers through some of the unfamiliar language used within the book. The study questions at the end of each chapter are intended to stimulate further discussion and to help review the material covered within the chapters. For teachers, the list of learning objectives may be helpful when assembling a syllabus. Clearly, it is impossible to capture all aspects of rehabilitation and engineering in a single book. Thus, we have only attempted to compile an introduction to provide an overview and to stimulate further study. We hope that you find this work beneficial and interesting.

Rory A. Cooper, Ph.D.
Pittsburgh, PA, USA

Douglas A. Hobson, Ph.D.
Pittsburgh, PA, USA

Hisaichi Ohnabe, Ph.D.
Niigata, Japan

Editors

Rory A. Cooper, Ph.D. is FISA and Paralyzed Veterans of America (PVA) Chair and distinguished professor of the Department of Rehabilitation Science and Technology, and professor of Bioengineering and Mechanical Engineering at the University of Pittsburgh. He is also a professor in the Departments of Physical Medicine and Rehabilitation and Orthopaedic Surgery at the University of Pittsburgh Medical Center Health System. Dr. Cooper is director and VA Senior Research Career Scientist of the Center for Wheelchairs and Associated Rehabilitation Engineering, a VA Rehabilitation Research and Development Center of Excellence.

Dr. Cooper has been selected for numerous prestigious awards, including the Olin E. Teague Award, the James Peters Award, the Paul M. Magnuson Award, and he is an inaugural member of the Spinal Cord Injury Hall of Fame. Dr. Cooper has authored or co-authored more than 175 peer-reviewed journal publications. He is an elected Fellow of the Rehabilitation Engineering and Assistive Technology Society of North America, the Institute of Electrical and Electronics Engineers, and of the American Institute of Medical and Biological Engineering. Dr. Cooper has been an invited lecturer at many institutions around the world, and was awarded "Honorary Professor" at The Hong Kong Polytechnic University and Xi'an Jiatong University.

Dr. Cooper has served as president of RESNA, a member of the RESNA/ANSI and ISO Wheelchair Standards Committees, IEEE-EMBS Medical Device Standards Committee, and as a trustee of the Paralyzed Veterans of America Research Foundation. In 1988, he was a bronze medalist in the Paralympic Games, Seoul, Republic of Korea. He has served as a member of the United States Centers for Medicare and Medicaid Services — Medicare Advisory Committee, Chair on the National Advisory Board on Medical Rehabilitation Research, National Institute of Child Health and Human Development, steering committee for the 1996 Paralympic Scientific Congress and as a member of the United States Secretary of Veterans Affairs Prosthetics and Special Disability Programs Advisory Committee.

Hisaichi Ohnabe, Ph.D. joined Ishikawajima-Harima Heavy Industries (IHI) in 1962 after graduation from the Japan Coast Guard Academy. Dr. Ohnabe was engaged to design a marine turbine for a huge tanker, the world's largest at that time. He was awarded a Fulbright Scholarship in 1968 and undertook an orientation program at the University of Kansas. He undertook a master of mechanical and aerospace engineering degree in 1971 as a Fulbright exchange student and then a Ph.D. at the University of Delaware in 1975 (in part as a Fulbright student). After coming back to Japan, he joined the Structure and Strength Department in the Research Institute of IHI. Dr. Ohnabe then collaborated in an international consortium of five nations (U.S., England, Germany, Italy and Japan) on the development of jet engines (V2500) for civil aircraft. He has also worked on ultra high temperature composite materials for the engines of a Japanese space shuttle.

Dr. Ohnabe was a professor in the Department of Biocybernetics, Faculty of Engineering, Niigata University from 1998 and also a professor in the Graduate School of Science and Technology at Niigata University. He then joined the Department of Rehabilitation Science and Technology, School of Health and Rehabilitation Sciences, University of Pittsburgh and Human Engineering Research Laboratories, VA Pittsburgh Healthcare System as visiting professor in April, 2004 and as adjunct professor in October, 2005. In the autumn of 2006 he returned to Japan and became a professor at Niigata University of Health & Welfare in March, 2006 and is now a professor there in the Department of Prosthetics & Orthotics and Assistive Technology, and continues as an adjunct professor, University of Pittsburgh.

Dr. Ohnabe is a member of the Japanese Society for Wellbeing Science and Assistive Technology, Rehabilitation Engineering Society of Japan (RESJA), the Japan Society for Welfare Engineering, the Japan Society of Mechanical Engineers, the Japan Society for Aeronautical and Space Science, and the Japan Society for Composite Materials. He is also a councilor of the Japanese Society for Wellbeing Science and Assistive Technology.

Dr. Ohnabe is interested in rehabilitation engineering, assistive technology and universal design. He has authored or co-authored more than 70 peer-reviewed journal publications and peer-reviewed proceedings publications.

Douglas A. Hobson, Ph.D. is associate professor-emeritus and associate director of the Rehabilitation Engineering Research Center (RERC) on Wheelchair Transportation Safety at the University of Pittsburgh, Department of Rehabilitation Science and Technology. Dr. Hobson began and directed the Rehabilitation Engineering Program at the University of Tennessee (UT), Memphis, TN from 1974 to 1990. It was recognized world-wide for its contribution to the field of specialized seating and wheeled mobility. From 1976 to 1981, he directed the NIHR-REC at UT that focused on research and development of wheelchair seating technology for children. Four seating products were developed, several of which are still being marketed by commercial suppliers. Many of the seating principles now being taught to clinicians and suppliers were developed and communicated by the UT staff. The International Seating Symposium, which is held every two years in the U.S., also had its genesis in the UT program. From 1992 to 2001, he served as managing director of the RERC on Wheeled Mobility (RERCWM) at the University of Pittsburgh. Until 2001, he served as chairman of the SAE, ISO, and the RESNA-SOWHAT standards committees related to wheelchair transportation safety. Both national (ANSI/RESNA, SAE) and international (ISO) industry standards for wheelchair securement and transport wheelchairs were completed during his tenure. Dr. Hobson also served as the president of RESNA during the period of 1991–1992. Dr. Hobson received the B.S. in Mechanical Engineering from the University of Manitoba in 1965 and the Ph.D. in Bioengineering from the University of Strathclyde, Scotland, in 1989.

Contributors

This book is the product of the efforts of many people. First, the editors thank all of those who contributed to the writing of this text; their contributions are invaluable. The authors are listed here in the order in which they appear in the book.

Rory A. Cooper
University of Pittsburgh and
U.S. Department of Veterans Affairs
Pittsburgh, Pennsylvania

Carmen Digiovine
University of Illinois
Chicago, Illinois

Douglas A. Hobson
University of Pittsburgh
Department of Rehab Science and Technology
Pittsburgh, Pennsylvania

Linda van Roosmalen
University of Pittsburgh
Department of Rehab Science and Technology
Pittsburgh, Pennsylvania

Hisaichi Ohnabe
Niigata University of Health and Welfare
Niigata, Japan
and
University of Pittsburgh
Department of Rehab Science and Technology
Pittsburgh, Pennsylvania

Jonathan Pearlman
University of Pittsburgh and VA Pittsburgh Healthcare System
Pittsburgh, Pennsylvania

Martin Ferguson-Pell
University College London
Institute of Orthopaedics and Musculoskeletal Science
London, United Kingdom

Mark Schmeler
University of Pittsburgh
Department of Rehab Science and Technology
Pittsburgh, Pennsylvania

Bengt Engstrom
Bengt Engstrom Seating
Stallarholmen, Sweden

Barbara Crane
University of Hartford
West Hartford, Connecticut

Rosemarie Cooper
University of Pittsburgh
Department of Rehab Science and Technology
Pittsburgh, Pennsylvania

David Brienza
University of Pittsburgh
Department of Rehab Science and Technology
Pittsburgh, Pennsylvania

Yihkuen Jan
University of Pittsburgh
Department of Rehab Science and Technology
Pittsburgh, Pennsylvania

Jeanne Zanca
University of Pittsburgh
Department of Rehab Science and Technology
Pittsburgh, Pennsylvania

Alicia Koontz
VA Pittsburgh Healthcare System
Pittsburgh, Pennsylvania

Brad Impink
University of Pittsburgh and VA Pittsburgh Healthcare System
Pittsburgh, Pennsylvania

Matthew Wilkinson
Carnegie Mellon University
Basking Ridge, New Jersey

Songfeng Guo
University of Pittsburgh and VA Pittsburgh Healthcare System
Pittsburgh, Pennsylvania

Karl Brown
University of Pittsburgh and VA Pittsburgh Healthcare System
Pittsburgh, Pennsylvania

Emily Zipfel
University of Pittsburgh and VA Pittsburgh Healthcare System
Pittsburgh, Pennsylvania

Yusheng Yang
University of Pittsburgh and VA Pittsburgh Healthcare System
Pittsburgh, Pennsylvania

Jue Wang
Xi'an Jiatong University
Research Center of Rehabilitation Science and Technology
School of Life Sciences and Technology
Xi'an, China

Tong Zhang
Xi'an Jiatong University
Xi'an, China

Elizabeth Leister
University of Pittsburgh and VA Pittsburgh Healthcare System
Pittsburgh, Pennsylvania

Patricia Karg
University of Pittsburgh
Department of Rehab Science and Technology
Pittsburgh, Pennsylvania

Gina Bertocci
University of Louisville
J.B. Speed School of Engineering
Louisville, Kentucky

Toru Furui
University of Pittsburgh and VA Pittsburgh Healthcare System
Pittsburgh, Pennsylvania

Dan Ding
University of Pittsburgh and VA Pittsburgh Healthcare System
Pittsburgh, Pennsylvania

Richard Simpson
University of Pittsburgh
Department of Rehab Science and Technology
Pittsburgh, Pennsylvania

Yoky Matsuoka
Carnegie Mellon University
Robotics Institute
Pittsburgh, Pennsylvania

Edmund LoPresti
AT Sciences
Pittsburgh, Pennsylvania

Diane M. Collins
University of Pittsburgh and VA Pittsburgh Healthcare System
Pittsburgh, Pennsylvania

Brad Dicianno
University of Pittsburgh and VA Pittsburgh Healthcare System
Pittsburgh, Pennsylvania

Amol Karmarkar
University of Pittsburgh and VA Pittsburgh Healthcare System
Pittsburgh, Pennsylvania

Paul F. Pasquina
Walter Reed Army Medical Center
Department of Physical Medicine and Rehab
Washington, DC

Rick Relich
Union Orthotics and Prosthetics, Inc.
Monroeville, Pennsylvania

Annmarie Kelleher
University of Pittsburgh and VA Pittsburgh Healthcare System
Pittsburgh, Pennsylvania

Michael Dvorznak
University of Pittsburgh and VA Pittsburgh Healthcare System
Pittsburgh, Pennsylvania

Kevin Fitzpatrick
Walter Reed Army Medical Center
Department of Physical Medicine and Rehab
Washington, DC

Thane McCann
Walter Reed Army Medical Center
Department of Physical Medicine and Rehab
Washington, DC

John Brabyn
Smith Kettelwell Eye Research Institute
San Francisco, California

Katherine D. Seelman
University of Pittsburgh, Department of Rehab Science and Technology
Pittsburgh, Pennsylvania

Sailesh Panchang
Deque Systems, Inc.
Centreville, Virginia

Elaine Mormer
University of Pittsburgh
Department of Communication Science and Disorders
Pittsburgh, Pennsylvania

Amanda Ortmann
University of Pittsburgh
Department of Communication Science and Disorders
Pittsburgh, Pennsylvania

Catherine Palmer
University of Pittsburgh
Department of Communication Science and Disorders
Pittsburgh, Pennsylvania

Stephanie Hackett
Health Information Management
University of Pittsburgh
Pittsburgh, Pennsylvania

Bambang Parmanto
University of Pittsburgh
Department of Health Information Management
Pittsburgh, Pennsylvania

Katya J. Hill
University of Pittsburgh
Department of Communication Science and Disorders
Pittsburgh, Pennsylvania

Bruce Baker
Semantic Compaction Systems and University of Pittsburgh
Pittsburgh, Pennsylvania

Barry A. Romich
Prentke-Romich Company and
University of Pittsburgh
Wooster, Ohio

Mary E. Buning
University of Colorado, Denver and
Health Sciences Center/Department of
Physical Medicine and Rehab
Denver, Colorado

Ian M. Rice
University of Pittsburgh and
VA Pittsburgh Healthcare
System
Pittsburgh, Pennsylvania

Shirley G. Fitzgerald
University of Pittsburgh and VA Pittsburgh
Healthcare System
Pittsburgh, Pennsylvania

Table of Contents

1 Introduction

Rory A. Cooper

CONTENTS

1.1 LEARNING OBJECTIVES OF THIS CHAPTER

Upon completion of this chapter, the reader will be able to:

Learn the basic principles of rehabilitation engineering
Learn about the basic methods used in rehabilitation engineering design
Understand the problems for which rehabilitation engineering and assistive technology are used
Gain knowledge about the research and development issues and opportunities in this field

1.2 INTRODUCTION

The future of rehabilitation engineering and assistive technology (AT) will depend on understanding and documenting the confluence of two critical factors represented by the paradigm of "functionality." This confluence is of medical restoration and consumer participation. Historically, medicine has attempted to "fix" or intervene on the injury or disease plateau, and the consumer simply "carried on." Now, the

International Classification of Function (ICF) has elevated therapy or technology to more strongly match the desired activity and participation of the consumer. Means of quantifying or qualifying participation are yet to be agreed upon, and so rehabilitation professionals are aspiring to harness their science toward this challenge.

Whether we are talking about a stroke, spinal cord injury, burn, or cognitive impairment, rehabilitation and habilitation need to address the ICF paradigm. To do so constructively, the field needs evidence from scholars, practitioners, and consumers, who can work together to successfully meet this challenge. The confluence of these new or realigned partnerships suggest the emergence of new endeavors, several of which will be described in this book. An important concept that will change future assistive devices is participatory action design (PAD).

Other factors that increase the usability of an AT device are the perceived advantages and disadvantages of AT devices. If the (perceived) benefits outweigh the (perceived) disadvantages, then there is a higher chance of that device being utilized. On the contrary, if the (perceived) disadvantages outweigh the (perceived) benefits of the device, then there is a higher chance that the device will not be used.

Personal factors including motivation, cooperation, optimism, good coping skills, and ability to learn or adapt to new skills work in combination for the functional independence of the user (Figure 1.1). In the older population, all the aforementioned factors may diminish gradually. Therefore, acceptance of AT devices can be a challenge in this population, which may result, in turn, in suboptimal use of AT devices for functional independence.

A variation in the number of AT devices used has been observed in people with varying disorders within the aging population. The overall trend indicated a positive relationship between the severity of disorders and the number of AT devices used by older people. The usability of AT devices also depends on environmental accessibility. The presence of environmental barriers can limit the acceptance of these devices. For instance, consider an elderly individual who lives in a two-storey house and has been prescribed a power wheelchair for functional mobility. There is a ramp to get in the house. The second floor, where the individual spends most of his time during a day, however, is not accessible. In this situation, the use of the power wheelchair inside the house will be limited as a result of an environmental barrier. Acceptance of AT in older adults is also determined to a large extent by views of society. An example of this is the higher acceptability of home modifications such as grab bars in the bathroom and a high-rise commode than that of a mobility device such as a cane or a walker. The latter are considered to be indicative of a significant disability.

The concept of person and society (P&S) must permeate all AT research, development, service delivery, and education efforts as a cross-cutting and integrated activity (Figure 1.2). The assimilation of the individual, social, policy, and privacy components of research, development, design, deployment, and education related to people with disabilities (PWD) and older people is among the more unique features of the domain of AT and rehabilitation engineering. Indeed, P&S form the foundation for AT innovation. P&S must create the pressure for the generation of research and development ideas, serve as a pathway through the research and development phases, and provide the consumer pull for the deployment of products and services.

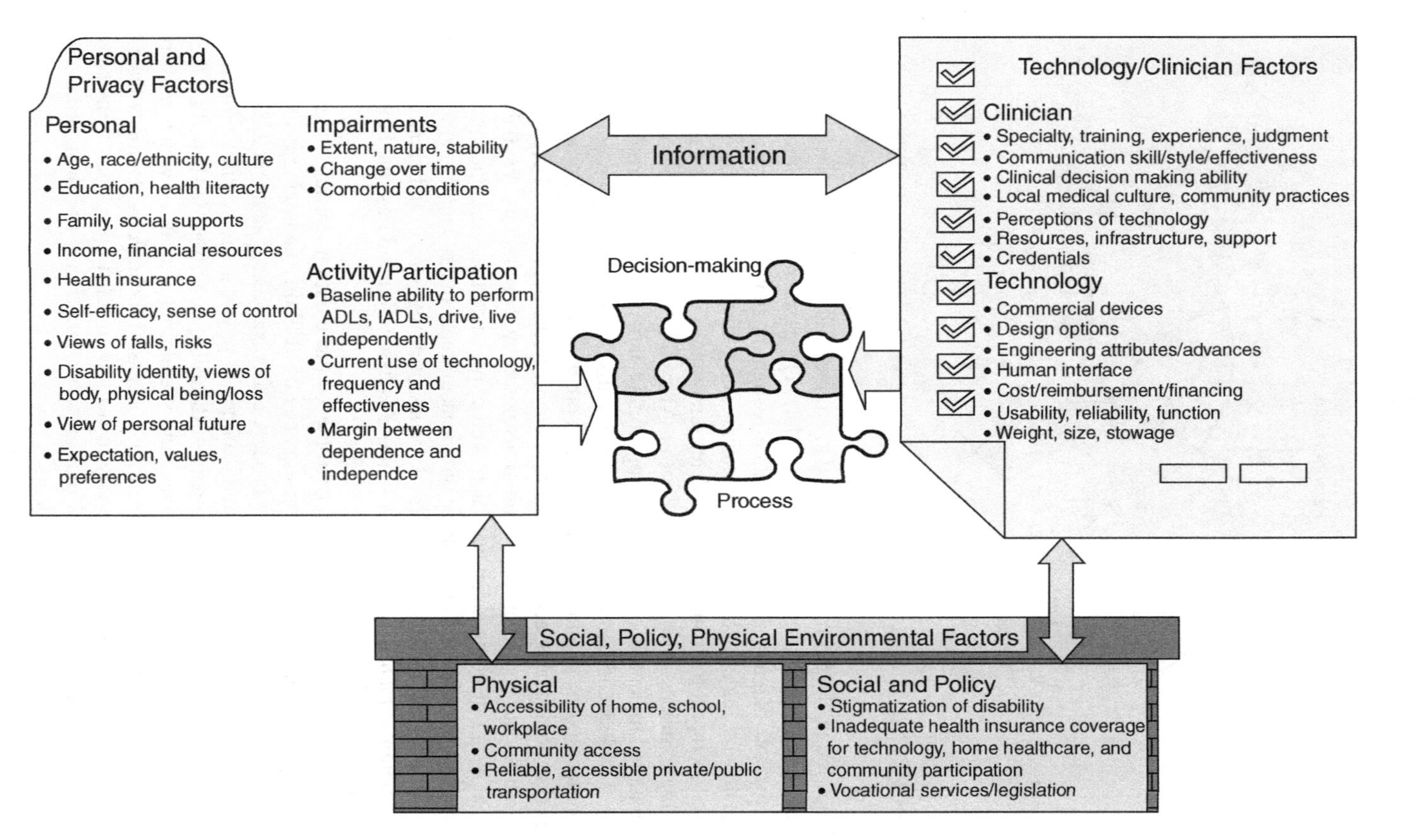

FIGURE 1.1 Illustration of person and society permeation in AT and rehabilitation engineering.

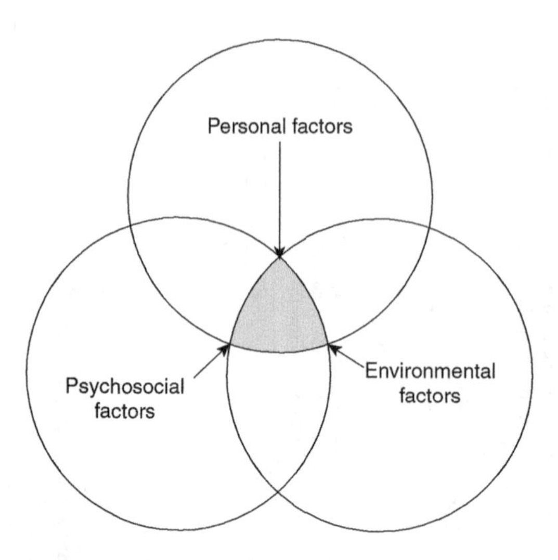

FIGURE 1.2 Interaction between different factors for acceptance of AT devices.

Knowledge of P&S must be imparted to engineering teams. Just as rehabilitation engineering cannot be done effectively in isolation from scientists and other clinicians connected with end users, social scientists investigating new technological directions need to be working effectively with engineers. The goal of rehabilitation engineering must be to support the concept of self-determination for older adults and people with disabilities. Success in the development of AT requires technical competence and imagination, but equally or even more importantly, it depends on a thorough appreciation and understanding of aging, older adults, disabilities, people with disabilities, environment, costs, regulations, policies, and other limiting factors. Some otherwise remarkable engineering accomplishments have been dismal failures because of lack of awareness and appreciation of limitations posed by these factors. We believe that a unique blend of complementary talent and experience that combines the best in engineering, science, and technology with consummate understanding and appreciation of the relevance of social, cultural, behavioral, regulatory, and economic and environmental dimensions and considerations is essential to address the burgeoning issues of aging and disability.

Independence, autonomy, and control over the environment are fundamental human motives (Figure 1.3). One consequence of disability and aging is that one's direct control over the environment is eroded, either because of acute medical events or because of more gradual age-related declines in physical and cognitive functioning. We further recognize that the impact of disabling conditions is shaped by their cause, timing, functional consequences, and social contexts, all of which must be considered in designing and implementing successful technology. Humans have a strong capacity to adapt psychologically to decline in functioning, but at the same time they will embrace technologies that preserve and/or enhance their abilities to maintain control over their environments.

Central to the future of rehabilitation engineering is the conduct of research and development in the "natural environment" of our extended community. This

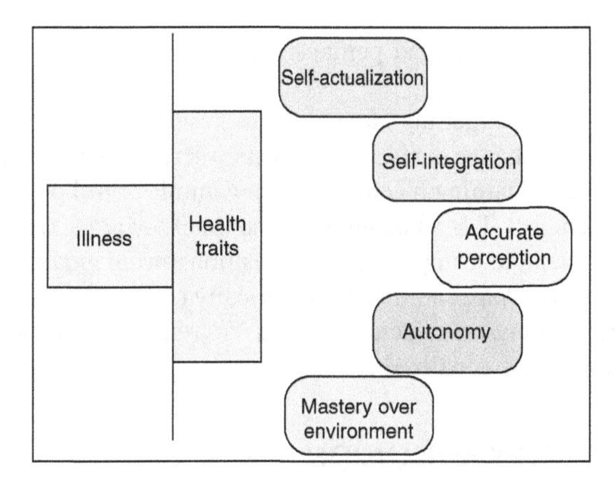

FIGURE 1.3 Illustration of human health motives.

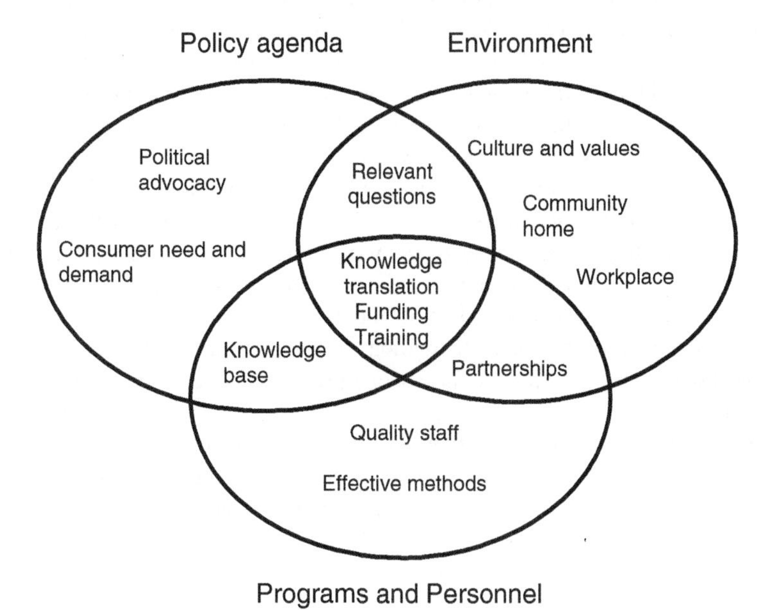

FIGURE 1.4 A new paradigm to address the physical, functional, social, behavioral, economic, regulatory, policy, and perhaps even spiritual dimensions of aging and disability.

"community" of human and organizational resources must be the "laboratory" for comprehensive study of aging and living with a disability and the issues associated with the development, transfer, and utilization of technology. Rehabilitation engineering research, development, education, and outreach activities must integrate expertise and experience in the areas of geriatrics, gerontology, engineering, computer science, AT, rehabilitation, behavioral and social sciences, policy study, education, participatory action research, and technology transfer. A new paradigm, as proposed in Figure 1.4, must be established to address the physical, functional, social, behavioral, economic, regulatory, policy, and perhaps even spiritual dimensions of aging and disability.

Assistive technology is getting "smarter." The availability and ubiquity of computing power are going to change the face of AT. This will result in better human–machine interfaces, adaptive or learning machines, more natural control of devices, and more capable devices. The iBOT is an example of the trend toward smarter, more capable technologies. AT manufacturing is going to transition toward greater customization or flexibility. Rapid prototyping, computer numerically controlled machines, computer-aided design/engineering/manufacturing, and Web-based systems are driving AT manufacturing toward personalized products.

1.3 PARTICIPATORY ACTION DESIGN

Technology plays a pivotal role in the lives of individuals with disabilities through promoting reintegration into community life. There is little doubt that as technological advances become available, PWD will live healthier, more productive, and longer lives. Some of the world's leading technological minds suggest that rehabilitation engineering will have a far greater impact on society than other fields of technology. In the largest survey of the US AT industry to date, 359 companies represented $2.87 billion in sales, with sales growth exceeding 20% from 1997 to 1999. The survey revealed that 60% of the respondents were firms of ten employees or less, and 11% of the surveyed companies accounted for 69% of the revenues. Of the companies surveyed, biomedical engineers/designers were reported to be in highest demand for the AT industry, followed by computer programmers and specialists in digital signal processing. Similarly, when asked about federal laboratory assistance, respondents reported a desire for transfer of technology related to electronic components and systems, board-level electronics, lasers or optics, and integrated circuits as their primary interests.

The PAD model describes a process of developing assistive devices (Figure 1.5). The process starts with identification of users' needs. There are several ways of doing this: through focus groups, where an open-ended discussion is moderated by a person

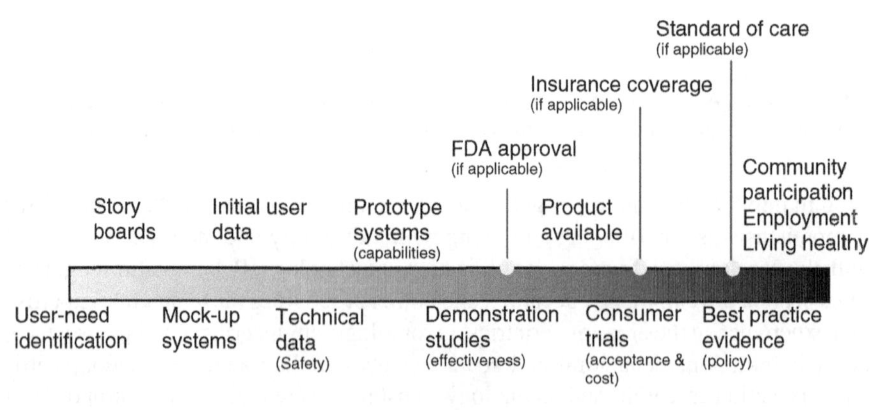

FIGURE 1.5 Illustration of the PAD process.

from a design team, although not one of the designers, or by getting feedback from users through surveys and questionnaires about specific requirements and possible solutions. All of this information is assembled, which helps to identify desirable features for prospective products. These data are helpful in comparing alternatives for a product and determining advantages or disadvantages of each. The next step includes development of a mock-up system, in which the devices' key features are incorporated on the way to a product design. All the features of the product are then compared to the benchmarks available, ensuring that the designed features are at par with industry standards. The prototype is constructed after this step, and it includes as many of the features as are feasible. With the prototype completed, a comparison is made with the standards for the product. Product efficacy is determined with appropriate tests. Product durability and reliability are often key aspects of consumers' desires. Durability testing typically determines the ability of individual components of a particular device to withstand repeated use by the end user. On incorporating the changes suggested by the efficacy testing, the results for medical devices, which include a range of assistive devices, are submitted to the Food and Drug Administration (FDA) for approval. The FDA approval process is an extensive procedure, with emphasis on ensuring safety to the end users. Clinical effectiveness of AT devices is best established in several phases. Typically, there are four phases of testing involved in determining clinical effectiveness. The first phase involves conducting a focus group of clinicians, end users, and manufacturers. These individuals provide their feedback on the benefits and disadvantages of the product. The second phase involves testing the product using an unimpaired population when appropriate. The third phase includes case studies, in which a small number of potential end users get tested on the device. The outcomes of determining clinical effectiveness could include physical capacity measures and/or functional performance measures. The fourth phase consists of testing a large group of potential end users, so that generalization can be made to the entire population, which will eventually be using the device. The most intricate step in this entire process is establishing insurance coverage, where potentially eligible, for a particular product. This involves identification or formulation of a common code for the device and establishing a fee schedule for the device. If the product meets the needs of end users and the approval of clinicians, it should become a part of the arsenal to ameliorate disability. With further clinical studies, the product may be incorporated into a clinical practice guideline.

Employment of the PAD process requires:

Rapid cycling from laboratory to clinical study, observation, and feedback.
Diverse environment from skilled-care institutions to independent home living and to community environments.
Broad range of participants, ranging from children to older adults, at home, school, and work throughout the community, and across a broad range of conditions and impairments.
Robust and expeditious regulatory training and *review processes* to be in place.

Employment brings social integration and pathways for upward mobility, and affords opportunities for improving quality of life. Through employment, PWD can obtain the discretionary income to purchase additional medical coverage and to purchase items to expand their range of activities (e.g., recreation and leisure). Work brings about the potential to exercise a voice through political and economic processes. Regrettably, most PWD remain unemployed and live on low incomes. Although US laws such as the Americans with Disabilities Act and the Rehabilitation Act have made headway in breaking down employment barriers, the social and attitudinal barriers remain a formidable, but not insurmountable, challenge.

1.4 POLICY, HUMAN, ACTIVITY, ASSISTANCE, TECHNOLOGY, AND ENVIRONMENT (PHAATE) MODEL

We have attempted to create a comprehensive conceptual model to represent the factors to be considered when designing AT or when developing a service delivery system. Our model is called the "PHAATE," pronounced fait, to incorporate policy, human (person), activity, assistance, technology, and environment (Figure 1.6). PWD are affected by public and often private/corporate policy. Reimbursement policy for AT and AT services is influenced if not determined by public and private (e.g., corporate policy, insurance carrier policy, and charitable organization). To underestimate the impact of policy on the design and service delivery process is to risk denial of reimbursement or even entry into the marketplace. Technology is used by PWD to accomplish some task or to perform an activity. Activity is a key factor in determining participation in society. Nearly all of us need assistance at some point in our lives. After even a little service delivery experience, people learn quickly that families and, in some cases, paid assistants are necessary for a significant number of people with disabilities. Moreover, AT must often interact with both the person with the disability and the persons providing assistance. People live in the "real world," and the environment impacts their functioning. It has been argued that the environment is what determines if impairment becomes a handicap. People perform activities in a variety of environments, and this needs to be given due consideration in the design,

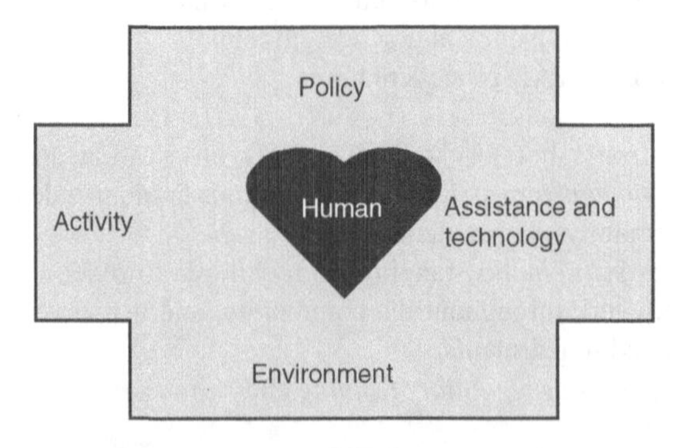

FIGURE 1.6 The PHAATE model incorporates policy, the person, activity, assistance, technology, and environment.

development, and service delivery processes. Last, rehabilitation engineers focus on technology for people, and any model would be incomplete without these components. Of course, the person needs to be at the core of the model and all of the process related to technology development and service delivery.

1.5 TRENDS IN AT SERVICE DELIVERY

Most PWD rely on a variety of different technologies: wheelchairs, adaptive driving aides, cushions, etc., for daily living and to participate in their communities. Assistive devices are among the few items that we can provide to PWD that make a profound difference. Part of the problem with AT provision lies in the paucity of AT outcome studies, somewhat because of the inadequate funding support for such studies and lack of understanding of the need for specialized clinical expertise, especially among insurers and nonrehabilitation medical professionals. Therapy and medicine residency programs spend too little time on AT training, and this is compounded by the limited focus on AT during accreditation. This results in inadequate prescriptions and specifications. Clinical education programs rely on fieldwork coordinators, who depend heavily on experience rather than evidence. AT is best delivered using a team approach, including an assistive technology practitioner (ATP), an assistive technology supplier (ATS), and a rehabilitation engineering technologist (RET) working in cooperation with a qualified physician. The Rehabilitation Engineering and Assistive Technology Society of North America (RESNA) provides the ATS and RET credentials to identify knowledgeable suppliers and engineers.

The AT service delivery field is becoming more professional. RESNA credentials are taking hold, as is necessary for the development of the field. It is essential that ATSs have training, practice standards, and credentials. If a hairdresser burns a client's scalp, there is a state board to appeal to, and the hairdresser can be sanctioned. However, a person can be injured because of an improperly provided wheelchair, and there are only the courts for recourse. Something as important and intimate as AT should require training and, eventually, licensure for providers. As clinical practice guidelines and clinical assessment tools take greater hold, the importance of highly skilled clinicians and suppliers is going to grow.

Individuals with disabilities and their families need to educate themselves to become better advocates. They should partner with an ATP, RET, and ATS (see www.resna.org) to help advocate for them. Also, finding a good physician is highly recommended. Physiatrists are trained in AT, and can be powerful advocates. Unfortunately, public policy on reimbursement for AT products and services has simply not kept pace with advances in service delivery (including rehabilitation engineers providing services) and the complexity of the products available. Creation and implementation of clinical practice guidelines is critical; although AT cannot be provided in a cookie cutter fashion, there is a need to create and translate scientific evidence into quality clinical practice. Tools for the guidance and assessment of clinical practice are core and as a whole we know too little about the appropriate provision of AT, and most decisions are made on very few visits. Devices, instruments, and other tools to collect better data to guide clinical decisions and the development of new devices are needed. Along similar lines, there is too little consumer (end-user) involvement in the AT design process.

1.6 NECESSARY EXPANSION OF AT CLINICAL SERVICE WORLDWIDE

In developed countries, AT has helped to maintain, increase, or improve the functional capabilities of PWD. In low-income countries, the availability of AT may be limited to some basic mobility and daily living aids. Although assistance from the spouse or family and skilled or unskilled attendants are sometimes available, AT services are still markedly limited or unavailable. If PWD are to maximally benefit from these devices and services, these need to be widely available throughout the world. There is an urgent need to collaborate with agencies abroad working in this field for making suitable AT for developing countries. One possible mechanism is to transfer technology that has been adapted for the culture and environment so that it can be indigenously manufactured at a low cost. Additionally, upgrading AT to improve upon already existing aids and appliances would help in making appropriate assistive devices.

An effective network of governmental and nongovernmental agencies working in this field with healthcare providers and consumers is required to create awareness in the community on how to access the already available government or nongovernmental facilities in this regard (both with regard to acquiring the assistive devices as well as to availing the government policies for getting soft loans or other financing options for the same), with a goal to optimize self sufficiency and independence for PWD.

Capacity building in developing countries must be encouraged. Whereas advances in technology for people with spinal cord injury (SCI) have notably improved the lives of PWD in the industrial nations, millions of people have been left behind in developing countries. We have an obligation to try and provide assistance to individuals with disabilities in developing countries. For example, this can be done by providing training to clinicians to take back to their homelands, transferring technology, educating consumers, and working with governments and international organizations to address the needs of individuals with disabilities.

1.7 TELEREHABILITATION

The potential of modern telecommunications and computing technologies as tools in the delivery and evaluation of AT has been discussed and has been termed *telerehabilitation*. The problems of providing AT in rural areas parallel the delivery of healthcare to rural areas, where the proportion of people with chronic illnesses is higher, and the means to pay for them is reduced. Large distances mean long travel times, increasing costs associated with any service delivery, and consuming valuable time that skilled professionals could be using to provide services elsewhere. The technology available for practicing telerehabilitation is significant and expanding at a rapid rate. Currently, plain old telephone systems (POTSs) and broadband videoconferencing equipment, the Internet and the World Wide Web, and embedded processor systems are most widely available. These technologies continue to evolve as along with emerging technologies such as wearable sensors that will have telerehabilitation applications. Issues of payment, safety, liability, and licensure need to be resolved because legislation lags the development of new technologies.

Rehabilitation engineering requires the creation of a fit between people and the environments in which they live, work, play, and otherwise participate in their communities. Recent advances in technologies to acquire dimensionally accurate, three-dimensional (3D) models of physical environments have presented an opportunity to enhance our ability to adapt physical environments to meet the needs of everyone. For example, 3D models of the home environment incorporating mobility simulations to make determinations of accessibility may optimize the selection of technology and home modifications. Support is also feasible of the decision-making process concerning appropriate AT selection and environmental modifications for specific people in particular environments and for accessibility analysis of public spaces relative to population norms. The use of this modeling technology in conjunction with advances in telecommunications and the Internet allows for remote use of such accessibility analysis systems via asynchronous or interactive communication links with a central resource.

1.8 RESEARCH AND DEVELOPMENT FUNDING

Communication with the professions that will change bioengineering is going to be a prerequisite for survival of the AT industry and for rehabilitation engineering. Computing, flexible manufacturing, nanotechnology, and automated controls are developing rapidly. It is only a matter of time before a disruptive technology creates a wave through the field of AT and rehabilitation engineering and changes the manner in which research, development, and business are conducted. The Internet certainly did this for the 1990s, and there are several candidate technologies looming just over the horizon.

There is too little federal, state, and private funding for research, development, and deployment. Technology has had tremendous impact on the quality of life and societal participation of people with disabilities. For example, pressure ulcers are one of the leading causes of death among people with spinal cord injury in developing countries, whereas seating technology has reduced the incidence significantly and almost eliminated the mortality in the U.S. PWD have little influence over funding priorities, decisions, and research outcomes. To maximize their impact, PWD need to form partnerships with medical rehabilitation professionals. The most effective means of influencing policy is to present compelling stories along with supporting scientific data.

Benefits from universal design for mainstreaming of products will likely result from the seeds of rehabilitation engineering research. Some types of impairments experienced by PWD essentially represent niche markets, and many assistive devices of use to them are not covered by insurers. Fortunately, some of the devices are or have the potential to benefit broader sectors of society, hence reducing their cost and increasing their availability. Some obvious examples are grab bars, door handles, and automatic door openers. It is not too difficult to envision other assistive devices becoming more mainstream; for example, secondary controls (sound systems, temperature controls, etc.) for automobiles and computer access systems (hands-free interfaces, accessibility operating systems, etc.). As a simple case in point, several popular computer operating systems offer accessibility features that are widely used by individuals without physical or sensory impairments who use these features simply to set software feature preferences. By finding a place in mainstream society or by being embedded in mainstream products, technologies that can benefits individuals with disabilities will become more available and make society more accessible. Of course, some products are not appropriate candidates for mainstreaming and private insurance, and governmental resources need to be directed towards these.

There are models to consider for improving the research environment. The Paralyzed Veterans of America (PVA) Research Foundation makes grant awards for assistive device design and development, basic science, and clinical science. The PVA Research Foundation also partners with other organizations to award additional grants. Partnerships are one means of having greater consumer involvement in the decision making process. The PVA Research Foundation also uses a two-tier review process. The Scientific Advisory Board reviews grants for their scientific content and forwards their recommendations to the foundation trustees. The trustees are members of PVA as well as professionals. The trustees consider both the scientific reviews and the relevance of the work to people with spinal cord injury or dysfunction when making award decisions.

Regenerative therapy has the potential to help new skin grow, to improve wound healing, to enhance bone regrowth, and to regenerate organs and cells. This area will combine engineering, biology, and medicine. It is not beyond expectation that future devices will see greater interaction between biological tissues and engineering materials. These advances will likely result in the greater integration of devices to individuals with SCI. For example, synthetic scaffolds can be used to form an artificial bladder. When such scaffolds are combined with human cells that are stimulated to grow into a bladder, the scaffold can serve as a bladder until the cellular bladder fully differentiates within the scaffold and takes over. Eventually, the scaffold could be absorbed by the body. Similar approaches can be applied to other organs or tissues to restore lost functions. Although these capabilities are not clinically available yet, it appears evident that they are on the horizon.

Disability and medical rehabilitation research is a quality-of-life issue, a technology issue, and a medical care issue. This is a public health imperative. We need increased federal, corporate, and foundation funding, an ample pool of well-trained investigators, mentors for young investigators, and tools to measure the outcomes of our interventions.

1.9 THE REHABILITATION ENGINEERING AND ASSISTIVE TECHNOLOGY SOCIETY OF NORTH AMERICA

The Rehabilitation Engineering and Assistive Technology Society of North America (RESNA) grew out of a series of conferences in the 1970s that were supported by the Veterans Administration (VA), the Department of Health, Education, and Welfare, and the National Academy of Sciences. With the Rehabilitation Act of 1973, the HEW Rehabilitation Engineering Centers in addition to the first VA Rehabilitation Research and Development Centers were established. These activities generated the first RESNA conference in 1980 and then the formal founding of RESNA in 1981. RESNA's structure can be viewed from multiple perspectives, and the organization appears different depending on the angle perspective. Let us first begin with the management of the organization. RESNA has a president elected every 2 years by the membership. The president serves a 6-year term — 2 years as president elect, 2 years as president, and 2 years as past president. This allows RESNA to have experienced leaders and a smooth transition of power.

From another perspective, it is an organization that promotes advancements in public policy, research, and clinical practice related to AT and rehabilitation engineering. Public policy is embedded in RESNA through multiple mechanisms — credentialing of professionals (ATS, ATP, and RET), work on coding for products and services for AT, legislative activities, and regulatory activities. The Government Affairs Committee and several ad hoc committees assist the RESNA board and staff with policy activities. Research was the basis for the formation of RESNA, and it remains a critical component. RESNA supports research through several key activities: *AT Journal*, Student Design Competition, Student Scientific Competition, Research Symposia, and policy initiatives. The third side of RESNA is the clinical practice focus. Since RESNA's founding, clinical practice has been an important aspect of the organization. Programs for clinicians include Professional Specialty Groups, Credentialing Program (ATS, ATP, and RET), Clinical Track at Conference, *AT Journal Clinical Practice Section*, RESNA News, and various educational activities (e.g., Fundamentals Course).

Another view of RESNA is as a collection of programs in AT and rehabilitation engineering. The major programs are the Conference, Standards, Credentialing, AT Journal, Educational Courses, Government Affairs, and Member Benefits. The RESNA conference is held each June, and typically has about 800 attendees, 25 different companies exhibiting, and 150 to 175 papers or posters presented. RESNA is the recognized standards body for AT by the American National Standards Institute (ANSI). As part of this arrangement, RESNA also has representation at the International Standards Organization (ISO). RESNA has supported AT standards development for 25 years. Currently, there are over 30 AT standards or test methods managed by RESNA. Standards are in the areas of wheelchairs, seating, recreational parks, and devices for people who are visually impaired. Credentialing is one of those areas that has far-reaching implications. The process of establishing the ATS, ATP, and RET credentials was started about 15 years ago, with the first people receiving their credentials about 10 years ago. There are over 2500 ATPs, nearly 1000 ATSs, and close to 50 RETs. The credential has become a symbol of a qualified professional, and there are state and federal agencies looking to adopt these credentials as requirements for prescribing or supplying certain types of AT (e.g., electric-powered wheelchairs).

International collaboration is becoming increasingly important to make engineering advances. Communication has become more rapid, and it is reasonably simple to share data across oceans with adequate security and speed. The World Wide Web has facilitated much larger studies of assistive devices and of device usage. Bridges to other countries are important to increase sample size, to investigate the effects of service delivery and payment models, and to include more representative technologies. With greater worldwide communication, the limited resources targeted towards research and development for PWD can be used more effectively, and it becomes likely that we will be able to benefit from advances in other areas of engineering.

1.10 THE FUTURE OF REHABILITATION ENGINEERING

The application of advances in power electronics, telecommunications, controls, sensors, and instrumentation have really only just scratched the surface of devices to assist people with disabilities. Advancing technology for PWD represents a significant career and business opportunity for engineers who wish to serve the public in a meaningful and tangible way. The AT industry has experienced about 5% growth per annum for 20 years, and there are no signs that this will not continue for the foreseeable future.

We envision intelligent systems, ranging from individual devices to comprehensive environments, which will monitor and communicate with people and understand their needs and task goals. The systems compensate for or replace diminished capabilities, as required, while adapting to the changing situation, so that tasks are performed safely, reliably, and graciously. A person's level of function is complex, comprising multiple determinants that have effects at many levels and involve various dimensions. These systems will especially impact those with partial loss of perception, cognition, and fine and large motor skills. Examples may include a future wheelchair that functions as a smart motion-and-manipulation compensator. Knowing the abilities of the rider, it would provide appropriate types and degrees of physical, navigational, and cognitive assistance to augment the rider's own mobility and manipulation capabilities, rather than merely being a power-assisted vehicle. A recognition-and-remembering coach would learn and know the person's daily activities, family, and friends, log their experiences, and relate them to current situations so that it provides reminders for taking medications, helps to recognize people, and aids in communication with other people. An assisted-living environment operating in a skilled-care or nursing home would continuously monitor residents' activities and behaviors in order to provide information to staff and reassurance to family. A smart public transportation system, detecting that a fragile person is near a bus stop or waiting for transportation, would minimize the time for her to wait in cold weather. Systems need to work daily in unstructured dynamic environments. They must work naturally with people and not be overpowering or overwhelming but should rather enable people to do what they want to do whenever and wherever possible. Systems must be safe and reliable, and users must be able to trust that their privacy is protected and modesty respected.

Environmental control systems are of tremendous importance for people with high levels of physical impairment. Current electronic devices offer some glimpse of future possibilities. Nearly any device that is operated via an infrared or wireless remote control can be controlled through an environmental control unit. This has made it easy for PWD to operate lights, doors, and home entertainment systems. Tasks such as meal preparation have remained the domain of personal assistants. However, the kitchen of the future may provide more independence for PWD while making life easier for everyone else. Such devices as appliance networks, smart appliances, and personalization algorithms combined with robot assistants all appear promising.

Leap forward to today, and the landscape has changed dramatically. PWD live to near the expected societal life spans, and other factors besides impairment dominate the causes for mortality. Return to home and community reintegration are the goal of rehabilitation. The aims of AT have changed tremendously in the past 50 years. Most of the federal agencies supporting assistive and rehabilitative technology research and development did not exist at that time. Indeed, the market-leading AT companies have all been established in that time period. There are multiple models of wheelchairs available, and the concept of fitting the wheelchair to the needs of the individual with disability is accepted practice. There are entire clinics dedicated to the appropriate delivery of coordinated AT services, and there are credentials in AT gaining momentum across the nation and serving as examples for many countries around the world.

1.11 STUDY QUESTIONS

Describe the PHAATE model.
Describe the key aspects of the PAD process.
Provide three sample uses of telerehabilitation.
What are two benefits of specialized credentials for AT service delivery providers (e.g., ATS, ATP, and RET)?
Why is the inclusion of PWD in the design process useful?
Briefly describe two ways the public policy affects AT service delivery?
Briefly describe how the environment impacts the choices of AT for an individual?
What is RESNA, and why was it established?
What are the three reasons for building international collaborations?
Why is it important to evaluate AT in natural environments?

BIBLIOGRAPHY

Applewhite A. and Kumagai J., Technology Trends 2004, *IEEE Spectrum*, January, pp. 14–19, 2004.

Boone D.A., Prosthetics Outreach Foundation Programme Evaluation of Automated Fabrication of Limb Prostheses in Vietnam, *Report of ISPO Consensus Conference on Appropriate Prosthetic Technology for Developing Countries*. Phnom Penh, Cambodia 5–10 June, 1995. Edited by Day H.J.B., Hughes J., and Jacobs N.A. — Copenhagen: ISPO. pp. 95–105, 1996.

Braddom R.L., Medicare Funding for Inpatient Rehabilitation: How Did We Get to This Point and What Do We Do Now? *Archives of Physical Medicine and Rehabilitation*, Vol. 86, pp. 1287–1292, 2005.

Brienza D., Chung K., et al., System for the Analysis of Seat Support Surfaces Using Surface Shape Control and Simultaneous Measure of Applied Pressures, *IEEE Transactions on Rehabilitation Engineering*, Vol. 4, pp. 103–113, 1996.

Bureau of Industry and Security, *Technology Assessment of the U.S. Assistive Technology Industry*, U.S. Department of Commerce, Washington, D.C., February, 2003.

Canadian Standards Association (CSA), *D435-02 Accessible Transit Buses*. Canadian Standards Association: Vancouver, 2002.

Cavuoto J., Neural Engineering's Image Problem, *IEEE Spectrum*, Vol. 41, pp. 32–37, 2004.

Chavez E., Boninger M.L., Cooper R., Fitzgerald S.G., Gray D., and Cooper R.A., Application of a Participation System to Assess the Influence of Assistive Technology on the Lives of People with Spinal Cord Injury, *Archives of Physical Medicine and Rehabilitation*, Vol. 85, pp. 1854–1858, 2004.

CIGNA Government Services DMERC, Region D DMERC Local Coverage Determination — DRAFT, www.cignagovernmentservices.com, September 16, 2005.

Clarkson J., Coleman R., Keates S., and Lebbon C. (Eds.), *Inclusive Design: Design for the Whole Population*, Springer-Verlag, London, 2003.

Committee on Prospering in the Global Economy of the 21st Century: An Agenda for American Science and Technology, *Rising Above the Gathering Storm — Energizing and Employing America for a Brighter Economic Future*, The National Academy of Sciences, 2005.

Cooper R.A., *Rehabilitation Engineering Applied to Mobility and Manipulation*, Institute of Physics Publishing, London, 1995.

Cooper R.A., Boninger M.L., Brienza D.M., van Roosmalen L., Koontz A.M., LoPresti E., Spaeth D.M., Bertocci G.E., Guo S.F., Buning M.E., Schmeler M., Geyer M.J., Fitzgerald S.G., and Dan Ding D., Pittsburgh Wheelchair and Seating Biomechanics Research Program, *Journal of the Society of Biomechanisms*, Vol. 27, pp. 144–157, 2003.

Cummings D., Prosthetics in the Developing World: A Review of the Literature, *Prosthetics and Orthotics International*, Vol. 20, pp. 51–60, 1996.

Department of Health and Human Services, Medicare Program; Conditions for Payment of Power Mobility Devices, Including Power Wheelchairs and Power Operated Vehicles; Interim Final Rule, Part VII, *Federal Register*, Friday, August 26, 2005.

Francis J., New QUERI Dedicated to Polytrauma and Blast-Related Injuries, QUERI Quarterly, *Newsletter of the Quality Enhancement Research Initiative*, September, pp. 1–2, 2005.

Frontera W.R., Fuhrer M.J., Jette A.M., Chan L., Cooper R.A., Duncan P.W., Kemp J.D., Ottenbacher K.J., Peckham P.H., Roth E.J., and Tate D.G., Rehabilitation Medicine Summit — Building Research Capacity, *American Journal of Physical Medicine and Rehabilitation*, Vol. 84, pp. 913–917, 2005.

Goldman M.A., Promises and Perils of Technology's Future, *Science*, Vol. 303, p. 629, 2004.

Goldstein H., A Dog Named Spot: When Microsoft Stops Imitating and Starts Innovating, Watch Out, *IEEE Spectrum*, January, pp. 72–73, 2004.

Heim S., The Work of GTZ, *Prosthetics and Orthotics International*, 20, 39–41, 1996.

Staats T.B., The Rehabilitation of the Amputee in the Developing World: A Review of the Literature, *Prosthetics and Orthotics International*, Vol. 20, pp. 45–50, 1996.

Institute of Medicine of the National Academies, *Responsible Research: A Systems Approach to Protecting Research Participants*, The National Academies Press, Washington, D.C., 2003.

Jones L.E., Does Virtual Reality Have a Place in the Rehabilitation World? *Disability and Rehabilitation*, Vol. 20, pp. 102–103, 1998.

Kanade T., Narayanan P.J., and Rander P.W., Virtualized Reality: Concepts and Early Results. Presented at IEEE Workshop on the Presentation of Visual Scenes, Boston, June 24, 1995.

Kintish E., Panel Calls for More Science Funding to Preserve U.S. Prestige, *Science*, Vol. 310, pp. 423–424, 2005.

Kiley K.C., Preserving Soldiers' Lives and Health — Anytime, Anywhere, *ARMY*, Vol. 55, pp. 121–124, 2005.

Kussman M.J., New Veterans and New Challenges, *FORUM Translating Research into Quality Health Care for Veterans*, September, pp. 1–2, 2005.

Lewis C., FDA Begins Product Approval Initiative, *FDA Consumer*, May–June, pp. 10–11, 2003.

Lightman A., Sarewitz D., and Desser C., *Living with the Genie: Essays on Technology and the Quest for Human Mastery*, Island Press, Washington, D.C., 2003.

LoPresti E.F. and Brienza D.M., Adaptive Software for Head-Operated Computer Controls, *IEEE Transactions on Neural Systems and Rehabilitation Engineering*; TNRE-12, pp. 102–111, 2004.

Moses H., Thier S.O., and Matheson D.H.M., Why Have Academic Medical Centers Survived? *JAMA*, Vol. 293, pp. 1495–1500, 2005.

National Research Council of the National Academies, *Preparing for the Revolution: Information Technology and the Future of the Research University*, The National Academies Press, Washington, D.C., 2002.

Nummelin L. and Lfstedt T., Finnish Red Cross Mulitprosthesis System, *Report of ISPO Consensus Conference on Appropriate Prosthetic Technology for Developing Countries*. Phnom Penh, Cambodia, June 5–10, 1995. Edited by Day H.J.B., Hughes J., and Jacobs N.A. — Copenhagen: ISPO. pp. 160–163, 1996.

Popovic D., Sinkjaer T., *Control of Movement for the Physically Disabled*, Springer-Verlag, London, 2000.

Prior S.D. and Warner P.R., Wheelchair-Mounted Robots for the Home Environment, *IEEE/RSJ International Conference on Intelligent Robots and Systems*, pp. 1194–1200, 1993.

Rados C., FDA Works to Reduce Preventable Medical Device Injuries, FDA Consumer, pp. 29–33, July/August, 2003

Scherer M.J., Virtual Reality: Consumer Perspectives, *Disability and Rehabilitation*, Vol. 20, pp. 108–110, 1998.

Scherer M.J. (Ed.), *Assistive Technology: Matching Device and Consumer for Successful Rehabilitation*, American Psychological Association, Washington, D.C., 2001.

Snelson R., Wings of Calvary. *Report of ISPO Consensus Conference on Appropriate Prosthetic Technology for Developing Countries*. Phnom Penh, Cambodia, June 5–10, 1995. Edited by Day H.J.B., Hughes J., and Jacobs N.A. — Copenhagen: ISPO. pp. 106–109, 1996.

Strax T., Clark G., Claypool H., Gamble G.L., Grabois M., Henrichs R.A., Lollar D., Thomas P.W., and Williams B. (Eds.), *Access to Assistive Technologies: Improving Health and Well-Being for People with Disabilities*, American Academy of Physical Medicine and Rehabilitation and Foundation for Physical Medicine and Rehabilitation, Chicago, IL, 2003.

Stemmler D.J., Freedom: No Longer Medically Necessary, *Mouth Magazine*, November–December, pp. 14–15, 2003.

Voelker R., Ramping Up Rehabilitation Research Urged as a "Public Health Imperative," *JAMA*, Vol. 294, pp. 2413–2416, 2005.

Wang J., Brienza D.M., et al., Biomechanical Analysis of Buttock Soft Tissue Using Computer-Aided Seating System, *International Conference of the IEEE Engineering in Medicine and Biology Society*, pp. 2757–2759, 1998.

Wang J., Brienza D.M., et al., A Compound Sensor for Biomechanical Analyses of Buttock Soft Tissue *In Vivo, Journal of Rehabilitation Research and Devices*, Vol. 37, pp. 433–439, 2000.

Wardell C., Here Comes Robo-Shop, *Popular Science*, November, pp. 20–21, 2003.

Wardell C., The Souped-Up Kitchen, *Popular Science*, March, pp. 30–40, 2004.

Wolfe C.E., Women in Medicine: An Unceasing Journey, *Archives of Physical Medicine and Rehabilitation*, Vol. 86, pp. 1283–1286, 2005.

2 Clinical Practice of Rehabilitation Engineering

Carmen Digiovine, Douglas A. Hobson, and Rory A. Cooper

CONTENTS

2.1 LEARNING OBJECTIVES OF THIS CHAPTER

Upon completion of this chapter, the reader will be able to:

Define the role of the rehabilitation engineering in a clinical setting
Define the different models of service delivery
Define the role of the clinical rehabilitation engineer (CRE) in each service delivery model
Analyze the key components of the service delivery process
Compare and contrast the differences among the principle clinical funding entities
Analyze evidence-based practice

2.2 INTRODUCTION

Clinical rehabilitation engineering (RE) may be considered a subspecialty of biomedical engineering. The three types of biomedical engineer are (1) the clinical engineer in healthcare, (2) the biomedical design engineer in industry, and (3) the research scientist. Clinical rehabilitation engineering parallels this model, in which we have (1) the CRE in direct service delivery (e.g., rehabilitation, vocation, and education) settings, (2) the design and development engineer in the rehabilitation technology and assistive technology (AT) industries, and (3) the research CRE or scientist working in rehabilitation technology and AT research.

Although rehabilitation technology and AT are often defined similarly, they represent two distinct concepts. Assistive technology has been defined as follows (1) any item, piece of equipment, or product system whether acquired commercially or off the shelf, modified, or customized that is used to increase, maintain or improve functional capabilities of individuals with disabilities; (2) devices and services that are used in the daily lives of people in the community to enhance their ability to function independently, examples being specialized seating, wheelchairs, environmental control devices, workstation access and communication aids; and (3) a broad range of devices, services, strategies, and practices that are conceived and applied to ameliorate the problems faced by individuals who have disabilities. Conversely, rehabilitation technology has been defined as follows (1) tool for remediation or rehabilitation rather than being a part of the person's daily life and functional activities; (2) technologies associated with the acute-care rehabilitation process; and (3) the segment of AT that is designed specifically to rehabilitate an individual from his or her present set of limitations due to some disabling condition, permanent or otherwise. Therefore, when discussing the roles of CREs, it is important to realize that their role may incorporate rehabilitation technology only, AT only, or both. For the purposes of this chapter, we will focus on AT as defined by Cook and Hussey, because a CRE typically interacts with AT as opposed to rehabilitation technology.

The focus of this chapter is on the CRE, who may work in a variety of settings, primarily concerned with the direct provision of AT services for persons with varying physical disabilities. The service delivery models of clinical rehabilitation engineering have their origins in the durable-medical-equipment-supplier field and in prosthetics and orthotics. The practice of the CRE has now further diversified into numerous service delivery arenas, most notably education and vocation. Therefore, this chapter will focus on the CRE in different service delivery models. To provide the reader with a broad understanding of the field, the service delivery models and the role of the CRE in each model will be discussed.

Rehabilitation engineering has been defined as follows (1) the application of engineering principles, technical expertise, and design methodology in the development and provision of assistive technology to help a person with a disability achieve his or her goals; (2) a total approach to rehabilitation that combines medicine, engineering, and related sciences to improve the quality of life of persons with disabilities; (3) the application of engineering concepts and techniques to understand, define, and solve problems associated with improving the quality of life for persons having chronic disabilities; and (4) the branch of biomedical engineering that is concerned with the application of science and technology to improve the quality of life of individuals with disabilities. The CRE has been described as being able to bring an organized approach to patient problem solving, through problem definition, analysis, synthesis, and application of solution.

The key point to note that is very unique to the CRE, as opposed to other biomedical engineering disciplines or other engineering disciplines, is the intimate interaction with the older adult or the individual with a disability. This presents a new set of challenges and rewards. The issues include the fact that the CRE must be acutely aware of the psychosocial and impairment issues. The rewards include the immediacy of the outcomes. This is one of the few fields in which the CRE can be an integral part of the outcome, enjoying the fruits of their skills, expertise, and knowledge. This reward is typically seen only in the service delivery setting, as opposed to the research or academic settings.

In terms of functional area, the breadth and depth of the CRE's knowledge and skill will depend on the setting, mission, and funding sources. An example of deep expertise in a single area is a CRE working in a seating-and-mobility clinic that is part of a comprehensive rehabilitation program. A broad knowledge of AT may be required in a state-agency-based program. Finally, a hybrid, where a deep knowledge of one area (e.g., home modifications) and a broad knowledge in all areas of rehabilitation technology may be necessary for consultative purposes, is seen at a university center.

The roots of clinical rehabilitation engineering date back to the early 1960s in Canada, following the "thalidomide tragedy," which occurred in the late 1950s. The drug thalidomide was prescribed to pregnant women in Canada, Europe, and Australia to alleviate morning sickness. Unfortunately, the drug affected the newly developing fetus and resulted in children being born with missing limbs or only rudimentary limbs. This resulted in the establishment of three research centers, in Winnipeg, Toronto, and Montreal, in which engineers and prosthetists were teamed with clinicians in leading rehabilitation facilities in an effort to create technical or therapy solutions for such children and their families. Emphasis was placed on developing high-tech prosthetic limb solutions, as well as devices to assist with toileting, feeding, and wheeled mobility. This model quickly expanded to other countries and soon involved other populations of children and adults with disabilities needing custom AT, such as specialized wheelchairs and wheelchair seating, augmented and augmentative communication (AAC) aids, environmental control, and computer access. Today, AT is a viable worldwide industry in which many companies are producing a vast array of commercial AT devices, and the design and fabrication of custom devices is the exception rather the norm. Now, the challenge for the CRE is to become knowledgeable about the features and capabilities of the commercially available AT products, how to modify, fit, or tune them to meet individual needs as may be necessary, and finally, design and construct custom solutions when the first two options do not suffice. Working with other rehabilitation professionals such as physicians, physical, occupation, and speech-language therapists, assistive technology suppliers (ATSs), teachers, and rehabilitation counselors offers CREs a network of colleagues in which their contribution will be unique and valued.

2.3 SERVICE DELIVERY MODELS

The role of the CRE in the rehabilitation technology and AT service delivery system has changed significantly over the years. No matter what the role, or the service delivery model in which they work, CREs must be well-versed in the goals and terminology of the other members of the team to effectively and efficiently provide services. The CRE needs to effectively understand and communicate with clinicians (e.g., the occupational therapist (OT), physiatrist, physical therapist (PT), speech and language pathologist, etc.), the funding sources, the certified rehabilitation technology supplier (CRTS), the manufacturer, and most importantly, the consumer. The communication abilities of the CRE are central to providing a successful outcome. To better understand the role of the CRE as a member of a multidisciplinary team, various service delivery models will be classified and detailed. The models can be classified in numerous ways. Given the type of administrative home base, eleven service delivery models exist (Table 2.1).

2.3.1 Rehabilitation Technology Supplier

The role of the CRE working in a rehabilitation technology supplier (RTS) model is an emerging practice area. The RTS as defined in the National Registry of Rehabilitation Technology Suppliers (NRRTS) code of ethics as a "specialist who is currently working for a medical equipment company that supplies rehabilitation equipment" (http://www.nrrts.org/body_frame_ethics.htm, accessed 23 May 2005). The role of the CRE working as an RTS is that of a certified rehabilitation technology supplier (CRTS), as accredited by the National Registry of Rehabilitation Technology Suppliers (NRRTS), and rehabilitation engineering technologist (RET),

TABLE 2.1
Service Delivery Models

Service Delivery Models
Rehabilitation technology supplier
Department within a comprehensive rehabilitation program
Technology service delivery center in a university
State-agency-based program
Private clinical rehabilitation engineering/technology firm
Local affiliate of a national nonprofit disability organization
Volunteer groups
Manufacturers
Technology service delivery center in the educational setting
Specialized rehabilitation technology supplier
Community-based or satellite services

as accredited by the Rehabilitation Engineering and Assistive Technology Society of North America (RESNA). CREs would have a very hands-on role in terms of the service delivery process and their interaction with the consumer and other members of the interdisciplinary team. Given the majority of today's funding systems, the CRE would be unable to make recommendations. Therefore, the CRE would play a consultative role, assisting a clinician (e.g., PT, OT, SLP) and the consumer when making recommendations that will be funded by a third-party funding source. The real value of the CRE as an RTS lies in the engineering background that the CRE brings to the table, thereby improving the final outcome of the service delivery process. The CRE would have direct access to manufacturer information, product specifications, technical training, and fabrication equipment, which are not typically available in most clinical settings. This intimate interaction with the consumer and manufacturer, plus the engineering expertise, significantly aids in the appropriate design and application of custom equipment that must be fabricated locally.

2.3.2 Department within a Comprehensive Rehabilitation Center

Rehabilitation services are most often provided within large rehabilitation centers located within major medical centers or as a division of a national for-profit organization. Clinical rehabilitation engineering services are usually located within an existing department, such as OT or PT. The role of the CRE in a department within a comprehensive rehabilitation center is typically on a consultative basis as an expert in rehabilitation technology design and application, which will complement the services provided by the therapy clinicians (e.g., OT, PT, SLP) and the RTS. When commercially available equipment, as supplied by the manufacturer, do not meet the needs of the individual, the CRE will provide the expertise required to modify the equipment or create customized equipment.

In this setting, it may be customary to include the CRE, even on a purely consultative role, in every case. This exemplifies the importance of the interdisciplinary team, with each member typically having his or her own way of looking at a situation. For example, the PT or OT will typically focus on the postural needs of the individual, the ATS on acquiring the equipment to meet the postural requirements of the individual as defined by the therapist, and finally, the CRE focuses on the effect of the selected components on the functionality and safety of the individual. Another example is a child who has grown to the point that the seating system on his wheelchair no longer fits. During a follow-up appointment, the therapist recommends increasing the seat depth of the wheelchair (given that this adjustment is possible). The ATS determines that the mobility base has a growth adjustment feature built into the wheelchair. The CRE can then verify that this adjustment will not make the mobility base unsafe and will not become unstable in the rearward direction as the clients' center of gravity would have shifted rearward. Using a few low-tech tools such as a scale and an angle finder, the CRE can directly calculate the change in stability based on shifting the back support rearward in order to increase the overall seat depth. This quantitative assessment can be completed prior to any final adjustments, allowing the family, the therapist, and the ATS to review the quantitative information in order to make an informed decision. Further, if the family has information on the angle of the typical ramps the child traverses, an even better estimation of the effect can be obtained.

Finally, the CRE has the technical training to quantitatively describe the effects of various pieces of equipment or modifications to the equipment. When two choices are equally feasible in terms of postural support, functionality and aesthetics, quantitative information may be the critical piece required to select one item over another.

2.3.3 Technology Service Delivery Center within a University Department

Typically, a university has three goals; teaching, service, and research, thereby providing the most diverse roles to the CRE. In the other models, the CRE will focus almost exclusively on service delivery.

Depending on the setting, the CRE may provide services on an as-needed basis to the different functional areas (e.g., AAC, home modifications, and computer access), or may become a specialist in a specific area of the service delivery process (described in Section 2.9) from the referral and intake processes to the follow-along processes. For example, the CRE may only be included when a custom mount and wiring is required to interface an augmentative communication device to a power wheelchair. In this case, the CRE will be working with the SLP and the PT to meet the specific needs of the consumer. In the second case, the CRE, as a specialist, will be involved in all aspects of the service delivery process for every client seen through a specific functional area (e.g., seating and mobility). Of course, the third option is a hybrid of the first two, in which the CRE is primarily responsible for a specific functional area, but is still available for consultative purposes on a case-by-case basis in other areas.

The unique opportunities that are available to the CRE when working in a service delivery center in a university setting are the opportunity to teach, typically at the graduate level to degree-seeking as well as continuing-education students, and the opportunity to be involved in various research projects. As a teacher, the CRE must remain informed on the current and future practices as well as current and future equipment. This requires constant review of the literature, technical training via manufacturer in-services, and attendance at national and/or international conferences. A knowledgeable CRE with direct links to consumer interaction and services is a highly valued asset to any research team which is developing or implementing a project that examines the role of AT in the lives of persons with disabilities. The CRE can often be the person who bridges the gap between the research setting and the clinical environment by clarifying and communicating the consumer's technical needs as well as by providing feedback on the outcomes of the evaluation of prototype developments in the home or community environments.

2.3.4 State-Agency-Based Program

The CRE in a state-agency-based program, such as vocational rehabilitation or developmental disabilities, tends to focus on the population that these services are provided to, primarily due to the funding restrictions imposed by state legislation. Depending on the location of the CREs and the ability of the consumer to come to a center-based location, CREs may be working by themselves as opposed to being the member of a team. This scenario is beginning to change as the ability to access other members of an interdisciplinary team becomes available via audio, video, and data conferencing. Traditionally, the CRE is involved only in areas of AT (e.g., those with computer access assessment) that are not already served by other professionals (e.g., OT or PT performing a seating and mobility assessment). That is, the CRE is most likely to be asked to solve the most difficult problems where prior attempts have not been successful. Therefore, the CRE needs to be to be proficient in overall AT knowledge and skills.

2.3.5 Private Rehabilitation Engineering or Technology Firm

The CRE, in this model, is not only concerned with the delivery of services but is also acutely aware of the business model and entrepreneurialism required to successfully function as a private entity. The firm may or may not include clinical staff. For any given project, the CRE may take on one of the roles described in the other models. In this case, CREs contract their services to meet a temporary need of the other service delivery model. For example, the CRE may contract with an RTS to fulfill a short-term need for CRE services. This may turn into a permanent position, where the independent CRE working as a consultant is hired as a full-time employee. In some cases, a private firm must have a licensed professional engineer on the staff when providing engineering services.

2.3.6 Local Affiliate of a National Nonprofit Disability Organization

Assistive technologies may be provided by programs located within national nonprofit advocacy organizations, such as United Cerebral Palsy or Independent Living Center.

The role of CRE in this model will greatly depend on the funding, population, and functional areas as defined in the mission statement. The CRE may take on the role of the ATS, depending on the availability of vendors to provide equipment in a specific service area. These organizations may incorporate large pools of loaned and/or demo equipment so that individuals may try equipment for extended periods of time prior to purchasing them. It often requires extensive modifications and/or repairs to customize each piece of equipment for the individual. Therefore, CREs may have to use their skills and knowledge to modify used and donated equipment to meet the needs of the consumer. In addition, the CRE will also be involved in the service delivery process. Depending on the population, the type of equipment, and the location, the CRE may be a single expert providing all services or the member of an interdisciplinary team. Once again, similar to the state-agency-based program, this is one group that will significantly benefit from the tele-presence of an interdisciplinary team as this communication resource becomes increasingly available in the future.

2.3.7 Volunteer Groups

The volunteer group model takes advantage of the vast engineering talent that usually resides within the industry and applies it directly to the resolution of technical problems for persons with disabilities. The difficulty is that, in most cases, volunteers have little knowledge of the AT industry and little appreciation of the process required to successfully select, provide, and maintain advanced technical equipment for this population. Therefore, the CRE is the ideal person to provide the interface between the two environments, being able to communicate in both engineering terms as well as having a good understanding of what is required to produce a successful clinical outcome.

Volunteer organizations are not as dependent on third-party reimbursement for custom equipment as the opportunity to develop custom equipment over a typically longer timeframe exists. The volunteer engineer must be cautioned against designing, developing, or implementing rehabilitation technology without fully understanding the amount of support that must be provided after deployment of the equipment. This level of commitment must be determined and conveyed to the consumer and volunteer group very early in the design process.

2.3.8 Manufacturer

The role of the CRE in a manufacturing firm is typically not included in service delivery models. In this setting, the engineer is involved in the design, fabrication, manufacturing, and troubleshooting of equipment in the manufacturing plant. However, the CRE employed by a manufacturer can play significant roles in the service delivery process. One role is as a field representative who works directly with consumers, clinicians, other CREs and ATS. Technical expertise and problem-solving capabilities are the primary assets when (1) providing solutions in the field that were not included in the original design of the equipment, (2) troubleshooting equipment while minimizing the downtime of the equipment, and (3) providing critical feedback to the design engineers, technical staff, and management on successes, failures, and future recommendations of the equipment. Furthermore, when field-testing prototype equipment, it is imperative that field representatives, with established relationships with consumers, clinicians, other CREs, and RTSs, are present. This will maximize the feedback obtained during field-testing and the future success of the AT.

A second role of the CRE, as the representative of a manufacturer, is as an educational advisor and trainer. It is critical that the CRE be able to synthesize the importance of functional goals with equipment features and relay that information to consumers, clinicians, technicians, and CREs. Once again, the ability of a CRE to "speak" technical, clinical, and consumer "languages" can be critical to a successful outcome, especially when recommending and implementing today's equipment as well as when introducing new equipment to the marketplace.

2.3.9 Technology Service Delivery Center in the School Setting

Assistive technology plays a very significant role in the education of children with disabilities, especially those with severe developmental disabilities. The technology ranges from low-tech adaptations for pens and pencils, making it easier to write, to high-tech applications such as computer access and power wheelchairs. Although the educational system may not fund all the technologies, the AT team is typically involved in the service delivery process as they must provide support for the equipment on a day-to-day basis. Traditionally, the AT team in an educational setting has not included the CRE. AT services are currently provided by the clinician with specialization in a particular area (e.g., ACC devices — SLP, mobility — PT, and computer access — OT). Given the clinician's limited amount of time in a given week, direct contact with the student is reduced so as to allow the clinician time to perform the research necessary to make appropriate equipment and training recommendations. However, a shift from this model to a model that is more similar to the university-based program described previously, where the CRE becomes an integral part of the AT team, would allow clinicians to maximize their direct contact time with the students. The one-on-one time would be increased by shifting the task of equipment research or development of custom equipment to the CRE. Furthermore, the incorporation of a CRE into the educational process would produce the value-added service of research and development of custom-fabricated AT when commercial equipment does not meet the educational needs of the student.

Another role of the CRE would be that of a consultant when integrating educational AT with that obtained through a health insurance funding source. Although the educational setting may not have high-tech fabrication equipment (e.g., drill press, band saw, and milling machine), a standard set of hand tools would increase the effectiveness of the AT and allow for on-the-fly adjustments to the system. The skills and knowledge of a CRE would be invaluable here. The school system has a significant advantage over the university system in that the AT team has the opportunity for long-term training of the student on consistent intervals and for observing the effects of minor adjustments on the abilities of the student. The inclusion of the CRE allows the teachers and clinicians (OT, PT, and SLP) to focus on educational objectives while the CRE addresses the technical issues and provides continuing education on AT. Continuing education programs related to AT are currently lacking for clinical staff, especially during days set aside for teacher in-services. Finally, the CRE would not only be available to address the concerns of individual students who are receiving AT services, but for addressing issues of universal design and accessibility of the entire school environment, as required by ADA, thus benefiting all students.

2.3.10 Specialized Rehabilitation Technology Supplier Environment

Assistive technology is shifting away from the realm of durable medical equipment to highly specialized equipment that requires extensive fittings and training in order to ensure its safe and effective use. A CRE working with traditional clinical staff (e.g., PT, OT, or SLP) will be able to provide these advanced technical services, thereby increasing the likelihood of a positive outcome. This new breed of specialized RTS will spawn another area in which the skills and knowledge of a CRE will add value to the rehabilitation process and ensure improved outcomes.

2.3.11 Community-Based or Satellite Services

Typically, the assistive services location is center based, and the consumers travel to the experts. However, individuals with disabilities may have difficulty traveling, and even if they can travel, they may have difficulty finding appropriate transportation. In this case, a satellite location, where the individual only has to travel a few miles as opposed to hundreds, or a community-based model, where the experts travel to the individual, may be more appropriate. The advantage of the satellite is that more caregivers may be involved in the evaluation, implementation, and training process. The obvious disadvantage is that the AT team must travel to the satellite location (loss of time) and bring all the demo equipment to a decentralized location. The community-based model is ideal because it typically includes the majority of stakeholders and allows the rehabilitation technology team to inspect the environment in which the individual will use AT as opposed to relying on the consumer's history. The disadvantages are the same as those of the satellite location, with the added disadvantage of seeing a much fewer number of clients. However, this must be weighed against the ability of the individual to receive services any other way.

In a center-based model, the CRE will typically rely on the RTS to specify the features on a standard piece of equipment, whereas in a satellite- or community-based model, the RTS may not be available to attend the assessment; therefore, the CRE must take on the added role of specifying each and every detail of both the custom equipment (typical) and the standard equipment (atypical).

2.4 TOOLS

The tools of the trade will vary based on the application and resources of the CRE, both in terms of time and money. For the purposes of this text, the tools will be divided into two groups: measurement/assessment tools, which we will simply refer to as measurement tools, and fabrication/adjustment/maintenance tools, which we will refer to as fabrication tools. Measurement tools can be further divided into low-tech, medium-tech, and high-tech equipment. *Low-tech tools* are defined as being low-powered, low-cost, portable, and of relatively low precision. Examples of low-tech tools are provided in Table 2.2.

Medium-tech tools are defined as being digital in nature, not requiring an external power source (e.g., outlet), of low-to-medium cost, portable, and of medium-to-high precision. Medium-tech tools, similar to low-tech tools, are commercially available and are not specific to the field of rehabilitation engineering. Examples of medium-tech tools are provided in Table 2.3.

High-tech tools are defined as being digital in nature, requiring an external power source (e.g., outlet), of medium-to-high cost, transportable to stationary, and of medium-to-high precision. High-tech tools are commercially available, but may require the CRE to assemble components from other fields, may require a computer to run, and may be specific to the field of rehabilitation engineering. Examples of high-tech tools are provided in Table 2.4

The tools listed in Tables 2.2 to 2.4 are not exhaustive and provide general guidelines as there will be some overlap among categories. The fabrication tools, on the other hand, consist primarily of standard hand tools that may be found in most home improvement/hardware stores. Overlap exists between the measurement tools and fabrication tools, especially for the low-tech measurement tools. The vast majority of fabrication, adjustment, and maintenance that a CRE will encounter in a service delivery setting can be accomplished using standard hand tools (e.g., screw drivers, wrenches, hex key, drills, and saws), irrespective of whether or not they are powered tools (battery or electrical line). A complete machine shop, including stationary power tools (e.g., drill press, lathe, table saw, band saw, sander, milling machine, and routing machine), is ideal but not required in all situations.

TABLE 2.2
Low-Tech Tools

Tool	Purpose	Example
Tape measure	*Measures linear distances*: seat-to-floor height on a wheelchair, doorway widths	Stanley steel tape measure
Analog angle finder	*Measures angles and slopes*: seat angle on a wheelchair and ramp slope	Craftsman magnetic angle finder
Goniometer	*Measures anatomical angles*: joint angles for active/passive range of motion	Prestige medical goniometer
Stopwatch	*Time*: time to complete an activity	Robic stopwatch
Bathroom scale	*Weight*: verify maximum weight capacity of a lift or ramp	Health-o-meter dial scale
Fish scale	*Force*: verify force to open/close door, force to lift an object	Mustad mechanical scale

TABLE 2.3
Medium-Tech Tools

Tool	Purpose	Example
Digital micrometer	*Linear distance*: tubing diameter (inner/outer)	Craftsman 6 in. electronic digital caliper
Ultrasound rangefinder	*Linear distance*: medium distances (<50 ft) with an end point (e.g., wall)	Craftsman laser-guided measuring tool
Laser range finder	*Linear distance*: long distances (>500 ft) with an end point (e.g., wall)	Leica DISTO laser instrument
Digital inclinometer	*Angle and slope*: seat angle on office chair and ramp slope	SmartTool electronic digital level
Digital force gauge	*Weight*: manual wheelchair *Force*: required to open door	Imada push/pull digital force gauge
Camera (analog/Polaroid/digital)	Documentation	Canon Sure Shot 35-mm zoom camera/Polaroid ONE 600 Pro Instant camera/Kodak Easy Share zoom digital camera

TABLE 2.4
High-Tech Tools

Tool	Purpose	Example
Wheelchair scale	Weight: wheelchair and occupant	Health-o-meter
Motion analysis system	Gait training, manual wheelchair propulsion training	Optotrak
Force plate system	Gait training	Kistler
Force measurement system for manual wheelchair propulsion	Wheelchair comparison, justification, funding, training	Three rivers
Datalogger for frequency, distance, time and duration	Mobility (lower limb prosthetics, wheelchairs); accessibility (reliability of lifts); AAC device	May be included in device software (e.g., AAC device and power wheelchair)
Pressure mapping system	Seating (wheelchair, office), prosthetics, orthotics, gait	Xsensor/FSA

2.5 THE REHABILITATION ENGINEER IN THE CLINICAL SETTING

Depending on the situation, as indicated in the descriptions of multiple service delivery models, the CRE can be a critical member in the service delivery process. As is the case for any piece of equipment, whether or not it is specifically designed for individuals with a disability or for the general population, the equipment will only meet the needs of a specific segment of the disability population. Even when incorporating the concepts of universal design, the unique nature of a physical impairment and disability as well as the context (environment) in which a piece of AT will be used requires the knowledge, skills, and expertise of an engineer trained in the field of rehabilitation engineering. One of the critical skills of a CRE, as opposed to a biomedical engineer who is designing an artificial organ, for example, is the ability to account for the psychosocial aspects of using AT. In particular, the CRE has a keen awareness of the aesthetics and outward appearance of the device as well as the its functionality. Therefore, with a number of professionals and the consumer already a part of the multidisciplinary team, why include a CRE in the clinical setting? The answer is twofold (1) the technical expertise that is required to customize a piece of commercially available equipment or to fabricate a custom piece of equipment that is not available from other disciplines and (2) the ability to account for the human, activity, and context in which the piece of equipment will be used. The inclusion of a CRE increases the likelihood of long-term success.

For the CRE to be an effective member of the service delivery process, and to increase the likelihood of long-term success, the CRE must have exceptional communication skills, both written and oral. Traditionally, communication has not been the strong-suit of many engineers; therefore, it is an area that the novice CRE must work on. The CRE must be able to communicate in layman's terms, in clinical terms, and in technical terms, depending on the audience. This includes both written and oral communication. An increasingly important asset is the ability to communicate in another spoken language, using sign language, or using Braille. Finally, and oftentimes most importantly, the CRE must be patient and have excellent listening skills as the consumer's rate of communication and clarity may not be within a range that the CRE is typically used to.

Although there is definitely a place for the CRE in the clinical setting, and the inclusion of the CRE will significantly increase the likelihood of an individual with a disability being able to perform a specific activity, there are significant differences between clinicians and CREs that must be acknowledged in order to achieve successful outcomes. The major difference is in the mindset that engineers can work in 20-min intervals, providing quick results. This has proved to be a strain not only on the relationship between engineers and clinical staff, but also on the CRE's ability to receive appropriate funding for engineering services.

The time commitment associated with the problem-solving process, in which the engineer is particularly well trained, is based on the size and scope of the project. Therefore, a relatively simple problem that has been previously solved may only take 10 min, while more difficult issues, which require numerous iterations of analysis, synthesis, and evaluation, may require much more time (hours, days, months, or even years). The sooner the clinical team members and funding sources realize the value of a project-based problem-solving methodology as opposed to a time-based scenario, the greater the perceived value.

2.6 THE CLINICAL REHABILITATION ENGINEER VS. THE CLINICAL ENGINEER

The field of clinical engineering in response to concerns about patient safety as well as about the rapid proliferation of clinical equipment was developing at roughly the same time as the field of CRE. Based on the definition of a clinical engineer provided by Bronzino, it does appear that clinical rehabilitation engineering and clinical engineering could be related. Potvin, Mercadante, and Cook indicate that the two may even be involved in similar activities such as "equipment design, procurement, evaluation, maintenance, use, calibration, and safety." The main difference, and this is critical, is that a clinical engineer deals only with the medical equipment that can be found in a healthcare setting (e.g., acute-care hospitals, sub-acute-care hospitals, rehabilitation hospitals, or out-patient clinics), with the patient as the focus of the medical model. Conversely, a CRE deals only with rehabilitation technology, with the individual as the focus, in any number of models and settings. Therefore, even though the histories of clinical engineering and clinical rehabilitation engineering have paralleled each other, and the term "clinical" is used to define the discipline of clinical engineering, a clinical engineer is very different from a CRE who works in a clinical setting.

2.7 INCORPORATING ENGINEERING INTO THE AT SERVICE DELIVERY PROCESS

The service delivery process, which never really ends once it has begun, is similar across service delivery models and functional areas. However, the role of the CRE is different from that of other clinicians (e.g., MD, OT, PT, and SLP), educators, counselors, and consumer. As discussed previously, whereas the service delivery process for the other team members has traditionally been defined by the amount of time of one-on-one interactions (i.e., 15- to 20-min intervals for billing purposes), the service delivery process for the engineer has been project based, which lends itself to the engineer's problem-solving skills. This difference is just one example of the challenges that has been presented to the multidisciplinary team. However, the practice of CRE is expanding to include more direct service delivery and also to include the individual tuning of complex systems. With the ever-growing usage of microprocessors in AT, features are being added to allow the software to be tuned for each client. Some of the more complex programming features require a CRE to be manipulated properly.

2.8 PRINCIPLES OF SERVICE DELIVERY

For the purposes of this chapter, the service delivery process described by Cook and Hussey and the principles of AT assessment described by Szeto will be used to develop the framework for incorporating rehabilitation engineering into the AT service delivery process. Both Cook and Hussey and Szeto describe the principles of the AT assessment and intervention processes. Both incorporate very similar concepts; however, for the purposes of this chapter, we will focus on the principles described by Cook and Hussey as they directly incorporate the Human Activity Assistive Technology (HAAT) model. These principles are repeated as follows:

1. AT assessment and intervention should consider all components of the HAAT model: the human, the activity, the AT, and the context components.
2. The purpose of AT intervention is not to rehabilitate an individual or remediate impairment, but to provide assistive technologies that enable an individual to perform functional activities.
3. AT assessment is ongoing and deliberate.
4. AT assessment and intervention require collaboration and a consumer-centered approach.
5. AT assessment and intervention require an understanding of how to gather and interpret data.

Examining each of these principles individually will provide insight into the role of the CRE in the service delivery process. The first principle addresses the importance of each component of the HAAT model to the overall success of the assessment and intervention processes. All too often, the focus of the CRE is on AT, especially if it is a technologically sophisticated device and/or is new to the market. This principle reminds the CRE that they are making recommendations based on the individual first, including the activity that will be performed and the context in which the device will function. The CRE is not looking for a person to fit the "cool" device; rather, they are making technical recommendations that fit the needs of the individual.

The second principle focuses on the need to enable an individual to perform a functional activity. Oftentimes, CREs have the point-of-view that they need to fix something. This is not necessarily the case. Rehabilitation engineering is about problem solving, specifically, enabling an individual to perform a specific task. This requires the CRE to examine the problem from many viewpoints, as there is usually more than one way to solve the problem. This is very similar to other engineering disciplines where the goal is to design a better method to perform a specific task, rather than just "fixing" the problem.

The third principle focuses on the fact that the assessment is ongoing and deliberate. It is important for the CRE to remember that the assessment process is ongoing. Previously, the interdisciplinary team could only request a CRE to join the team when an issue arose. The shortcoming with this scenario was that the CRE only saw a small component of the service delivery process. The CRE should be involved in, or at least informed of, all aspects of the assessment process to properly define the issue, the characteristics of the ideal solution, and possible solutions, given the activities that the individual wants to perform and the context in which those activities will occur. From a CRE point-of-view, one often thinks that once the issue has been addressed and the individual is able to perform a specific task, the job is done. This is not the case. Due to the often-intimate connection between the AT and the individual, even if the AT is not custom or customized, there is a constant need for the CRE to reassess its effectiveness.

The fourth principle focuses on the collaboration among peers, which specifically includes the consumer. From a traditional engineering point of view, it may be easy to define your client as the clinicians; however, the clinicians are really your colleagues in the interdisciplinary framework, whereas the consumer is the client. The consumer must always be included as a collaborator. It is rather simple to rely on the clinicians who are working with the consumer to provide the engineer with the issue at hand, the specifications of what is required, and the desired end result. The engineer can then solve the problem and provide a recommendation to the clinician without ever interacting with the consumer. However, this scenario will rarely lead to a successful outcome. The key to successful implementation, training, and continued use of AT is the inclusion of the consumer and other stakeholders in all aspects of the assessment process. Often, there are specific questions, measurements, and visual cues that can only be addressed by a one-on-one interaction with the consumer. Furthermore, the focus on consumers should not be limited to the consumers alone, but to their activities and performance. There is a maximum of information that can be transmitted through a clinician, via verbal or written notes, photographs, and videos. Oftentimes, the success of the recommendations depends on the CRE's ability to interact with the consumer, ask the consumer questions, listen to the consumer's answers as well as his or her wants and needs, and record appropriate performance measures. The importance of this principle of focusing on and interacting with the consumer has been brought to light by the limited success, at least to this point in time, of computer-based recommendation algorithms (e.g., recommendation is based on answering the yes/no questions found in hierarchal answer tree) and telerehabilitation. These methodologies are limited by the amount of information that can be recorded and transmitted.

The fifth and final principle focuses on the ability of the CRE to gather and interpret data. Traditionally, a CRE is well trained for this endeavor. The CRE's analytical and quantitative training allows him or her to determine what can be measured, what should be measured, and what tools are necessary when measuring the parameters. Furthermore, they typically have the skills necessary to analyze the information. The methods used to gather and interpret outcome data fall under the heading of AT outcome measures. Outcome measures can be used to develop group characteristics; however, the CRE in the service delivery setting is typically concerned with outcome measures that are designed to assess the abilities of an individual and how those abilities change or remain the same with the intervention of AT services.

2.9 SERVICE DELIVERY PROCESS

An overview of service delivery in AT was provided by Cook and Hussey and has been summarized in the following, with some revisions to emphasize the incorporation of the engineering contribution. This will provide the generic foundation for the service delivery process, independent of the functional area (i.e., the type of equipment).

Cook and Hussey's diagram depicting the service delivery process has been revised to address both the service delivery process and the role of the CRE (Figure 2.1). The diagram incorporates the majority of components without providing excessive detail. The detail is left out because the service delivery process is generalizable to any AT equipment and any professional. Conversely, Szeto does not cover the entire service delivery process, but nicely provides much more detail on the role of the

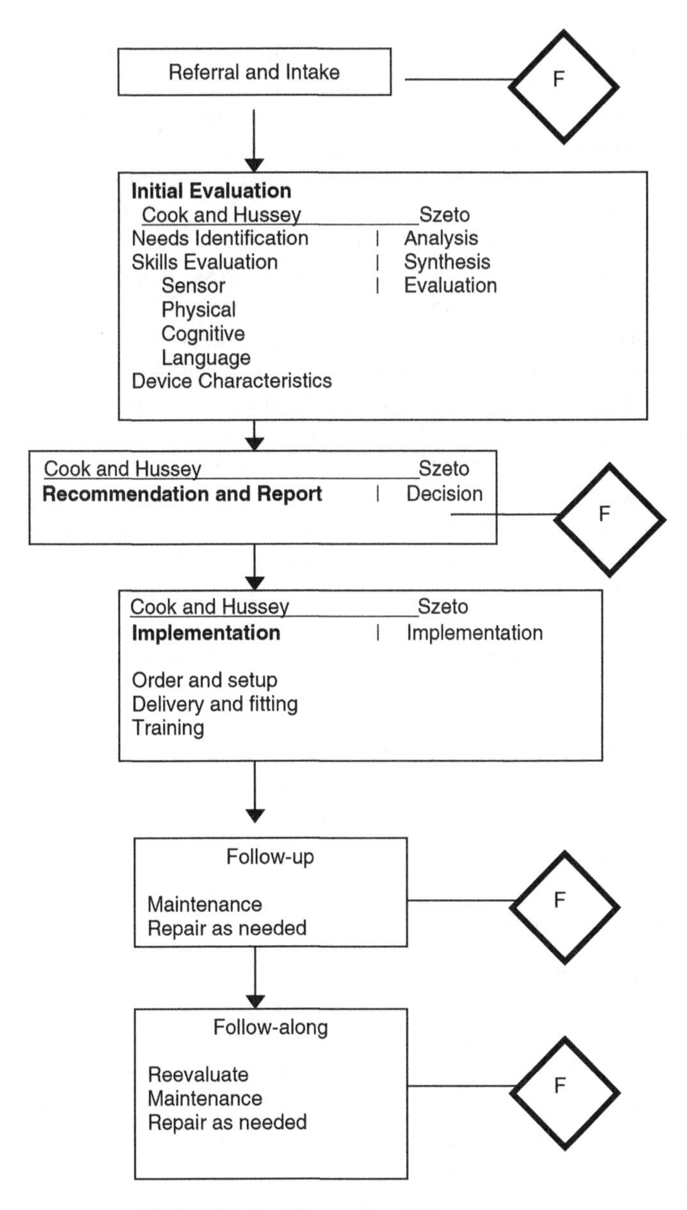

FIGURE 2.1 The service delivery process.

CRE. The purpose of this section is to integrate the concepts described by both Cook and Hussey and Szeto, while filling in any gaps related to the engineering contribution.

2.9.1 Referral and Intake

Cook and Hussey begin the service delivery process with the referral and intake. Szeto does not address this component, which makes sense, as this does not traditionally include the engineer. However, there is a great opportunity to increase referral rates by including the CRE. This is where the communication skills of CREs are critical, as they will have to be able to clearly and succinctly articulate the services they can provide and how these services are better than another organization that does not have access to a CRE. By articulating the services to referral entities (e.g., individuals, agencies, and clinics), CREs promote themselves along with the organization they work for, thereby increasing referral rates.

The CRE should be involved in the intake process, if not directly on a case-by-case basis, then indirectly through the development of the intake questionnaire. The intake questionnaire is critical to streamlining the service delivery process as this will provide the information necessary to begin the analysis process. Essentially, a well-worded intake questionnaire guides the interdisciplinary team into the "ball park," making it possible to have the appropriate demonstration equipment and tools available. This forces the rehabilitation professionals as well as the consumer to focus on the real issues at hand, and should limit the possibility of the consumer and/or rehabilitation professional from going off on a tangent. This is especially important for the CRE who has the tendency to potentially overanalyze the issues and/or recommend the most technologically advanced equipment.

The other critical component that the Cook and Hussey model addresses during the referral and intake process, as well as other points throughout the service delivery process, is the funding. This directly impacts the CRE's ability to provide services, as will be noted in following sections. The CRE needs to ensure that adequate reimbursement will be received for the services provided. To guarantee that reimbursement for services will be received, the funding issue must be addressed early in the process. This will alleviate any confusion and potentially awkward situations later on in the service delivery process.

2.9.2 Initial Evaluation

The initial evaluation closely parallels the analysis, synthesis, and evaluation components of the principles of rehabilitation engineering, as described by Szeto. The analysis process begins during the intake process and continues in full force in the initial evaluation. The CRE begins to describe, in quantitative terms, the current situation. The analysis process is the most important step in the process and, oftentimes, an overlooked step, especially by newcomers to the field of rehabilitation engineering. Possible solutions to the issue cannot be generated without a well-defined problem with well-defined operational features and performance specifications.

Using the HAAT model as a frame of reference, the CRE can begin to describe the individual (e.g., height, weight, reach), the activity (e.g., how fast, how far, how often, with what accuracy and precision), the AT (e.g., dimensions, weight, durability, failures), and the context (e.g., where, when, quantitative information about environment). The key difference and the primary strength of the CRE, at this point, is that the CRE will be defining each component with quantitative measurements when possible. This information is critical in making informed comparisons among the multiple options, especially when justifying the selection to a funding source. The consumer will make the final selection, and the quantitative information will aid in the decision process. Szeto lists five questions that the CRE could use to begin to define the issues in the analysis process. These questions are as follows:

1. What is the environment or task situation?
2. How have others performed the task?
3. What are the environmental constraints (cost, size, speed, weight, location, physical interface, etc.)?
4. What are the psychosocial constraints (user preferences, support of others, gadget tolerance, cognitive abilities, and limitations)?
5. What are the financial considerations?

Since the fourth question is typically new to the engineer, the CRE should deliberately focus on these psychosocial issues, particularly the user's preferences. An endpoint of the analysis process would be a list of the operational features or performance specifications of the "ideal" solution. Many of these ideal features will come directly from the consumer. Once again a key requirement of a successful CRE will be communication skills, primarily listening skills, when recording the user's preferences. The communication skills will also be important when describing in real-world terms the consequences and/or compromises, based on engineering principles, of selecting a given piece of equipment. This will lead the CRE to the next component of the service delivery process, the synthesis process.

The synthesis process requires skills that are not always readily available based on a traditional engineering curriculum. Whereas, the strength of most engineers and CREs, in this case, is applying mathematical principles to a specific process, the synthesis process requires the CRE to describe the issues in writing. Therefore, the synthesis process requires both mathematical and writing skills to adequately define the issue and begin to provide solutions for the issue. As potential solutions are defined during the synthesis process, the analysis process may need to be revisited, specifically to acquire more quantitative information. Multiple solutions will become apparent, each with its set of pros and cons. Typically, none of the solutions will meet all of the requirements of the "ideal" solution. The synthesis process should end with sketches and technical description of each solution. This may include specifications for commercially available equipment, stating how the equipment will interact with each other; customized equipment, stating how the commercially available equipment should be modified and/or customized to meet the individual's specific needs; or custom-fabricated equipment, providing specific dimensions and performance specifications for the equipment. At this point, the CREs have the best understanding of the technical requirements, the ideal solution, and the possible solutions. Therefore, they are best prepared to describe the pros and cons, or the compromises that will have to be made to arrive at a successful outcome for the consumer.

The evaluation process, which follows the analysis and synthesis processes, usually requires taking the two or three best solutions and evaluating these solutions via field trials, computer simulation, and/or detailed drawings. The degree of evaluation will depend on the complexity of the problem as well as the time and funding constraints. In the service delivery model, the most common evaluation is the field trial where a mock-up and/or simulation of the possible solution is directly evaluated by the consumer. All stakeholders, particularly the consumer, should be involved in this part of the process. Once again, the principles of the HAAT model should be revisited as was the case in the analysis section. The role of the CRE is critical at this point in order to gather the quantitative data that will be necessary to compare the possible solutions to the original system. The quantitative data is also necessary to determine if the possible solutions meet or exceed the operational functions and performance characteristics of the "ideal" solution. Finally, new solutions may develop from this evaluation process.

It is important to remember that the *initial evaluation*, specifically the subprocesses of *analysis*, *synthesis*, and *evaluation*, which specifically address the role of the CRE, may be performed during a single appointment, or may require multiple appointments with the consumer in order to provide the time and energy that is necessary to develop possible solutions. Therefore, even though the initial evaluation in Figure 2.1 is given a single discrete box, it may be a looping function until a final solution is agreed upon on by all the stakeholders.

2.9.3 Recommendation and Report

The third component of the service delivery process as defined by Cook and Hussey is the recommendation and report. Szeto defines this as the decision process in which the choice of the final solution is made. Given that the initial evaluation process, including the analysis, synthesis, and evaluation, was preformed correctly and repeated as necessary, the decision-making process should be a relatively simple endeavor. Based on the final evaluation, particularly via field trials and the collection of quantitative data, one or two final solutions should become readily apparent; then it simply becomes the choice of the consumer as to which solution meets his/her intangible needs. The role of the CRE at this point is careful articulation of the quantitative information that went into making the final decision. This information will need to be clearly presented in the final report.

The content of the report will depend on the CRE's contribution to the evaluation process and the requirements of the funding source. It is important to realize that two or more entities may be involved in funding the evaluation process, the implementation process, and the equipment. For example a state agency may fund the evaluation; a private health insurance company may pay for the medically necessary AT, while a school system may pay for the AT that meets educational goals. All three sources, along with the requirements of best practice, must be addressed when writing the final report. In general, funding sources require either an in-depth report requiring information on the individual, the individual's current system, the intended use of the equipment, and the rationale for each item, or simply the vocational, educational, or medical justification without significant background and technical information. The differentiation of the report type is most critical for the CRE working in a private rehabilitation engineering/technology firm, as the type of report will depend on the client and the client's goals. A report addressing multiple activities and multiple pieces of equipment in multiple environments may be on the order of 20 to 30 pages, while a report describing the rationale for a single piece of equipment (e.g., manual wheelchair) for use in a single environment (e.g., employment setting) may be less than five to ten pages. Also, depending on the relationship between the funding source for the evaluation (i.e., the entity paying for the service of generating a report) and the CRE, the report may be narrative in nature, form based (e.g., check-boxes), or tabular (e.g., short descriptions). The report must clearly and concisely describe the technical information and pricing necessary to accurately implement, customize, or fabricate the equipment at the implementation, and repair/replace the equipment during the follow-along process. This may require technical drawings and a bill of materials.

2.9.4 Implementation

Once funding has been secured, the CRE can proceed with the actual implementation of the recommendations. The role of the CRE is to acquire any raw materials and commercial equipment that is necessary to fulfill the final solution. In most cases, the CRE will work with an equipment specialist and/or machinist to fabricate any custom equipment and assemble or customize any commercially available equipment. The final and most important component is the actual installation of the equipment. Ideally, this should take place in the context that the individual will typically use the equipment, so the individual can verify that the AT will allow him/her to perform the desired activity. Oftentimes, minor tweaking of the equipment is necessary, so a general assortment of hand tools may be necessary. Finally, the CRE will be involved in training the individual and any other stakeholders on the proper use of the equipment.

Individuals new to the CRE field often think that the service delivery process ends once the final payment for equipment and services have been secured. However, this is just the initial phase, as the follow-up and follow-along phases are just beginning. The amount of maintenance and repairs that the CRE will need to provide is dictated by the degree of customization and modification required in the final solution. For a completely custom system, the CRE will be responsible for all maintenance and repair issues. For commercially available AT provided by a separate RTS, the CRE will facilitate maintenance and repair issues, while relying on the RTS to repair and maintain the equipment.

The CRE's role may be minimal in the follow-up process, especially if the recommendation primarily included commercial equipment. However, the CRE will have a significant role in the follow-along process. The follow-along process may proceed concurrently with the follow-up process. The primary roles of the CRE are to provide training if needed, provide maintenance as deemed appropriate, and, most importantly, constantly reevaluate the needs of the individual. The reevaluation will include the same processes that were included in the analysis process, constantly examining the quantitative information for significant changes in an individual's ability to perform the desired activities. The end for quantitative information in the analysis process is accentuated at this point, as good information will allow the CRE to easily document any changes, and facilitate the process of recommending new equipment if changes have occurred or if the AT no longer meets the consumer's functional requirements. If changes occur to any component of the HAAT model, then a new evaluation may need to be initiated, starting the process all over again.

2.10 REIMBURSEMENT

One of the most difficult areas is the reimbursement of the services that are provided by CRE. Historically, clinical rehabilitation engineering services in the clinical setting have not been reimbursed for their clinical services by some medically based third-party payers (e.g., CMS, private health insurance). The CRE has been able to recoup some of the service delivery costs via the charges associated with custom-fabricated or customized commercial equipment. However, this does not adequately cover the value of the CRE in the service delivery process. There has been a transition for private health insurance to reimburse for direct client-to-engineer interaction. However, this is limited to direct contact time and does not account for the research, design, and fabrication times. Once again, this fundamentally limits the CRE's potential funding sources. Recall that the CRE's problem-solving skills flourish in a project-based paradigm rather than in a time-based one. A limited number of funding sources provide reimbursement on a service fee schedule basis, where the funding source pays for a complete evaluation. This will incorporate the time and expertise of the multidisciplinary team. This fee-for-service model is in constant flux as health insurance companies attempt to contract out services in order to minimize costs. Furthermore, due to health insurance probability and accountability act (HIPAA) and protected health information (PHI), all medical funding sources are moving to the codes currently generated by the healthcare procedure coding system (HCPCS) for equipment reimbursement and current procedural terminology (CPT) for service delivery. Within the CPT codes, there are new codes that directly address AT and are available for use by the CRE. Even though the codes exist, and may be applied by CRE, this does not guarantee that the funding sources will pay for these services. Medicare has yet to reimburse CRE for these services; however, it remains to be seen if individual states will reimburse for these services via the individual Medicaid programs. Furthermore, it remains to be seen if private health insurance companies will reimburse CREs for these services. It is anticipated that if more private medical insurance companies recognize clinical rehabilitation engineering services, then Medicare will also recognize CRE.

Worldwide demographics indicate that the proportion of older people is increasing in the general population, and that people with disabilities will generate a much larger need for CRE services in an interdisciplinary team framework. Public and private insurance funding constraints could significantly limit access to CRE services, particularly for economically disadvantaged individuals who would be unable to pay "out-of-pocket" for these services.

2.11 CREDENTIALING

All clinical professionals hold either licensure or independent credentials. Independent validation of education, experience, and knowledge is critical to ensuring the minimal skill set and for the safety of the consumer. The CRE for many years did not have access to licensure or credentialing. To partially alleviate this problem, RESNA created the Rehabilitation Engineering Technology (RET) credential. The RET credential requires education as an engineer, supervised experience in CRE, and passing an examination offered by RESNA. Only engineers who have the Assistive

Technology Provider (ATP) or who take the examinations concurrently may sit for the RET examination. The ATP examination tests for the fundamentals of AT and service delivery processes. The RET tests for engineering knowledge and the application of engineering knowledge to the analysis, design, delivery, and training in the use of AT. The RET is currently the only clinical credential available to the CRE. Currently, the number of certified RETs is limited, however the numbers are steadily growing. The rate of growth will likely increase dramatically once the number of insurance companies paying for CRE services improves. Individuals with the RET and ATP credentials must provide proof of participating in continuing education to retain their credentials.

In the US, engineers may become licensed professional engineers in the state in which they reside and in which they practice. The licensure is provided by the states, and those engineers who have the experience, recommendations, and examination results may call themselves *professional engineers*. Unfortunately, there is no professional engineering (PE) examination for rehabilitation engineering, and this credential in the US is not intended for clinical service delivery. Other countries (e.g., UK, Australia) have actually shown greater foresight and incorporated rehabilitation engineering into their professional engineering credentialing programs and incorporated the appropriate clinical training as part of the process.

2.12 EVIDENCE-BASED PRACTICE

Science is constantly changing as are the technologies available to assist people with disabilities. It is important for the CRE to remain current with the scientific and engineering literature. Often, the other members of the clinical team look to the CRE to introduce new technologies at the opportune time and to inform them of the potential benefits that these new technologies may bring for their clients. There is a paucity of scientific data in many areas of AT but, in some areas, there is sufficient data and even clinical practice guidelines. Clinics and clinicians should strive to base their practice on the best available data and, wherever possible, to add to the body of scientific knowledge. The US National Library of Medicine manages a Web site and database called Pub-Med that contains a large number of abstracts and full-text journal articles. This is a very useful tool to search for studies on specific topics or to learn about research in more general areas. There are a growing number of scientific databases that are open to the public, and increasingly more countries are requiring that research supported with public funds be disclosed in public databases. Knowledge of public databases and in searching the World Wide Web for information is critical for the CRE.

It is important for CREs to document the outcome of their work. This can be done in a number of valid ways. All devices that are custom made need to be documented and should meet pre-specified criteria even if only critical dimensions. Outcome measures relate to both services and equipment. Services are measured using tools to track services and surveys to collect the opinions of consumers. Some of the measures to track are:

Time for appointment
Time to prepare report

Time for delivery of devices
Quality of the reports/documentation
Consumer satisfaction
Consumer device usage or abandonment
Changes in activity-, community-participation-, and health-related factors

Rehabilitation engineers should also be specialists in quantitative measurement. This expertise lends itself well when delivering devices to consumers. The CRE can check and record dimensions, control settings, system preferences, and user effectiveness with the device. Specialized tools are more frequently becoming available. Pressure mapping systems are useful for estimating the seated interface pressures. The numerical and graphical display helps to identify areas at risk for pressure ulcer and permits comparison of different seat cushions. Similar information exists for manual wheelchair propulsion and input devices for electric-powered wheelchair. Data logging devices provide important data about devices that are actually used, allowing for optimal tuning and providing useful information for improved designs. As quantitative data become more available, computer files can be used to conduct comparisons and to support improvements in practice.

2.13 SUMMARY

There are a number of exciting settings for CREs to practice within. No matter where people work, it is critical to remain current with the scientific and clinical literature. The CRE is often the member of the team to whom the others turn when seeking to learn about new technologies or when they have questions about the attributes of something that they are not familiar with. The opportunities for the CRE should grow as more insurers recognize their unique contributions.

BIBLIOGRAPHY

Bauer S. M., Lane J. P., Stone V. I., and Unnikrishnan N., The voice of the customer — Part 2: Benchmarking battery chargers against the Consumer's Ideal Product, *Assist Technol*, vol. 10, pp. 51–60, 1998.

Beaumont-White S. and Ham R. O., Powered wheelchairs: are we enabling or disabling? *Prosthet Orthot Int*, vol. 21, pp. 62–73, 1997.

Bergamasco R., Girola C., and Colombini D., Guidelines for designing jobs featuring repetitive tasks, *Ergonomics*, vol. 41, pp. 1364–83, 1998.

Berry B. E. and Ignash S., Assistive technology: providing independence for individuals with disabilities, *Rehabil Nurs*, vol. 28, pp. 6–14, 2003.

Biomedical Engineering: A Historical Perspective, in *Introduction to Biomedical Engineering*, J. D. Enderle, S. M. Blanchard, and J. D. Bronzino, Eds. New York: Academic Press, 2001, pp. 1–28.

Bleck E. E., Rehabilitation engineering services for severely physically handicapped children and adults, *Curr Pract Orthop Surg*, vol. 7, pp. 223–45, 1977.

Bronzino J. D., Clinical engineering: Evolution of a discipline, in *The Biomedical Engineering Handbook*, vol. 2, J. D. Bronzino, Ed., 2nd ed. Boca Raton, FL: CRC Press LLC, 2000, pp. 167-1–167-7.

Brown R. and Wright D. K., An integrated approach to rehabilitation engineering education: the development of a new masters programme at Brunel University, *Biomed Sci Instrum*, vol. 28, pp. 75–80, 1992.

Comments on Warren's cost effectiveness and efficiency in assistive technology service delivery, *Assist Technol*, vol. 5, pp. 66–73, 1993.

Cook A. M. and Hussey S. M., Delivering Assistive Technology Services to the Consumer, in *Assistive Technologies: Principles and Practice*, 2nd ed. St. Louis, MO: Mosby, Inc., 2002, pp. 91–142.

Cook A. M. and Hussey S. M., Introduction and Overview, in *Assistive Technologies: Principles and Practice*, 2nd ed. St. Louis, MO: Mosby, Inc., 2002, pp. 3–33.

Diez C., More than just credentials: the personal and financial rewards of certification, *Biomed Instrum Technol*, vol. 37, pp. 69–70, 2003.

Enderle J. D., Blanchard S. M., and Bronzino J. D., *Introduction to Biomedical Engineering*. New York: Academic Press, 2001.

Foort J., Comments for a new generation of rehabilitation engineers, *J Rehabil Res Dev*, vol. 22, pp. 2–8, 1985.

Foort J., Hannah R., and Cousins S., Rehabilitation of engineering as the crow flies. Part I — Development of the biomechanics clinic team, *Prosthet Orthot Int*, vol. 2, pp. 15–23, 1978.

Foort J., Hannah R., and Cousins S., Rehabilitation engineering as the crow flies. Part IV — Criteria and constraints, *Prosthet Orthot Int*, vol. 2, pp. 81–5, 1978.

Foort J., Hannah R., and Cousins S., Rehabilitation engineering as the crow flies. Part V — a problem-solving method for rehabilitation engineering, *Prosthet Orthot Int*, vol. 2, pp. 157–60, 1978.

Glanville H. J., Rehabilitation and engineering, *Biomed Eng*, vol. 10, pp. 297–9, 310, 1975.

Goodman G., The profession of clinical engineering, *J Clin Eng*, vol. 14, pp. 27–37, 1989.

Hobson D. A., Rehabilitation engineering — a developing specialty, *Prosthet Orthot Int*, vol. 1, pp. 56–60, 1977.

Hobson D. A. and Trefler E., Rehabilitation Engineering Technologies: Principles of Application, in *The Biomedical Engineering Handbook*, Vol. 2, J. D. Bronzino, Ed., 2nd ed. Boca Raton, FL: CRC Press LLC, 2000, pp. 146-1–146-9.

Hutchins B. M., The slow, slow death of clinical engineering, *Biomed Instrum Technol*, vol. 35, p. 289, 2001.

Iles G., Rehabilitation engineering, *N Z Nurs J*, vol. 75, pp. 7–9, 1982.

Jedeloo S., De Witte L. P., Linssen B. A., and Schrijvers A. J., Client satisfaction with service delivery of assistive technology for outdoor mobility, *Disabil Rehabil*, vol. 24, pp. 550–7, 2002.

Johnson G. R., Special issue on rehabilitation engineering, *Proc Inst Mech Eng [H]*, vol. 215, p. i, 2001.

Kenworthy G. and Simpson D. C., The provision of a service in rehabilitation engineering, *Biomed Eng*, vol. 9, pp. 515–6, 1974.

Kizer K. W. and Norby R. B., Internal practice barriers for non-physician practitioners in the veterans healthcare system, *J All Health*, vol. 27, pp. 183–7, 1998.

Kohn J. G., LeBlanc M., and Mortola P., Measuring quality and performance of assistive technology: results of a prospective monitoring program, *Assist Technol*, vol. 6, pp. 120–5, 1994.

Kohn J. G., Mortola P., and LeBlanc M., Clinical trials and quality control: checkpoints in the provision of assistive technology, *Assist Technol*, vol. 3, pp. 67–74, 1991.

Lane J. P., Usiak D. J., Stone V. I., and Scherer M. J., The voice of the customer: consumers define the ideal battery charger, *Assist Technol*, vol. 9, pp. 130–9, 1997.

Langbein W. E. and Fehr L., Research device to preproduction prototype: a chronology, *J Rehabil Res Dev*, vol. 30, pp. 436–42, 1993.

Logan G. D. and Radcliffe D. F., Supporting communication in rehabilitation engineering teams, *Telemed J*, vol. 6, pp. 225–36, 2000.

Lowe P. J., Richardson W., and Smallwood R. H., Physical disability: the role of the physical scientist in the health service. A report of the Institute of Physical Sciences in Medicine, *Clin Phys Physiol Meas*, vol. 9, pp. 81–4, 1988.

Mann R. W., Selected perspectives on a quarter century of rehabilitation engineering, *J Rehabil Res Dev*, vol. 23, pp. 1–4, 6, 1986.

Mann W. C., Goodall S., Justiss M. D., and Tomita M., Dissatisfaction and nonuse of assistive devices among frail elders, *Assist Technol*, vol. 14, pp. 130–9, 2002.

Mann W. C., Ottenbacher K. J., Fraas L., Tomita M., and Granger C. V., Effectiveness of assistive technology and environmental interventions in maintaining independence and reducing home care costs for the frail elderly. A randomized controlled trial, *Arch Fam Med*, vol. 8, pp. 210–17, 1999.

Meanley S., Rehabilitation engineering and leprosy in Nepal: a personal view, *J Med Eng Technol*, vol. 16, pp. 165–9, 1992.

Milner M., Naumann S., Literowich W., Martin M., Ryan S., Sauter W. F., Shein G. F., and Verburg G., Rehabilitation engineering in pediatrics, *Pediatrician*, vol. 17, pp. 287–96, 1990.

Montan K., Rehabilitation engineering — a growing part of the rehabilitation services, *Prosthet Orthot Int*, vol. 2, pp. 111–13, 1978.

Nosek M. A. and Krouskop T. A., Demonstrating a model approach to independent living center-based assistive technology services, *Assist Technol*, vol. 7, pp. 48–54, 1995.

O'Dea T., Clinical engineering without clinical engineers, *Biomed Instrum Technol*, vol. 35, pp. 225–6, 2001.

Perlman L. G. and Enders A., *Rehabilitation Technology Service Delivery: A Practical Guide*. Washington, DC: RESNA, 1987.

Potvin A. R., Mercadante T. C., and Cook A. M., Skill requirements for the rehabilitation engineer: results of a survey, *IEEE Trans Biomed Eng*, vol. 27, pp. 283–8, 1980.

Reswick J. B., Rehabilitation engineering, *Annu Rev Rehabil*, vol. 1, pp. 55–79, 1980.

Reswick J. B., Technology — an unfulfilled promise for the handicapped, *Med Prog Technol*, vol. 9, pp. 209–15, 1983.

Reswick J. B., How and when did the rehabilitation engineering center program come into being? *J Rehabil Res Dev*, vol. 39, pp. 11–16, 2002.

Ring N., Rehabilitation engineering, *Nurs Mirror Midwives J*, vol. 140, pp. 61–5, 1975.

Riva G. and Gamberini L., Virtual reality as telemedicine tool: technology, ergonomics and actual applications, *Technol Health Care*, vol. 8, pp. 113–27, 2000.

Rowley B. A., Mitchell D. F., and Weber C., Educating the rehabilitation engineer as a service provider, *Assist Technol*, vol. 9, pp. 62–9, 1997.

Scherer M. J. and Lane J. P., Assessing consumer profiles of 'ideal' assistive technologies in ten categories: an integration of quantitative and qualitative methods, *Disabil Rehabil*, vol. 19, pp. 528–35, 1997.

Seelman K., Aging with a disability: views from the National Institute on Disability and Rehabilitation Research, *Assist Technol*, vol. 11, pp. 84–7, 1999.

Selwyn D., Rehabilitation engineering: new hope for the permanently disabled, *J Am Soc Psychosom Dent Med*, vol. 22, pp. 114–28, 1975.

Selwyn D., Tandler R., and Zampella A. D., Engineering as a clinical tool for geriatric rehabilitation, *Med Instrum*, vol. 16, pp. 259–60, 1982.

Shaffer M. J., A new clinical engineering curriculum, *Biomed Instrum Technol*, vol. 29, pp. 448–9, 1995.

Shuster N. E., Addressing assistive technology needs in special education, *Am J Occup Ther*, vol. 47, pp. 993–7, 1993.

Smith R. O., Chapter one: Models of service delivery in rehabilitation technology, in *Rehabilitation Technology Service Delivery: A Practical Guide*, L. G. Perlman and A. Enders, Eds. Washington, DC: RESNA, 1987, pp. 7–25.

Staros A., Rehabilitation engineering and the growth of prosthetics/orthotics practice, *Int Rehabil Med*, vol. 6, pp. 79–84, 1984.

Story M. F., Mueller J. L., and Mace R. L., *The Universal Design File: Designing for People of All Ages and Abilities*. Raleigh, NC: NC State University, The Center for Universal Design, 1998.

Szeto A. Y. J., Rehabilitation Engineering and Assistive Technology, in *Introduction to Biomedical Engineering*, J. D. Enderle, S. M. Blanchard, and J. D. Bronzino, Eds. New York: Academic Press, 2001, pp. 905–41.

Szeto A. Y. J., Rehabilitation Engineers in Government Service, *IEEE Eng Med Biol Mag*, vol. 23, pp. 8–9, 2004.

Thacker J. G. and Kauzlarich J. J., Rehabilitation engineering education at the University of Virginia, *J Clin Eng*, vol. 7, pp. 329–34, 1982.

Trachtman L. H., A review of practices among information resource programs on assistive technology, *Assist Technol*, vol. 3, pp. 59–66, 1991.

Vanderheiden G. C., Service delivery mechanisms in rehabilitation technology, *Am J Occup Ther*, vol. 41, pp. 703–10, 1987.

Vasa J. J., Electronic aids for the disabled and the elderly, *Med Instrum*, vol. 16, pp. 261–2, 1982.

Vitalis A., Walker R., and Legg S., Unfocused ergonomics?, *Ergonomics*, vol. 44, pp. 1290–301, 2001.

Volinn E., Do workplace interventions prevent low-back disorders? If so, why?: a methodologic commentary, *Ergonomics*, vol. 42, pp. 258–72, 1999.

Warren C. G., Cost effectiveness and efficiency in assistive technology service delivery {see comments}, *Assist Technol*, vol. 5, pp. 61–5, 1993.

Winters J. M., Rehabilitation engineering training for the future: influence of trends in academics, technology, and health reform, *Assist Technol*, vol. 7, pp. 95–110, 1995.

3 Universal Design

Linda van Roosmalen and Hisaichi Ohnabe

CONTENTS

3.1 LEARNING OBJECTIVES OF THIS CHAPTER

Upon completion of this chapter, the reader will be able to:

Have an understanding of universal design (UD)
Know the difference between UD and orphan technology (OTech)
Understand how UD can benefit products and the built environment
Select and use tools to apply UD to products and the built environment
Evaluate products and the built environment for compliance with principles of UD
Know where UD fits within existing design standards

3.2 INTRODUCTION

Look around us at the products and systems shaping our society — how many of these products or systems have been designed with the elderly individual in mind? How many of these products and systems can be used by an individual with a hearing impairment or by an individual using a wheelchair?

This chapter explains universal design (UD), and how a UD approach can guide product designers, engineers, and companies in the design, development, and launch of products and systems for a population that includes individuals with a variety of physical characteristics, sensory abilities, and cognitive abilities. This chapter also gives an explanation of tools that can be used to help design products that meet the principles of UD.

3.3 BACKGROUND

Ron Mace, the founder of UD and an architect by profession, stated in 1985 that "universal design is an approach to design that incorporates products as well as building features which, to the greatest extent feasible, can be used by everyone." Another leader in the field of UD is Patti Moore. For the purpose of her gerontological research twenty years ago, she went around the country dressed as an elderly woman so as to experience the difficulties that this part of the population encounters with products, the built environment, and the public attitude regarding aging individuals.

When designing products and our built environment to make them usable by everyone, we have to define who is meant by "everyone." McNeil found that an estimated 54 million individuals within the US alone have some type of limitation. This estimate comes from the Survey of Income and Program Participation (SIPP) data, and includes individuals who are at least 15 years old and meet any of the following criteria:

1. Use a wheelchair, a cane, crutches, or a walker.
2. Have difficulty performing one or more functional activities (seeing, hearing, speaking, lifting/carrying, using stairs, walking, or grasping small objects).

3. Have difficulty with one or more activities of daily living (getting around inside the home, getting in or out of bed or a chair, bathing, dressing, eating, and toileting).
4. Have difficulty with one or more instrumental activities of daily living (going outside the home, keeping track of money and bills, preparing meals, doing light housework, taking prescription medicines in the right amount at the right time, and using the telephone).
5. Have one or more specified conditions (a learning disability, mental retardation or another developmental disability, Alzheimer's disease, or some other type of mental or emotional condition).
6. Have any other mental or emotional condition that seriously interferes with everyday activities (frequently depressed or anxious, trouble getting along with others, trouble concentrating, or trouble coping with day-to-day stress).
7. Have a condition that limits the ability to work around the house.
8. Have a condition that makes it difficult to work at a job or business.
9. Receive federal benefits based on the inability to work.

The art of product design has always existed, and goes as far back as the first hand tools designed by cave dwellers. Up to this day, people always seem to invent new products. With improved healthcare services and the initiation of Americans with Disabilities Act (ADA), the past couple of decades show a growing percentage of elderly individuals and those who have disabilities. Enabling integration of people with disabilities and elderly into our society requires the development of "adaptive" or "accessible" products in the marketplace. Examples of adaptive design or accessible design are lifts or entrance ramps designed to provide access to people who use wheelchairs into vehicles and/or buildings. Handrails in restrooms provide stability for elderly individuals and individuals who need to transfer from a wheelchair to a toilet. Making products accessible for the disability market using an adaptive or accessible design approach is not necessarily cost-effective, nor does it result automatically in successful solutions. Products and systems that are altered especially for people with disabilities can be stigmatizing and poorly integrated, resulting in rejection of the product or system by the target population. A human-centered design approach needs to be used to design an all-inclusive environment that meets the needs of all potential users, including the elderly and those with disabilities.

In 2001, George W. Bush made a push for UD during the introduction of his "Freedom Initiative." This initiative has secured funds for research and development programs that are aimed towards making assistive technology available as well as to increase availability of universally designed technologies. This initiative identified six research priorities. One priority aim is to increase "accessibility to and mobility within the physical environment." The physical environment includes public and private buildings, tools and objects of daily use, as well as roads and vehicles. To make our environment accessible to all, the priority suggests the use of a process that incorporates UD.

3.4 ORPHAN TECHNOLOGY

Although it is a good initiative to design most publicly used products according to the UD approach, not all products are suitable for this design approach. Some products that have been designed for a specific application, for example, to prevent a person from developing pressure sores, would not necessarily end up as a mainstream product, and would not necessarily be designed using the UD approach. This type of product will stay in the "disability" market, and will fall under the so-called orphan technology (OTech).

An OTech or product is one that has been designed especially for a population with a specific disability. OTech is the opposite of mainstream technology. Also, the position that OTech takes with respect to UD is that products that fit into the OTech category are difficult to make "universal." Examples of OTech are therefore products that are custom-made for people with disabilities, such as a wheelchair seat cushion, an augmentative communication device, or a watch that also adjusts a hearing aid.

Seelman concludes that the implementation of UD and the availability of OTech (technology solely designed for use by people with disabilities) are both necessary to allow full participation of people with disabilities in our society. Factors external to technology development are also important for the success of design, such as attitudes, education, markets, and government incentives and mandates.

3.5 UNIVERSAL DESIGN

3.5.1 Origin of Universal Design

Originating in the United States in the 1990s, UD is a design approach that, when properly implemented throughout the product development process, results in the creation of new products that meet the needs of all potential consumers including those with disabilities and the elderly. There are a number of other design streams that have emerged throughout the world that encompass similar ideas as those contained in UD. Kyoyo-Hin originated in Japan in the 1990s. Instead of designing from the ground up, Kyoyo-Hin redesigns existing products and systems to enable usability by a larger user population, including people with disabilities and the elderly. Inclusive design and design for all originated in Europe, and barrier-free design originated in the United Nations (Figure 3.1). What these definitions have in common is that they focus on the design of a product and/or system. Additionally, their approach focuses on "all people," meaning that the target population includes people with disabilities and the elderly.

We can conclude that UD, Kyoyo-Hin, inclusive design, design for all, and barrier-free design all focus on creating products or systems that keep the abilities and disabilities of all potential users in mind in the design process.

3.5.2 The Seven Principles of Universal Design

When North Carolina State University was awarded the RERC on UD (funded by NIDRR), performance measures were developed to help consumers, product

Time line	United Nations/ ISO	USA	Europe	Japan
1950			Normalization	
1960				
1970	Barrier-free design →		Barrier-free design	
1980				
1990		Universal design	Design for all-inclusive design	Kyoyo-Hin
2000	Accessible design →		Accessible design	
2001	ISO/ IEC Guide 71			

FIGURE 3.1 International variations to universal design vs. time.

designers and marketers apply UD. The evaluation of the performance measures was completed in 1999, and the measures are now available for use by the public. Based on the research conducted at North Carolina State University, seven principles of UD have been developed. These seven principles are listed in Table 3.1. The principles of UD are to be used to design, improve, or make decisions regarding products and systems that include as many people as possible.

3.5.2.1 Principle One: Equitable Use

An example that explains "equitable use" is the EasyGrip™ kitchen tools that feature a rubbery thick handle so that they are easy to grasp. Although initially the "elderly" were targeted for using this type of handle, the design appeared to be comfortable and attractive to a much larger group of users.

3.5.2.2 Principle Two: Flexibility in Use

The lever-type door handle is a type of handle design that can be used by a variety of people in a variety of ways. Elbows, wrists, fists, single fingers, or even an item held in someone's hand can be used to open a lever-type door handle. When we compare equitable use of the spherical shaped door handles commonly used in the United States, the round door handles are less easy to grasp and rotate with less flexibility.

3.5.2.3 Principle Three: Simple and Intuitive Use

Airports have a large number of visual signs to guide travelers to the correct location. These signs are often in the form of single-color pictograms and are therefore easy to

TABLE 3.1
Principles of Universal Design (North Carolina State University, 1999)

Universal Design Principles	Guidelines
Principle one: equitable use The design is useful and marketable to people with diverse abilities	1a. Provide the same means of use for all users: identical whenever possible, equivalent when not 1b. Avoid segregating or stigmatizing any users 1c. Provisions for privacy, security, and safety should be equally available to all users 1d. Make the design appealing to all users
Principle two: flexibility in use The design accommodates a wide range of individual preferences and abilities	2a. Provide a choice in methods of use 2b. Accommodate right- or left-handed access and use 2c. Facilitate the user's accuracy and precision 2d. Provide adaptability to the user's pace
Principle three: simple and intuitive use Use of the design is easy to understand, regardless of the user's experience, knowledge, language skills, or current concentration level	3a. Eliminate unnecessary complexity 3b. Be consistent with user expectations and intuition 3c. Accommodate a wide range of literacy and language skills 3d. Arrange information consistent with its importance 3e. Provide effective prompting and feedback during and after task completion
Principle four: perceptible information The design communicates necessary information effectively to the user, regardless of ambient conditions or the user's sensory abilities	4a. Use different modes (pictorial, verbal, or tactile) for redundant presentation of essential information 4b. Provide adequate contrast between essential information and its surroundings 4c. Maximize "legibility" of essential information 4d. Differentiate elements in ways that can be described (i.e., make it easy to give instructions or directions) 4e. Provide compatibility with a variety of techniques or devices used by people with sensory limitations
Principle five: tolerance for error The design minimizes hazards and the adverse consequences of accidental or unintended actions	5a. Arrange elements to minimize hazards and errors: most used elements, most accessible; hazardous elements eliminated, isolated, or shielded 5b. Provide warnings of hazards and errors 5c. Provide fail-safe features 5d. Discourage unconscious action in tasks that require vigilance
Principle six: low physical effort The design can be used efficiently and comfortably and with minimum fatigue	6a. Allow user to maintain a neutral body position 6b. Use reasonable operating forces 6c. Minimize repetitive actions 6d. Minimize sustained physical effort
Principle seven: size and space for approach and use Appropriate size and space is provided for approach, reach, manipulation, and use, regardless of user's body size, posture, or mobility	7a. Provide a clear line of sight to important elements for any seated or standing user 7b. Make reach to all components comfortable for any seated or standing user 7c. Accommodate variations in hand and grip size 7d. Provide adequate space for the use of assistive devices or personal assistance

see and understand by people with a variety of language skills and levels of knowledge. The use of pictograms also benefits individuals with a low concentration levels.

3.5.2.4 Principle Four: Perceptible Information

Cell phones now feature a range of options not only to personalize a phone but to also make the phone useful for people with disabilities. A phone's direct dial option is commonly used by the majority of phone users to speed-dial an often-dialed phone number. This feature is also useful for people who are cognitively impaired and cannot remember phone numbers easily. Also, a visual signal in combination with audio and vibration can alert the phone user in a noisy or dark environment. Other features may include a speaker that is adjustable to the noise level of the environment.

3.5.2.5 Principle Five: Tolerance for Error

Products that reduce the risk of errors are, for example, the temperature limiters that are available on some shower faucets. When adjusted correctly, a temperature limiter prevents the water from getting too hot and burn the user if the user accidentally turns the faucet to hot instead of cold.

3.5.2.6 Principle Six: Low Physical Effort

In the 1960s, power steering was introduced in motor vehicles to make them easier to maneuver. Another example that shows a decrease in physical effort is the introduction of the elevators in addition to stairs. In many buildings, stairs are only used for emergency situations and elevators are the main mode of vertical travel.

3.5.2.7 Principle Seven: Size and Space for Approach and Use

Compared to the commonly used hinged doors, sliding doors take up less space to use and construct. They are also easier to use and navigate for wheelchair-seated individuals. Less back and forth maneuvering is needed when traversing a sliding door compared to the opening and closing movements related to a hinged door.

3.5.3 Benefits of Universal Design

3.5.3.1 Societal Benefits

For the longest time, our society has been providing rehabilitation according to the disability model. More recently, a social model, putting participation of individuals with disabilities in the environment on the forefront, has become increasingly popular. Whereas the outdated disability model focused on "treating (rehabilitating) the patient," the social model puts the individual in the center and focuses on full functional participation of that individual in the environment. The disability model and the social model are, therefore, competitive models and difficult to integrate. Rehabilitation engineering plays a role within this social model and enables participation of individuals with disabilities in society.

One of the objectives of rehabilitation engineering is its contribution to "participation" in society. One of the roles of rehabilitation engineering is to integrate the "disability model" and the "social model." Although rehabilitation engineering is the key to integrate the disability model and social model, UD plays an additional role to meet this objective. Rehabilitation engineering alone will not enable full participation and full function of people with disabilities into society. In order to have all people (including those with disabilities and the elderly) participate successfully in our society, both the environment needs redesigned and rehabilitation technology needs further development within a user-centered design framework. Only a user-centered design framework, including a UD approach, can enable full integration of the disability model and the social model, resulting in a safe, useful, effective, and comfortable environment for all.

3.5.3.2 Personal Benefits

Most products in our built environment have been designed for the general public. The focus has been primarily on designing for the healthy, average male user who has good sensory, mobility, and cognitive skills. To make products and the built environment usable for people who have disabilities, most often adaptations are made to existing products.

Some example people who can be benefited by a UD approach include:

- Individuals with a disability (i.e., cognitive, mobility, or sensory limitation)
- Older people who have limited sensory, mobility, and cognitive abilities
- Individuals who use assistive technology such as wheelchairs, canes, walkers, or crutches for mobility
- Short people and children
- Tall and/or heavy people
- People with visual impairments using guide dogs
- People with disabilities who are accompanied by a caregiver
- Parents with strollers
- Individuals who are visually impaired
- Pregnant women
- People who speak a different language
- A left-handed person

Some products are so well integrated into our society that not only people with disabilities benefit from their existence but also other potential users. Think of the recently developed low-floor public buses that "kneel" at the bus stop. This development has allowed people to enter the bus without having to step up or step down, making it safer and easier for not only people with disabilities and the elderly but all passengers. Other examples are curb cuts on sidewalks and on street corners. These environmental changes were originally implemented to create accessible routes for people in wheelchairs who use the sidewalk. However, families with small children and strollers also benefit from these developments.

By developing products and systems in our society that are "universally designed," people with disabilities and the elderly will benefit by being more independent. With the availability of products and systems that people with disabilities can use to get around in their environment, there is no need for costly interventions and support systems to be in place for this population. Additionally, when products are designed to "blend in" with mainstream products, they are less stigmatizing to use, and product acceptance is therefore increased. For example, among the wheelchair-using population, motorized scooters have gained much popularity among the elderly population. One of the reasons for this increased popularity is that wheelchairs are associated with medical devices and scooters are considered mobility devices.

3.5.3.3 Benefits to Industry

There are several reasons why the industry can be reluctant to adapt UD in their operations. The most common reason is that changing one's development process costs money. Another reason is a lack of knowledge of who the potential end user is and what their characteristics and capabilities are. Companies may be confused by the complexity of UD and where it fits into the design process.

When the industry has access to knowledge how to apply UD and who to design for, the benefits of UD are clear. When applying the UD approach throughout the development process, the benefits can be one or more of the following:

Products are more effective in meeting consumer needs.
More competitive products.
Increase consumer acceptance of product.
Larger market due to increased number of potential buyers (existing consumer group plus people with disabilities and the elderly).
Increased production can result in lower production costs and more profit.
Increased cost-effective production process.

In the following section, an explanation is given on how to design a product that complies with the principles of UD.

3.5.4 Barriers to Implementing Universal Design

Although the benefits of UD appear imminent, there exist several barriers preventing UD from being implemented. Managements of companies think that additional costs are involved with implementing UD and that adoption of UD requires retooling of machines and extensive training of development personnel. Furthermore, UD may slow down the time to market because the initial thought process that includes user research requires additional time prior to product development. Industry often shows a lack of knowledge on how to solve access problems or practice UD. Companies often do not have the knowledge of people with limitations owing to a lack of time to learn about diverse populations and how to design for them. Also, many companies currently do not seem to use a systematic product design structure so that it is difficult for them to implement UD. Companies think UD may add cost to their production process and are unsure if UD will pay off. Finally, companies may be unsure of the potential user group that may be interested in UD products.

Many of the barriers listed to implementing UD involve a lack of knowledge of a design process and a lack of knowledge of what people with disabilities can do and how we should design to include their needs. The following subsection will deal in depth on how to implement UD into the product development process.

3.5.5 The Universal Design Matrix

A few products in our society are widely known to be "universally designed" and have therefore successfully entered the mainstream market (e.g., EasyGrip and Microsoft Tools). For the most part, however, individuals who meet the "disability criteria" listed by McNeil remain limited in their participation in society because of inaccessible or unusable products in our built environment.

The core reason that mainstream products (furniture, motor vehicles, or packaging) remain inaccessible to the disability population is because the (special) needs of this population have not been included in the design process during which these products or systems were developed. Excluding the capabilities and needs of people with disabilities or the elderly from the design process can result in exclusion of these individuals from participating in our society.

The UD matrix will help categorize types of products and users. The UD matrix will also assist in what products or systems are more open to UD and what product or systems are more difficult to design according to the UD principles. Figure 3.2 shows in the horizontal direction the type of user, including an individual user, a family user, or a public user. In the vertical direction, there are three classifications including consumer goods (quick use), durable goods (use over a certain period), and social or capital goods (long-term use).

Examples of consumer goods for personal use (a1) are clothes or office supplies. Durable goods for personal use (a2) are, for example, assistive technology,

Duration	Personal	Family	Public
Social goods	a3	b3 Housing	c3 Public building Road Station
Durable goods	a2 Assistive technology Computer Cell phone Chair Bicycle	b2 Refrigerator Washing machine Car	c2 Train Bus station
Consumer goods	a1 Clothes Dishes	b1	c1 Trash can Bus stop
	Personal	Family	Public
	Use		

FIGURE 3.2 Universal design matrix.

Duration		Personal	Family	Public
	Social goods		Housing	Product group 3 Public building Road Station
	Durable goods	Assistive technology Computer Cell phone Chair Bicycle	Product group 2 Refrigerator Washing machine Car	Train Bus station
	Consumer goods	Clothes Dishes Product group 1		Trash can Bus stop
		Use		

FIGURE 3.3 Representation of the expansion of universal design throughout various product groups.

a computer, cell phone, chair, or bicycle. Durable goods for family use (b2) are, for example, a refrigerator, washing machine, or car. Social or capital goods for family use (b3) include housing, whereas consumer goods for public use (c1) include public trash cans or bus stops. Finally, durable goods for public use (c2) include a train, bus, or bus station, whereas social or capital goods for public use (c3) would be a public building or a highway.

Figure 3.3 represents how universally designed products can expand within the various product-type categories and various user types. There are generally three types of product groups that seem responsive to a UD approach:

Product group 1: There are many personal product selections made by an individual user in the area of consumer and durable goods. Up to now, readymade products (e.g., clothes or cell phones) have been designed to fit only young and healthy individuals and not people with disabilities or the elderly. To expand UD into the category of consumer goods and durable goods, readymade products and systems should also be developed towards the population with special needs.

Product group 2: This product group consists of durable goods and public goods for use by individual users and families. To expand UD into this product category, special features can be included into existing product designs to make them usable and accessible to all. Features in durable products can result in a reduction of resources over time, because these types of products, when universally designed, have the tendency to grow with the user (e.g., accessible housing and adjustable car seat).

Product group 3: To allow public goods to be suitable for use by a wide variety of users, they need to be designed for multiple purposes. This product group includes buildings that are likely to see all types of users. UD from the ground up is very important in this product category.

Universal design is the key to ensure independence, safety, and active engagement in daily activities and improved quality of life of persons with disabilities. By widely incorporating a UD approach, the development of possibly more expensive OTech can be reduced, because OTech is now often needed because our society and its products are inaccessible.

3.5.6 Applying Universal Design

The next question we need to ask ourselves is how we can design products that meet the needs of all users, including those with disabilities and the older people. Many products have been developed solely for people with disabilities. When designing for people with disabilities, the following two methods are commonly used:

1. Designers take the "accessible design" approach to serve the needs of individuals with disabilities. A solution to meet the needs of people with disabilities is sought long after the environment has been designed. This design approach has the disadvantage that the solution often negatively affects the aesthetics and the environment. Environments are often not equipped or suited for an addition or modification to ensure accessibility. An example of accessible design is an addition of a ramp structure next to the stairs of an existing building to allow wheelchairs to enter the building.
2. Another design approach is "adaptive design." Existing products or systems are made usable for people with disabilities by after-market add-on components to make the product or system usable also for people with disabilities. The advantage of adaptive design is that products and systems can be customized to people with disabilities. The disadvantages are that installing adaptations can be costly and the use of add-on products by people with disabilities can be stigmatizing. An example of adaptive design is the addition of a special wheelchair lift in a public bus to enable a wheelchair user to enter the bus.

On the contrary, there is the UD approach that supports all potential users and not only those with disabilities. Products designed for this user group are according to the functional and cognitive abilities of a broad population. Although accessible design and adaptive design are user-driven design processes, they have a different definition of who that user actually is. Accessible and adaptive design considers only people with disabilities being the potential user, whereas UD includes besides people with disabilities also able-bodied individuals.

Think user-oriented. When designing for "all users," the key point is to adopt a structured design process when developing a product or system for human use. Next is to include a representative group of users or the personas of these users and the recognition of their sensory, mobility, and cognitive limitations early on in the development process.

When designing, for example, a display on a car dashboard, a designer would previously focus on the average reach capabilities of an adult user and not necessarily include the reach capabilities of an 85 year old. When including only those physical capabilities of an average adult user, a potential elderly user population may be compromised and unable to use the product effectively, comfortably, and safely.

Consumers of products consist of a very diverse population. And to develop successful products for a wide range of users, we need to be able to place ourselves in the shoes of the various end users in order to meet their needs. To "design with a universal approach," it is important to be able to conceptualize broadly and to think about the potential limitations your user population may have. Prior to even putting the first sketch on paper, the users or users being designed for must be given full consideration. Additionally, think also about those users being excluded when setting up the list of design requirements.

3.6 DESIGN AND HUMAN ABILITIES

To organize human abilities, a user matrix can be used (Table 3.2). The horizontal axis in Table 3.2 represents the four ability categories of people. The vertical axis represents seven aspects of interventions to consider.

By using the matrix, relations can be made between abilities, a proposed product, and the stage the product is in. If we design, for example, an automated teller machine, factors to be considered are related to sensory, physical, and cognitive capabilities of the user, which results in design requirements that include lighting, glare, color/contrast, size/style of font, loudness/pitch, clear language, and ease of handling.

3.6.1 PERSONA

There are several techniques to observe a user population and to uncover their problems, needs, and cognitive and physical abilities. We can interview people and ask a variety of potential user questions about what their needs and abilities are. We can

TABLE 3.2
User Matrix Consisting of User Abilities and Product Life Aspects

	Human abilities			
Product life aspects	**Sensory**	**Physical**	**Cognition**	**Allergy**
	Seeing	Dexterity	Intellect/memory	Contact/food/
	Hearing	Manipulation	Language/literacy	respiratory
	Tough	Movement		
	Taste/smell	Strength		
	Balance	Voice		
Information				
Packaging				
Materials				
Installation				
User interface				
Maintenance, storage/disposal				
Built environment				

also observe a site that is related to our problem, and we can document people's habits, actions, and concerns. Another useful tool that places you in the skin of the user is "persona" — the creation of an imaginary "family" of potential users. This family consists, for example, of the following individuals:

1. Cheryl, a 16-year-old girl, who uses a power wheelchair for daily mobility.
2. Robert, Cheryl's father, is 43 years old. He is a diabetic and weighs over 250 lb. Robert has lost part of his hearing due to his work with heavy machinery.
3. Mary-Beth, Cheryl's grandmother, is 83 years old and lives in a nursing home. She is short of memory and recently had a stroke, which affected the right side of her body, reducing the function of her right arm, hand, and leg.
4. Finally, there is Brian, Cheryl's uncle, who has poor vision and uses a guide dog for navigation.

In addition to the personas listed here, there are obviously many other limitations that users may have with more or less severity:

Low vision or no vision
Hard of hearing or no hearing
Loss of smell, taste, or senses
Physical challenges; lower extremity and/or upper extremity limitations
Cognitive challenges: difficulty with language, reading, understanding, memorizing
Poor dexterity or sensory limitations

When using personas that have few limitations, design ideas can be evaluated based upon effectiveness, safety, fit, comfort, and ease of use.

3.7 USE OF A DESIGN PROCESS

A structured design process can be used to solve an existing problem and systematically design a successful solution. Such a design process can, for example, include the steps listed in Table 3.3. To successfully implement a UD approach in a design process, the management as well as designers must be involved and accept the UD approach to be beneficial.

The steps in this design process are a combination of a variety of existing design processes. However, none of the existing design processes describe clearly how and where in the process UD is needed. Also, not many other design processes describe how and where to incorporate the potential limitations of users.

3.8 STANDARDS RELATED TO UNIVERSAL DESIGN

There are a variety of existing standards that incorporate the UD approach to make products and our built environment more friendly and safe for everyone. The U.S. has

TABLE 3.3
Universal Design Approach Implemented in the Design Process

Problem definition
List those people who have a problem and list in detail what the problem entails
Describe the functional and cognitive limitations that may relate to the problem

Identify stakeholders
List all people who are related to the stated problem
List the functional and cognitive limitations of the people related to the problem

Describe stakeholder needs
With input from stakeholders, list their needs and prioritize them in order of importance
List in detail the functional and cognitive needs of stakeholders

Describe product requirements and technical specifications
For each stakeholder need, list several ways a product needs to comply with to specifically address this need

Brainstorm on ideas
Generate ideas based on other similar problems and their solutions
Generate ideas based on each product requirement or a set of product requirements

Select feasible ideas
Evaluate ideas upon their compliance with the product requirements and their potential feasibility, usefulness, and innovation
Evaluate the ideas upon compliance with the seven principles of universal design
Where possible, replace bad design features with good ones

Refine ideas
Detail ideas and create mockups of best concepts

Concept evaluation
Have a cross section of potential end users evaluate the best concepts upon compliance with the stakeholder needs, product requirements, and the universal design principles

Final design
Based on stakeholder input, select the best concept, make modifications where necessary, and create a product prototype

Final product evaluation
Evaluate the final product based on the stakeholder needs, product requirements, and the universal design principles

the legislation and regulations, the Americans with Disabilities Act (ADA) for accessible design and various ergonomic standards such the ANSI-117 and the FCC 255 part 308.

The Canadian Standards Association's (CSA) CAN/CSA-B651-95 Barrier-Free Design Standard (hereinafter the CSA Barrier-Free Design Standard), which was first published in 1990, contains requirements for making buildings and other facilities accessible to persons with various physical and sensory disabilities. The CSA

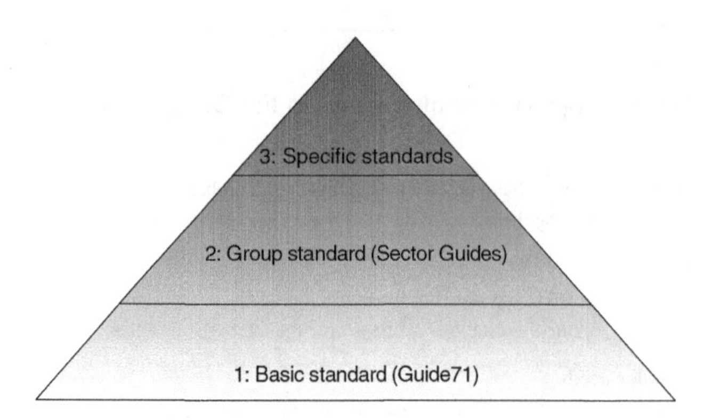

FIGURE 3.4 Guide 71 standard tiers.

Barrier-Free Design Standard states that it was developed to fulfill an expressed need for a national, technical standard covering a broad range of building and environmental facilities, and can be referenced in whole or in part by a variety of adopting authorities.

The ISO/IEC Guide 71 standard consists of three tiers of standards (Figure 3.4): the overarching basic standard (ISO/IEC Guide 71), the group standard (Guide Sectors), and the specific standards. Each tier can be described as follows:

Tier 1: General guidelines including ISO/IEC Guide 71, CEN/CENELEC Guide 6, and JIS Z8071. Internationally, the ISO/IEC Guide 71 is used, which is a guideline for standards developers to address the needs of older persons and persons with disabilities. This standard provides direction to writers of international standards on products, services, and environments as to how to address the needs of older persons and persons with disabilities. Europe uses the equivalent of the ISO/IEC Guide 71, called the CEN/CENELEC Guide 6.

The standard compilation consideration guide ISO/IEC guide 71 is the equivalent of the Japanese Industry Standard (JIS) Z8071 for older people and people who have disabilities. JIS Z8071 includes guidelines for older persons and persons with disabilities on information and communication equipment, software, and services. This standard consists of JIS Z8071-1 Common Guidelines, JIS Z8071-2 Information Processing Equipment, and JIS Z8071-3 Web Content.

Tier 2: Guidelines for specific sectors. This tier includes, for example, the JIS X8341-1 Standard, which has been used in Japan since 2004. It includes guidelines to design for older persons and persons with disabilities. It has UD guidelines for information and communications equipment, software, and services.

Tier 3: Individual standards. This tier includes, for example, JIS X8341-2, related to information processing equipment, and JIS X8341-3, related to Web content.

3.9 EXAMPLES OF UNIVERSALLY DESIGNED PRODUCTS

This section presents some case studies to illustrate UD in products and the built environment.

3.9.1 Living

Designs for various activities have begun in Europe; design guidelines and actual prototypes were developed for a public restroom that can be used by all. This project was made possible by collaboration among various universities, user organizations, and design studios through funds from the Belgian government. Focus groups, user evaluations, and usability studies ensured an end product that would meet most users' needs.

Another approach was taken by the Japanese. Technology of a rehabilitation or assistive device (a special toilet seat for people with disabilities) was integrated in a mainstream product, resulting in a universally designed product that is easy to use by everyone. An adjustable toilet seat for use by people with disabilities was integrated in a regular toilet seat. Figure 3.5 shows the spiral type of development process of this rehabilitation device into a mainstream, universally designed product.

3.9.2 Packaging

A Japanese manufacturer added a tactile strip to the side of a shampoo bottle to make it easier for people with visual limitations to distinguish the shampoo bottle from the conditioner bottle (Figure 3.6). This design can be introduced with minimal cost to producers of shampoo bottles.

Although the tactile strip was intended for individuals with visual impairments, it was found that the tactile strip made the bottles easier to distinguish for people who shut their eyes while washing their hair.

Another example of tactile identification on packaging is the example in Figure 3.7. The Braille dots indicate that the can contains alcohol instead of soda.

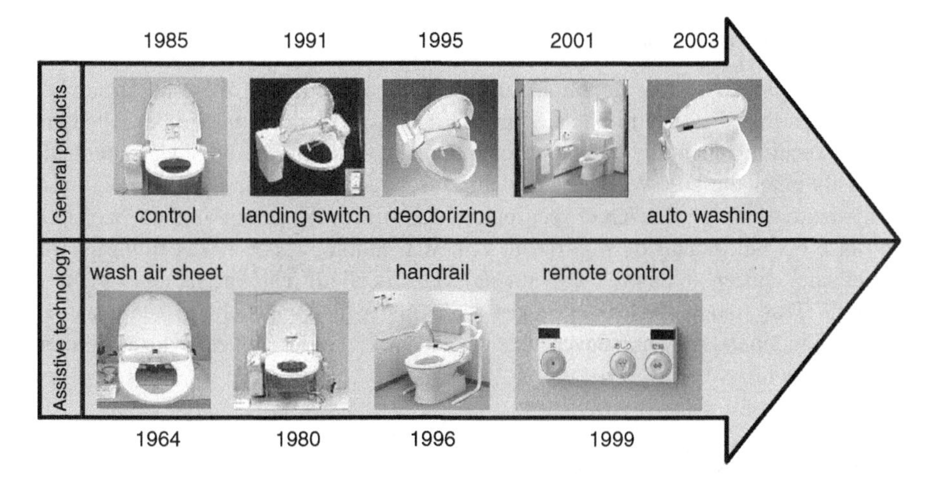

FIGURE 3.5 Spiral-up development of assistive technology into mainstream technology (Photograph courtesy of TOTO Ltd.).

FIGURE 3.6 Shampoo bottle with integrated tactile bars.

FIGURE 3.7 Braille text indicating that the can contains an alcoholic beverage instead of soda.

3.9.3 Transportation

When wheelchair-seated individuals use public transportation, there are a number of features that they use to be safe during transit. First, many public buses feature lifts or kneeling technology to allow both people in wheelchairs and older individuals to easily enter the bus. When a wheelchair user enters the bus, a special section is reserved in the bus (wheelchair securement station) for the person in the wheelchair to place his wheelchair during transit. A new technology that makes riding public buses safer, easier, and more independent is the so-called "rearward facing passenger station." This station consists of a clear space into which wheelchair seated passengers can independently maneuver their wheelchairs to reduce wheelchair movement during travel.

3.10 SUMMARY

Designing products for people with disabilities and older people is very important; however, UD has the potential to reduce costs and benefit a larger number of people. UD is a process that needs to begin at the conceptualization phase of device design in order to achieve maximum benefit. International standards that describe methodologies for implementing UD processes have been approved.

3.11 STUDY QUESTIONS

1. What is universal design?
2. List the steps of a design process and indicate where universal design is integrated.
3. What is the difference between universal design and adaptive design?
4. What are the societal benefits of universal design?
5. What are the personal benefits of universal design?
6. What are the industry benefits of universal design?
7. Name four barriers that have prevented implementation of universal design.
8. Name two commonly used tools that assist designers in understanding the user.
9. Name three examples of a universally designed products or systems.
10. What is ISO/IEC Guide 71?
11. What is orphan technology?
12. What is the purpose of a "universal design matrix"?

BIBLIOGRAPHY

Department of Education. NIDDR Long Range Plan for Fiscal Years 2005–2009. Federal Register: February 15, 2006, 71(31).

Giannini M.J. (2003). *How the Federal Government is Working to Tear Down Barriers for Persons with Disabilities, and How Can We Use the ICF to Ensure Maximum Impact?* Draft Remarks. Presented at the Ninth Annual NACC Meeting on ICF June 16, St. Louis.

ISO 159 http://www.iso.org/iso/en/CatalogueDetailPage.CatalogueDetail?CSNUMBER=40933&scopelist=PROGRAMME.

Kose S., et al., *What is Universal Design, Beyond Barrier Free, Toshi Bunnka Sha.* 1998: Japan.

Kyoyo-Hin Foundation, ISO/IEC Guide 71 Application Method, NIKKEI, 2002.

Mace R. (1990). Definitions: Accessible, Adaptable, and Universal Design, Fact Sheet #6, The Centre for Universal Design.

McNeil J. (2001). Americans with disabilities. Current Population Reports: Household studies, no. 1997. http://www.census.gov/prod/2001pubs/p70-73.pdf.

Moore P. (1985). *Disguised: A True Story*. New York: Word Publishing Group.

Ohnabe H., Cooper R.A., Wheelchairs in the Aging Society, *The 19th Japanese Conference on the Advancement of Assistive and Rehabilitation Technology*, 2004, pp. 215, 216.

Seelman K.D. (2005). Universal design & orphan technology: do we need both? *Disability Studies Quarterly*, 25.

Story M.F. Maximizing usability: the principles of universal design. *Assistive Technology* 10 (1988): 4–12.

4 Technology Transfer

Rory A. Cooper and Jonathan Pearlman

CONTENTS

4.1 LEARNING OBJECTIVES OF THIS CHAPTER

Upon completion of this chapter, the reader will be able to:

- Have an understanding of technology transfer
- Know the difference between various technology-transfer mechanisms
- Understand how intellectual property can benefit products
- Know how to select and use methods to apply universal design (UD) to transfer technology
- Have an understanding of the supply-push and demand-pull processes
- Have some knowledge of reimbursement mechanisms

4.2 INTRODUCTION

Technology is often developed within laboratories or centers or by individuals, and they want to see their invention distributed to people who can benefit from them. There are a number of ways to accomplish this, as will be briefly described in this chapter. In addition, technologies available in one country could benefit people in other countries as well. Transfer of technology within a global climate is important for raising the quality of products available to all people with disabilities.

From a business perspective, assistive technology (AT) has grown to be a business worth $5 billion annually worldwide. Although this is clearly not as large as many sectors of the global economy, it is certainly worthy of notice. In reality, the AT marketplace is probably much larger than current measures can predict. This is for two notable reasons. First, most AT is simply purchased by people for their personal use without utilizing a healthcare system, and many of these consumers would not identify themselves as having a disability. Second, there are tens (if not hundreds) of millions of people, on a worldwide basis, who currently do not have access to AT or the means to purchase it. Because of their large populations of potential AT users, this is likely to change as the economies of China and India continue to grow.

Assistive Technology can reach consumers through number of market channels — some through commercial mechanisms and others through altruistic means. There are a number of issues that need to be considered when attempting to transfer a technology.

4.3 THE SUPPLY-PUSH PROCESS

When an inventor has an idea for a product, process, or service but is either not interested or capable of bringing it to market without support from manufacturers or distributors, then a supply-push process is helpful. Devices, processes, or services that will have significant impact on people with disabilities or for medical rehabilitation is an ideal candidate for development through a supply-push process, which typically consists of a five-step process. The first step is the development of a prototype device, process, or service that must be significant enough to develop a business or to become a notable enhancement to an established company. It is not unusual for devices to be developed to meet the needs of an individual with the possibility that it could be of benefit to a much larger group of people. Hence, during the second phase, new inventions need to be viewed in terms of the competitive market, the market size, market growth potential, and opportunities for the technology under development. The third step involves studying alternative designs, investigations of standards and regulations, and the gathering of consumer input. Consumer input is often obtained through focus groups intended to identify areas for improvement and items of particular strength. This step also includes developing detailed specifications, drawings, charts, and prototypes. Testing for compliance with standards and regulations are often required. The invention may go through multiple iterations during this phase. The fourth phase involves finding a licensing partner or forming strategic alliances for manufacture and distribution of the invention. The fifth phase involves bringing the product to the market.

4.4 THE DEMAND-PULL PROCESS

Research and market analysis are sometimes needed to identify the need for a product or service. The demand-pull process is a means of describing the manner in which a product is developed to meet a specific market need. The first step in the process is to conduct a review of the market sector of interest. This is helpful in identifying underdeveloped or underserved areas of the business sector. The second step is to hone in on specific areas for potential development. This can be accomplished through consulting with stakeholders to further identify and to prioritize unmet needs. The third step is to develop technical and performance specifications. These can be in the form of problem statements and/or design criteria. The fourth step is a cost and feasibility analysis. During this phase, a review of standards, regulations, and payment methods is conducted. Supply chains, manufacturing models, and distribution methods need to be carefully examined, as well as the appropriate business model. The fifth step involves developing a business plan or technology-transfer agreement. The business plan or technology-transfer agreement needs to be presented to potential collaborators or financiers. The final phase involves bringing the product to the market.

4.5 INTELLECTUAL PROPERTY

Inventors create intellectual property (IP) — fruit of their labors. IP is essentially an invention or method created by the inventors. IP is not necessarily a product, nor will it become a product by itself, but it is often the first step. Once IP has been created, the inventor may choose three paths. The first and most common path is to apply for a patent or copyright. Copyrights are used for written documents, regardless of the form in which they are published, and for some software. Patent is a formal recognition by a government body that the inventors have created unique IP and that they have exclusive use of the IP for the term of the patent. Individuals can files patents for devices (e.g., a unique wheelchair design) or for methods (e.g., a process to make something such as to sequence a gene). The rules for applying for patents vary from country to country. Obtaining a patent typically requires the assistance of a patent attorney. The cost of filing a patent can be high, but can be reduced by conducting a thorough online patent search. In the United States, inventors typically have 1 year from initial disclosure of the invention to apply for a patent. However, the filing must take place before disclosure for international patents; otherwise, the patent will be disallowed. Another notable difference is related to the date of invention. For U.S. patents, the date of invention traces back to the earliest documentation of the concept leading to the patent; in other countries (notably European countries), the filing date for the patent constitutes the date of invention. Hence, in many countries, inventors sometimes race to file a patent to protect their intellectual property, whereas in the United States, the filing tends to occur when the decision has been made to bring a product to market or to attempt to license a technology or method. To license a patent is to assign the rights to one or more individuals or organizations. There are literally millions of patented assistive devices and methods related to AT. Students should familiarize themselves with the search engine of the U.S. Patent and Trademark Office as well as similar sites for other countries. Often, inventors do not publish their work in traditional academic journals, and significant contributions to AT would be missed without conducting a patent search. Judgment needs to be exercised before filing for an AT patent because markets for some devices are small and a patent could be a hurdle to bringing a product to the people who need it. Some companies simply protect their products by maintaining trade secrets, that is, their processes or technologies are sufficiently complex that their competitors are not likely to be able to copy them or the cost of engineering a similar product would be prohibitive.

4.6 U.S. FEDERAL FUNDING ASSISTANCE

4.6.1 SBIR/STTR Program

A popular and effective mechanism for funding the development of a new AT within the United States is to apply for small business grants offered by the U.S. governmental agencies. The two main granting mechanisms are the Small Business Innovative Research (SBIR) and the Small Business Technology Transfer Research (STTR). The SBIR grant program has two programs. The primary SBIR program provides funds in the form of grants to small businesses to develop a technology with the goal of bringing it to the market. The STTR program is designed to support the transfer of technology from academic institutions to small businesses. By definition, STTR is a university and industry collaboration, whereas SBIR requires no university partner. However, in reality, applicants to both programs typically have a university partner simply because few companies have the in-house capability to conduct the research component of the SBIR.

The amount of funds available for STTR and SBIR grants vary with the agency that provides the funding. However, all of the programs fund Phase One and Phase Two grants. Phase One grants are to develop preliminary prototypes and to collect pilot data. Promising projects move on to Phase Two, where more advanced prototypes are developed and extensive research is conducted. Phase Three represents commercialization phase, which is funded and performed solely by the small business. The strength of the SBIR program is that it provides funding for innovative new AT, without burdening the small business with paying back the research and development costs. This has allowed companies to take some risks and to invest in assistive technologies for smaller markets.

4.6.2 Small Business Loans or Grants

The Small Business Administration and several states have programs that provide loans and grants to help small businesses get started or to tide over difficult times. Some AT companies have started using these programs. Debt is often difficult for small companies to manage, especially when trying to launch a new product. Grants do not need to be repaid as along as the specified terms of the agreement are met. Most small businesses need to have a line of credit to cover expenses that are not allowed under grants. Small companies tend to be more successful if they can focus their loans on the end stage of bringing a product to the market or to prop up operating expenses while a product gains traction in the marketplace. Ultimately, sales need to generate enough revenue to cover operating expenses and loan payments.

4.7 PRIVATE FUNDING

4.7.1 Charitable Assistance

Charitable organizations typically do not assist new companies to get established in the United States, but they play an important role in other countries, especially developing countries. There are several organizations that provide AT free or at low cost to people with disabilities in developing countries. There are several models of charitable business development. *Micrograms* are small amounts of money provided to an individual to start a business. These grants provide as little as hundreds of dollars or as much as a few thousand dollars. The goal is to help the individual develop a small self-sustaining business to help bring them out of poverty. Some charitable organizations work to build small factories that would produce assistive devices for the local population. Whirlwind Wheelchairs International, in San Francisco, CA (www.whirlwindwheelchair.org) is an organization, for example, that creates small factories in developing countries. In some cases, the goals of a charitable organization may result in the development of business opportunities for AT manufacturers. For example, The Wheelchair Foundation has purchased over 1,000,000 wheelchairs for distribution to people in developing countries. This demand has resulted in a factory focusing on their needs in China.

4.7.2 Foundation Funding

Foundations are organizations set up to provide specific programs or services. There are numerous foundations worldwide that have been established to help build small businesses or to provide advice and consulting for small businesses. Often, there is an application process requiring a business plan, budget, and preliminary market analysis. Some foundations provide grants or services with no expectation of a return on their investment. However, it is more common that, if actual funds are invested by the foundation, then usually some sort of equity or royalty arrangement is required. Foundations are a good source of support to help start an AT business or to help a small business that is experiencing some financial difficulty.

4.7.3 Individual Investors

Individual investors are often called *Angels*. They typically give funds with few strings attached, the exception being the common request for substantial equity in the company often in excess of 50% and sometimes over 90%. Angels expect to see a large return on their investment in a short period of time, typically a few years. Inventors often overestimate the value of their idea to the business. In reality, the idea or prototype frequently represents about 10% of the potential value of the business. Therefore, individual investors will often insist upon selecting a management team to assist with bringing a product to the market, unless the individual starting the company has had previous success in starting a business.

4.7.4 Venture Capital

Venture capitalists are wealthy individuals or organizations that invest in business start-ups in return for equity. As with Angels, venture capitalists will ask for a substantial, if not complete, control of the company. Venture capitalists require a more formal process involving a detailed business plan, market analysis, description of the IP protection, and background information on the principles of the company. It is very competitive to obtain venture capital funding, but if one manages it, the assistance provided in the form of funding and business management can be of tremendous help in making the company successful.

4.8 PAYMENT MODEL

There are two basic models for the payment of AT. The most common and largest in total expenditures is consumer sales. Direct sales to the customer are often overlooked, because they typically involve products that have largely been integrated into stores serving the general population, and many of these consumers do not consider themselves to have a disability. For example, eyeglasses are worn by millions of people, yet few would consider them to be assistive devices. Other products, such as grab bars or software to vary text size have become mainstream. This has had the effect of lowering cost and increasing availability, but at the same time many of these ATs are no longer reimbursed by third-party payment systems.

Reimbursement for assistive devices is often covered by insurance agencies, and various federal, state, and local governments, summarily called third-party payers. The products that are most commonly thought of as assistive devices, such as wheelchairs and prosthetic limbs, are mostly paid for by third-party payers. Reimbursement for assistive devices through third-party payers impacts the market in three important ways. First, devices must be registered and qualified for payment, which often results in guidelines set for eligibility for reimbursement as well as restrictions on the amount to be paid. Second, reimbursement by third-party payers often alters the consumer–supplier relationship, making consumer choice and opinion subservient to the rules set forth by the third-party payer. Third, reimbursement tends to slow development because third-party payers often impose barriers to entry into the marketplace. Naturally, reimbursement is essential for people with disabilities to receive the products and services that they truly need; this is the foundation of the insurance industry and of civilized nations.

Both reimbursement and purchases by consumers are needed to ensure that the products and services beneficial to people with disabilities are available and that they continue to improve. However, it is a constant struggle to strike the appropriate balance. Any new venture needs to study both sales models and determine which is best for their product.

4.9 MANUFACTURING

There are two basic models for manufacturing assistive devices. The traditional model relies upon creating a manufacturing plant with the resources to create and assemble all or most of the parts necessary to produce the product. This often requires considerable capital investment, and can be a significant hurdle for a start-up company. However, as a means of exercising greater control over the process and thus to control costs, some companies transition to this model as their sales and distribution grow.

Virtual companies and outsourcing have become the norm for small companies introducing an AT. As there are companies around the world that specialize in the manufacture of components and assemblies for other companies (sometimes referred to as virtual manufacturing); it has become increasingly easy to outsource most, if not all, of the manufacturing processes. Electronic printed circuit boards (PCBs) and populating the PCBs can often be done in days by communicating purely via the Internet and telephone with high-quality and reliable results. Because of the widespread use of computer-aided design (CAD) programs and computer-aided manufacturing (CAM), it is no longer necessary for the design engineer and manufacturing engineer to be located in close proximity. The process from design to manufacturing has been further facilitated through rapid prototyping, which allows working models to be made directly from CAD files through machines such as three-dimensional printers. Because of these technologies, design cycles are much quicker and innovations can reach the marketplace faster, often years before reimbursement mechanisms respond.

4.10 DISTRIBUTION MODELS

There are a number of distribution models that have been used effectively. Most traditional assistive devices that are reimbursed through third-party payers use a network of dealer representatives. An interesting aspect of the AT marketplace is that there are few exclusive dealerships. Most AT suppliers (i.e., dealers) represent multiple products and, in many cases, competing products. This provides the supplier the advantage of presenting clients with multiple options to identify the product that best meets the client's needs. Some manufacturers have established exclusive distributor networks that represent a single product line. This allows the supplier to stay focused on serving their manufacturing partner. It is also simpler for the supplier to gain and maintain in-depth knowledge of the product line represented.

The direct-sales model is similar to exclusive distributor networks, the main difference being the ownership of the distributor. In a direct-sales environment, the product representative is an employee of the manufacturer, whereas in an exclusive distributor relationship the distributorship is independently owned and operated. Direct sales may also be performed by the manufacturer without face-to-face contact with the consumer. This can be accomplished through catalogs, the Internet, and telephone sales. Direct sales is most effective for commodity items, standardized products, and replacement components. Only very knowledgeable consumers should consider buying complex items, such as wheelchairs, without the assistance of a professional during an in-person visit.

Many manufacturers use a combined model of nonexclusive dealerships and factory sales representatives. Factory sales representatives have regional sales territories. Within their territories, they provide support and education for their affiliated suppliers and maintain contacts with the clinicians as well. Factory representatives are most effective when they have in-depth product knowledge and clinical expertise. Much of their job is based on educating suppliers and clinicians about the benefits of their products and potential customers.

There is no formula for determining the most appropriate distribution model. Companies and products have their unique marketing needs. It takes careful study of the market, payment mechanisms, and business resources to optimize distribution channels. Of course, the approach can vary from product to product and over the time course of the company. Flexibility and responsiveness are the keys to success.

4.11 A TECHNOLOGY-TRANSFER FRAMEWORK

To compare each of the models based on factors relevant to the success of a small AT business, the following broad definitions describing the important factors in the technology-transfer process may be used:

Input: The financial and/or technical requirements necessary to perform the project.

Sustainability: The ability of the business, after initial funding, to have established production that does not require persistent external input. This could also include the economic value to the local community generated because of the business or technology (e.g., developing jobs, income into a region, educational benefit by training workers, etc.).

Appropriateness: A metric describing how well the AT fits the need of the consumer in his/her cultural, physical, and psychosocial environment. This could be measured by the average number of years the device functions properly, abandonment rates, changes to community participation, and/or maintenance rates.

Impact: The number of devices produced and or delivered in a given period of time and the changes that they make to the users' lives.

This language is used because each of the four factors (input, sustainability, appropriateness, and impact) have been deemed important both in measuring the success of an AT business (impact, appropriateness, and sustainability) and evaluating whether one should be started (input). By combining these metrics, one could, for example, compare the cost per wheelchair delivered (impact $\div$ input) between businesses. Similarly, one could develop a metric for local impact (sustainability $\div$ input, i.e., number of jobs developed $\div$ input cost).

4.12 SUMMARY

Technology transfer and business development are complex processes that require business acumen and, often, technological expertise. It is best to consult with business, marketing, and manufacturing experts while exploring the avenues to bringing a new technology to market. As there are many avenues for transferring concepts to products, the best choices may not be obvious to the inexperienced.

4.13 STUDY QUESTIONS

1. Briefly describe the supply-push process.
2. What are the key differences between the supply-push process and the demand-pull process?
3. Name three means of protecting intellectual property?
4. Describe the attributes of three different mechanisms for funding the transfer of a technology to a commercial product?
5. Why are different payment models employed for AT, and how does that affect the consumer and distributor relationship?
6. What is virtual manufacturing?
7. Describe a distribution model for selling an assistive device.
8. What is the role of a manufacturer's representative?
9. Why does the U.S. government invest in developing and transferring AT?
10. Describe the four basic steps of a technology-transfer framework.

BIBLIOGRAPHY

ALIMCO, Artificial Limbs Manufacturing Corporation of India 2003 Annual Report, Indian Government: Kanpur, India, 2003.

Boonzaier D.A., Ten Years of Rehabilitation Technology in a Developing Country: A Review 1980–1990, *RESNA 13th Annual Conference*. Washington, D.C., 1990.

Donald A., Political Economy of Technology Transfer,*BMJ*, 319: pp. 1–3, 1999.

HHIM, Hope Haven International Ministries, 2004.

Hof H., Hotchkiss R., and Pfaelzer P., Building Wheelchairs, Creating Opportunities: Collaborating to Build Wheelchairs in Developing Countries, *Technology and Disability*, 2, 1993.

Hotchkiss R., Independence through Mobility, *Assistive Technology International*, 1984.

Hotchkiss R., Ground Swell on Wheels: Appropriate Technology Could Bring Cheap, Sturdy Wheelchairs to Twenty Million Disabled People, *The Sciences*, 1993.

Hotchkiss R., Putting the Tools in the Hands That Can Use Them: Wheelchairs in the Third World, *RESNA 10th Annual Conference*, San Jose, CA, 1987.

Hotchkiss R. and Knezevich J., Third World Wheelchair Manufacture: Will It Ever Meet the Need? *RESNA 13th Annual Conference*, Washington, D.C.: RESNA, 1990.

Hotchkiss R. and Pfaelzer P., Measuring Success in Third World Wheelchair Building, *RESNA International*, 1992.

Jeserich M., Building Appropriate Chairs for the Developing World, *AT Journal*, 2003.

Kim J. and Mulholland S.J., Seating/Wheelchair Technology in the Developing World: Need for a Closer Look, *Technology and Disability*, 11: pp. 21–27, 1999.

Krizack, M., Discussion Regarding Estimated Number of Wheelchairs Produced by Whirlwind Wheelchair Affiliated Workshops, Editor: Pearlman J., Pittsburgh, 2004.

Lane J., Understanding Technology Transfer, *Assistive Technology*, 11: pp. 5–19, 1999.

Leavitt, R.L., *Cross-Cultural Rehabilitation: An International Perspective*. London; Philadelphia: W.B. Saunders, p. 413, 1999.

McCambridge M., Coordinating Wheelchair Provisions in Developing Countries, *RESNA 2000*, 2000.

Pfaelzer P., Whirlwind Network News: Advancing Wheeled Mobility Worldwide 98, *Wheelchair Production in Developing Countries: Some Observations*, Whirlwind Wheelchair International: San Francisco State University, 1998.

Prabhaka M. and Thackker T., A Follow-Up Program in India for Patients with Spinal Cord Injury: Paraplegia Safari, *The Journal of Spinal Cord Medicine*, 27: pp. 260–262, 2004.

WCF, Wheelchair Foundation Website. The Wheelchair Foundation, 2003.

Werner D., Thuman C., and Maxwell J., *Nothing About Us Without Us*, Palo Alto: HeathWrights, p. 350, 1998.

WFH, Wheels for Humanity Web site, 2004.

WFW, Wheels for the World Web site, 2004.

5 Standards for Assistive Technology

Douglas Hobson, Rory A. Cooper, and Martin Ferguson-Pell

CONTENTS

5.1 LEARNING OBJECTIVES OF THIS CHAPTER

Upon completion of this chapter, the reader will be able to:

Have an understanding of the history of assistive technology (AT) standards
Know the current status of several AT standards
Understand the practical application of AT standards
Understand how AT standards are developed
Have an understanding of the engineering contributions to AT standards
Have some knowledge of future needs for AT standards

5.2 INTRODUCTION

5.2.1 RATIONALE FOR INDUSTRY STANDARDS

Products for persons with disabilities, especially those on which users rely to carry out their daily activities, assume a high level of expectation in terms of design, technical performance, cost–benefit, reliability, and safety. Prior to 1980, there was no systematic way for wheelchair users, product prescribers, or healthcare insurers to determine, in advance of purchase, what products actually met these higher needs and expectations. Reports of poor reliability, user injury, and sometimes fatalities were documented. Large bulk purchasers, such as the Department of Veterans Affairs in the United States and national healthcare providers in Europe were also in need of objective information that would guide their decisions related to choice of products for inclusion on their provider listings of approved products for government payment.

For example, despite many years of research on wheelchair seat cushions that incorporate design features intended to prevent the formation of pressure sores and maintain tissue integrity, there were no objective tests or established criteria that could help differentiate the efficacy of different products. Since pressure sores can be very debilitating, costly, and, in some cases, life-threatening, the performance of these products takes on an important medical as well as functional dimension. In the absence of industry standards that provide objective test information, any manufacturer can claim superior performance capabilities for their cushion products. Objective presale information is critical for clinicians and users to aid in their product selection decision-making process. Healthcare insurers also need objective test information upon which to base their payment codes that reflect a justifiable relationship between the quality of product performance and reimbursement level.

In the late 1970s, working under the auspices of the International Standards Organization (ISO) and inspired by European countries with large national healthcare programs (Sweden, Netherlands, and UK), a Subcommittee on Wheelchairs (SC-1) was established within Technical Committee (TC-173) to develop voluntary industry standards for wheelchair products.

Rationale or purpose of standards — the rationale for the development of voluntary industry standards may be summarized as follows:

- To provide a common minimum benchmark of quality and safety that all manufacturers should attain
- To promote improved safety for areas in which problems have arisen or may arise with existing products
- To provide standardized product information, based on objective test information, that can be used for decision making by service providers, product users and insurance agencies
- To facilitate barrier-free trade of AT products on a the worldwide scale
- To consolidate technical, scientific, and clinical knowledge so as to advance the quality and safety of AT products worldwide

5.2.2 Historical Overview of U.S. Involvement in National and International Standards Development

In the late 1970s, in the United States, the Health Industry Manufacturers Association (HIMA) served as the Technical Assistance Group (TAG) for the American National Standards Institute (ANSI) in terms of representing the US wheelchair manufacturer's interests within the newly formed ISO TC173-SC-1 structure. Keith Rodaway, then head of engineering at Everest and Jennings, Inc., attended the early ISO TC-173/SC-1 meetings as a representative of HIMA. HIMA was unable to provide the necessary ongoing resources for development of a multidisciplinary U.S. national working group on wheelchair standards, which is essential for meaningful involvement in ISO activities. In 1981, the lead author was approached by Rodaway and asked to set up a forum within Rehabilitation Engineering and Assistive Technology Society of North America (RESNA) for developing voluntary industry performance standards for wheelchairs so that U.S. wheelchair manufacturers could officially participate in ISO standards activities that had begun in 1978. The RESNA board approved this proposal, and, a few years later, RESNA became designated by ANSI as the U.S. standards development body for disability products and also the official U.S. designee (TAG) to participate in related ISO activities (TC-173/SC-1).

Since that time, many very positive activities have occurred. For example, with direct financial support from the Veteran's Health Administration (VHA) and the Paralyzed Veterans of America (PVA) and with indirect support via industry and federal funding to research institutions, mainly from the National Institute on Disability and Rehabilitation Research (NIDRR) of the Department of Education, we now have in excess of 30 RESNA voluntary standards for wheelchairs and wheelchair seating. This series covers a wide range of design and performance requirements, test methods, and information disclosure requirements. Most wheelchair products now being manufactured and marketed in the United States conform to the majority of the requirements of the RESNA and/or ISO standards. The U.S. standards, to a large extent, have been harmonized with Canadian and ISO equivalent standards so that barriers to import and export for both consumers and manufacturers, respectively, are minimized. Table 1 in Annex A contains a listing of the ISO standards that have been published or are under development to date.

Occupied wheelchairs are used as seats in some motor vehicles. In partnership with the Society of Automotive Engineers (SAE)–Adaptive Devices Committee, a SAE recommended practice document, SAE J2249-Wheelchair Tie-down and Occupant Restraint Systems (WTORS) was published in 1996. This document sets out the requirements for WTORS intended for use with occupied wheelchairs during transport in a motor vehicle. Virtually all WTORS marketed in North America now conform to the SAE J2249 recommended practice. Because of RESNA's prior work in wheelchair standards and its long-standing liaison with ANSI and ISO, this SAE work was transferred to RESNA and established as the Subcommittee on Wheelchairs and Transportation (SOWHAT), later named the Committee on Wheelchairs and Transportation (COWHAT).

In May 2000, SOWHAT completed the first voluntary standards for wheelchairs that are intended for use as a seat in a motor vehicle (ANSI/RESNA WC-19 and ANSI/RESNA, 2000). This effort was made possible as a result of combined private, wheelchair industry, school bus associations, and federal grant support, all managed and supported by a consortium of research institutions. This 4-year effort to develop the WC-19 standard cost approximately $350,000. Also, this effort has permitted active participation in ISO WG-6, in which an equivalent standard, ISO 7176-19, was completed in 2001.

In June 1998, the RESNA Technical Guidelines Committee authorized the formation of the Committee on Wheelchair Seating Standards and three related working groups to begin work on voluntary wheelchair seating standards. A parallel effort in ISO, WG-11, was established shortly thereafter.

For reasons that will be expanded upon later in the chapter, it is reasonable to say that no other activity or development has contributed more to the advancement of quality and safety of wheelchair-related products on a worldwide scale than the voluntary industry standards program.

5.2.3 Voluntary vs. Regulatory Industry Standards

Unlike Federal Motor Vehicle Safety Standards (FMVSS) in the United States, for example, that have mandatory compliance requirements, all the wheelchair and wheelchair seating standards development efforts to date have been based on the voluntary participation by industry. The main advantage of this approach is that it is easier to reach a consensus on contentious issues as compliance is not mandated. The disadvantages are that manufacturers may or may not opt to comply with the full requirements of standards.

Depending upon the laws and regulations in each country, government agencies at various levels can decide to make compliance a requirement. In Europe, Japan, and the

United States, wheelchairs are considered medical devices. In Europe, this subjects wheelchair products to European Union (EU) laws requiring compliance to requirements that govern all medical devices, including CE marking. In most cases, the ISO voluntary industry standards will form the basis for these requirements of wheelchair related technology, thereby making compliance to the EU-selected parts of ISO standards mandatory by law. The irony is that, if government agencies make selective use of the standards to suit their needs for regulation or device classification and cost containment, then that drives a payment system that can stifle future product development. In its worst case, it can lead to inferior clinical outcomes rather than an improvement. In an environment that depends on the voluntary compliance by industry, such as the United States, those standards and related test methods that have been made selectively mandatory for coding and payment purposes by a government agency will, in all likelihood, become the pseudo industry standard, with the real risk that the remainder of the standards will be being largely ignored.

In the United States, the FDA has selectively recognized the ANSI/RESNA and/or ISO standards as proof of safety compliance for 510k approvals on new wheelchair products. More recently, the federal government's Centers for Medicare and Medicaid Services (CMS) has decided to selectively reference test methods to classify wheelchairs and wheelchair seating products into categories for which funding codes for reimbursement are being assigned. To the extent that a manufacturer wishes to have a wheelchair or seating product made eligible for reimbursement through the CMS system, compliance with those test methods referenced by CMS is mandatory. While encouraging, it may prove problematic, given the limitations and application of the test data (Raflo, 2002).

Although there is little experience to date with these new policy decisions, potential problems are on the horizon. For example, in the case of wheelchair seat cushion products, CMS has published guidelines for classifying cushions into six billing code categories. The same document also specifies the eligibility of different groups of users for these billing codes. Although the introduction of the new codes is widely supported, there is concern regarding the manner in which CMS has elected to selectively use the draft standards and regarding the fact that the resulting clinical outcome (and future product development) will be driven by the funding formula (codes). Unfortunately, CMS has elected to narrowly and prematurely use relative peak pressures test data as a means to classify seat cushions. In later versions of the draft standard, these tests were withdrawn owing to difficulties in achieving repeatable results between laboratories and comparable results between different pressure-mapping systems. What is required is a balance across a wider range of cushion parameters that are applicable to a wider range of users. As intended by the standards, this approach will then provide the disclosure of the technical characteristics necessary to allow the judgment of skillful judgment clinicians working with their client to arrive at the desired clinical outcome.

Similar steps have been taken by CMS to selectively reference RESNA test methods from industry standards for the coding of powered wheelchairs. Among several concerns and for reasons that are not clear, CMS has violated one of the key tenets of interlaboratory testing by allowing the use of any size test mass in the wheelchair as may be specified by the manufacturer. Standard test masses have historically been required so the test results could be compared between products — information critical to both wheelchair prescribers and users. Also, testing to some worst-case mass is not typically done in standards testing. For example, cars are not tested with largest possible occupant crashing into a barrier at the highest possible speed. Furthermore, the wheelchair industry may now primarily focus on the more limited CMS required tests and largely ignore the remainder of the RESNA voluntary standards. Of course, manufacturers may do this at some peril to their bottom line, as litigation that can effectively demonstrate that injury has resulted from lack of full compliance to a nationally recognized voluntary industry standards can be costly. The point being that selective adoption of industry standards to meet the needs of national government agencies should be done with extreme caution; otherwise a great deal of standards development progress can be quickly destroyed (DMERC, 2001).

5.3 ORGANIZATION OF NATIONAL AND INTERNATIONAL INDUSTRY STANDARDS

Acquiring an understanding of the structure and processes involved in taking a new standard through the various review and voting approval stages can be a daunting task. This section attempts to simplify that task.

5.3.1 Brief Description of the Standards Development Process

Because the ISO standards are the most universally applicable and referenced standards by most national bodies, the following description will be limited to the ISO standards. The following briefly outlines the six stages that each new standard progresses through from a preliminary new work item (PWI) to the end product — a published international standard (IS). To progress from one stage to the next requires a 70% approval response by the national member bodies that have elected to participate in a particular standard development process.

Preliminary work item (*PWI*): The early stage in which a work group informally consolidates its ideas and prepares a working draft of a new standard. The main goal at this time is to seek agreement on the scope (purpose) of the standard. To initiate this activity, a proposal for a preliminary work item is made to the responsible subcommittee (SC) for approval.

New work item (*NWI*): This is the first formal stage in the process in which a proposal for a NWI is circulated to all participating member countries, complete with a working draft of the proposed standard for approval. Upon approval, the ISO clock begins to run and timetables are established for each stage of advancement, with a completion target of 36 to 40 months for the Final Draft International Standard (FDIS) vote.

Committee draft (*CD*): Assuming voting approval as an NWI, the earnest work of the working group now begins. Test methods must be fully developed and validated, terms and definitions agree upon, a multitude of editorial revisions made, and graphics and text all formatted to strict ISO rules, in preparation for the CD review and vote by the participating national bodies. At this point in the process, the standard takes its initial form, and test methods are often evaluated by multiple test laboratories. Once the working group members are satisfied that the test methods and requirements are reliable and reasonable, the CD is prepared for national body voting.

Draft international standard (*DIS*): Assuming voting approval of the CD version, this stage usually results in many pages of suggested revisions that result from the comments associated with the CD voting process. The work group must act on each comment and, if rejecting a comment, explain the reason. Upon revision, the document is then sent out again for national body voting as a proposed draft international standard (DIS). At this level, the voting process is managed by the ISO central secretariat in Geneva, Switzerland. At any stage to this point, if the voting comments are largely technical in nature (as opposed to only editorial), even though there are sufficient positive votes to warrant advancement to the next stage, a decision can be made to continue the development process at the same stage. Also, if the timeline has not been adequately adhered to, the process will be terminated by ISO and a request must be made by the working group to restart the process.

Final draft international standard (*FDIS*): Assuming a successful vote at the DIS level, the voting comments should be much fewer and suggestions for change reserved to editorial comments, as the document has now reached the penultimate stage of the process. National voting at this stage is only a "yes" or "no" vote as comments are not solicited. At this stage, the work of the working group has been completed and it is removed from their list of work assignments if the FDIS passes.

International standard (*IS*): This is the final stage of completion in which final approval must be granted by an internal ISO committee. The standard is then finally edited into ISO format in Geneva, translated into French, published in both English and French as an international standard, and posted on the ISO Web site. This final process typically adds an additional 6 months to the FDIS level process. The working group that prepared the standard never sees a copy of the published (IS) standard unless they purchase it from ISO.

5.3.2 Anatomy of a Typical ISO Industry Standard

As a standards-setting body, ISO has very detailed formats for the preparation and publishing of its documents. The following briefly walks through a typical layout and explains what content appears in each of the major sections of a typical ISO standard. Examples have been excerpted from the existing ISO 7176 and ISO 10254 series on wheelchair-related technology.

Foreword: This contains mainly information about ISO, intellectual property, collaboration with other international standards bodies, ISO rules that have been applied, edition being replaced, technical committee (TC) and SC responsible for work, and a listing of all the other parts of the ISO 7176. A sample foreword is presented in the following:

> *Foreword:* ISO (the International Organization for Standardization) is a worldwide federation of national standards bodies (ISO member bodies). The work of preparing International Standards is normally carried out through ISO technical

committees. Each member body interested in a subject for which a technical committee has been established has the right to be represented on that committee. International organizations, governmental and non-governmental, in liaison with ISO, also take part in the work. ISO collaborates closely with the International Electrotechnical Commission (IEC) on all matters of electrotechnical standardization. International Standards are drafted in accordance with the rules given in the ISO/IEC Directives, Part 3. Draft International Standards adopted by the technical committees are circulated to the member bodies for voting. Publication as an International Standard requires approval by at least 75% of the member bodies casting a vote. Attention is drawn to the possibility that some of the elements of this part of ISO 7176 may be the subject of patent rights. ISO shall not be held responsible for identifying any or all such patent rights. International Standard ISO 7176-2 was prepared by Technical Committee ISO/TC 173, *Technical systems and aids for disabled or handicapped persons*, Subcommittee SC 1, *Wheelchairs*.

This second edition cancels and replaces the first edition (ISO 7176-2:1990), which has been technically revised.

ISO 7176 consists of the following parts, under the general title *Wheelchairs*: (lists all other parts of 7176 series) (ISO, 2001).

Source: ISO 7176-2, Determination of static stability for powered wheelchairs, 2001.

Introduction: This is a brief overview, less than a page in length, of the problems encountered in the field that the standard has been designed to address. Also, a brief overview is provided of the direction taken in the standard and what specific technology is covered by the standard. An example follows.

Introduction: The provision and selection of wheelchairs and associated seating supports relies on clear communication of information relating to these devices. Over time, many terms and definitions have evolved. Unfortunately, this process has resulted in a lack of clear meaning for some terms and duplication of other terms (sometimes with conflicting messages). For example, the terms tilt and recline are sometimes used interchangeably, but usually have quite distinct meanings. If used inappropriately, an entirely inappropriate wheelchair may be specified or purchased. The purpose of this part of ISO 7176 is to provide a nomenclature of terms and their definitions to form the basis of clear communication across the field of wheelchair and associated seating and to eliminate confusion from duplication or inappropriate use of terms. The nomenclature is drawn from surveys of the literature and language used by experts in this field. It excludes, however, terms which are adequately defined in the everyday language of English, medicine, and technology. The standard recognizes that there are a number of terms in use which, because of duplication in adequacies of meaning, should be replaced by terms from this nomenclature. To help people move towards a common vocabulary, these deprecated terms are included along with a reference to the preferred term from the nomenclature.

The development and application of wheelchair standards is particularly dependent upon clear and consistent terms and definitions. Hence, a major proportion of this part of ISO 7176 includes terms and definitions used in more than one of the ISO standards specifically related to ISO Wheelchair Standards. These include the ISO 7176, 10542, and 16840 series, and ISO 7193. In the future standards in these series will cite this do comment for definition of terms wherever possible, thus ensuring consistency of definitions.

This part of ISO 7176 is intended purely as a means of specifying terms and definitions. It does not attempt to classify wheelchairs and associated seating into any classification of device groupings as this is the purpose of ISO 9999. Annex A provides a standard set of descriptors for characterizing wheelchairs (ISO, 2004).

Source: ISO DIS 7176-26-Wheelchairs — Part 26: Vocabulary, 2004.

Scope: This is probably one of the most important sections in a standard, as it is intended to concisely specify what the standard applies to, and in some case, when it does not apply. A sample "scope" is presented in the following:

Scope: This international standard applies to all manual and powered wheelchairs, including scooters, which, in addition to their intended function as mobility devices, are also intended for use as forward-facing seating by adult occupants of motor vehicles. It also applies to wheelchairs with add-on components designed to meet one or more of the requirements of this standard.

This standard specifies wheelchair design and performance requirements and associated test methods, as well as requirements for wheelchair labelling, presale literature disclosure, user instructions, and user warnings. These requirements are applicable to wheelchairs that are designed to be secured by any type of wheelchair tiedown that complies with ISO 10542-1 and any other applicable parts of 10542 (ISO, 2000).

Source: ISO 7176-19: Wheelchairs — 7176-Part 19: Wheeled Mobility Devices for Use in Motor Vehicles, 2000.

Normative references: This section references any other standards that need to be considered when applying the standard. An example follows.

Normative References: The following normative documents contain provisions, which, through reference in this text, constitute provisions of this part of ISO 7176. For dated references, subsequent amendments to, or revisions of, any of these publications do not apply. However, parties to agreements based on this part of ISO 7176 are encouraged to investigate the possibility of applying the most recent editions of the normative documents indicated below. For undated references, the latest edition of the normative document referred to applies. Members of ISO and IEC maintain registers of currently valid International Standards (ISO, 2000).

Source: ISO 7176-19: Wheelchairs — Part 19: Wheeled Mobility Devices for Use in Motor Vehicles, 2000.

Terms and definitions: These provide definitions for any terms used in the standard that are not terms in common usage. An example follows.

Term and definitions: For the purposes of this document, the terms and definitions given in ISO 6440 and the following apply:

3.1 Running brake: Means to stop or to slow the wheelchair.
3.2 Control device: Means by which the user directs an electrically powered wheelchair to move at the desired speed and/or in the desired direction of travel.
3.3 Parking brake: Means to keep the wheelchair stationary (ISO, 2002).

Source: ISO/FDIS 7176-3 Wheelchairs — Part 3: Determination of effectiveness of brakes, 2002.

Design requirements: These are the minimum design features that must be exhibited by the product to be in compliance. These are used when the committee believes that safety or performance warrants specific requirements for the design of the device or potential product. Design requirements may restrict creativity and limit the ability of manufacturers in the design of devices; therefore, they must be specified requirements only when necessary. An example follows.

Design requirements: The wheelchair shall be designed to:

4.1.1 Provide for forward-facing securement in a motor vehicle by one or more types of wheelchair tie-down systems that conform to ISO 10542.
4.1.2 Have a minimum of four securement points, two at the front and two at the rear that conform to the specifications set forth in Annex B (ISO, 2001).

Source: ISO 7176-19: Wheelchairs — Part 19: Wheeled Mobility Devices for Use in Motor Vehicles, 2000.

Identification, information, and instruction requirements: These specify what must be provided by the manufacturer regarding permanent product labeling, user information, and, if applicable, installation instructions. An example follows:

Identification and labeling:
WTORS and replacement parts shall be permanently and legibly marked with:

a) Manufacturer's name or trademark,
b) Month and year of manufacture, and any other identification necessary to clearly identify a WTORS in the event of a product recall, and
c) A mark showing that the WTORS conforms to ISO 10542-1.

Instructions for installers:

Manufacturers of WTORS shall provide written instructions for the installer in the principal language(s) of the country in which it is marketed.

The instructions shall include statements that:

a) The WTORS should be installed for forward-facing wheelchairs,
b) Identify the number of separate packages containing WTORS components,
c) The WTORS conforms to ISO 10542-1,
d) In order to fit low across the pelvis and/or over the upper thighs and thereby reduce the possibility of the belt loading the abdomen (ISO 2000).

Source: ISO 10542-1: Wheelchair tiedown and occupant restraint systems — Part 1: Requirements and test methods for all systems, 2000.

Performance requirements: Performance requirements specify how the product must perform when tested in accordance to the test methods contained in the standard. They are intended to set a minimum performance level to ensure the safety of the device user or to ensure adequate performance during normal usage. An example follows:

Frontal impact test:

The wheelchair shall be dynamically tested in accordance with Annex A, using a four-point strap-type tie-down that conforms to ISO 10542-2. It may also be dynamically tested using other methods of securement.

The following requirements shall be met during and after each test conducted. During the test:

a) The horizontal excursions of the assistive technology devise (ATD) and the wheelchair with respect to the impact sled shall not exceed the limits in Table 3.
b) The knee excursion shall exceed the wheelchair Point P excursion as follows:

$$X_{knee}/X_{wc} \geq 1.1$$

Note: Compliance with this requirement reduces the potential for the wheelchair to apply large horizontal loads to the wheelchair occupant.

c) The rearward excursion of the head of the ATD shall not exceed the limits shown in Table 3 (ISO, 2000).

Source: ISO 7176-19: Wheelchairs — Part 19: Wheeled Mobility Devices for Use in Motor Vehicles, 2000.

Test report: The test report section specifies what information must be contained in a test report and kept on file by the manufacturer. Only the information in the information disclosure section needs to be available in the manufacturer's presale literature. However, regulatory bodies typically examine all aspects of testing results. An example follows:

The test report shall contain the following information:

a) A reference to this part of ISO 7176;
b) The name and address of the testing institution;
c) The name and address of the manufacturer of the wheelchair;
d) The date of issue of the test report;
e) The wheelchair type and any serial and batch numbers;
f) The size of the dummy used or, if a person is used, the mass of the driver and weights;
g) Details of the setup of the wheelchair as specified in ISO 7176-22, including equipping and adjustments;
h) A photograph of the wheelchair equipped as during the test;
i) Description of the parking brake(s) tested including method of operation such as finger/hand/foot control, manual, electrical, automatic, etc.;
j) If preparation of the wheelchair requires measurement of the brake operating force (as specified in 6 b), the force, in newtons, required to operate the brakes during the tests;
k) The results of the parking brake tests as determined in 7.2 (ISO, 2002).

Source: ISO /FDIS 7176-3 Wheelchairs — Part 3: Determination of effectiveness of brakes, 2002.

Information disclosure: The information disclosure section specifies what test information that must be disclosed in the manufacturer's presale literature intended for use by clinicians and users. An example follows.

In addition to the requirements in 7176, Part 15, the wheelchair manufacturer's presale literature shall include:

a) A statement that the wheelchair is designed to be forward facing when used as a seat in a motor vehicle and that it complies with the requirements of ISO 7176/19-20XX,
b) A description of the types of tie-downs that are suitable for use with the wheelchair (i.e., four-point, strap-type, clamp systems, specific type of docking system, etc.),
c) A statement that ease of access to, and maneuvrability in, motor transit vehicles can be significantly affected by wheelchair size and turning radius, and that smaller wheelchairs and/or wheelchairs with a shorter turning radius will generally provide greater ease of vehicle access and maneuverability to a forward-facing position,

d) A statement of whether the wheelchair provides for, and has been tested with, any wheelchair-anchored occupant restraint belts (ISO, 2000).

Source: ISO 7176-19: Wheelchairs — Part 19: Wheeled Mobility Devices for Use in Motor Vehicles, 2000.

Test methods: Laboratory tests are used to verify that design and performance requirements have been met. A normative test means that it is required and must be followed to comply with the standard. An informative annex is for information or guidance only.

Annex A (normative): Method for determining brake lever operating force
A.1 Test method

a) Select the part of the lever through which the force is to be applied from the following (see Figure A.1), with precedence for selection given to the earliest in the sequence below:
   1) If the lever is fitted with a generally spherical knob, apply the force through the centre of the knob;
   2) If the lever is tapered, apply the force through the point where the largest cross section intersects the centerline of the lever;
   3) If the form of the lever is such that the lever is gripped by the whole hand, apply the force through the centerline of the lever, 15 mm from the end;
   4) If the brake is operated by pushing or pulling a bar or pad, apply the force to the centroid of the bar or pad;
   5) If the lever is parallel or any shape other than those above, apply the force through a point on the centerline of the lever, 15 mm below the top;
   6) If the lever is telescoping or is supplied with an extension handle, apply the force 15 mm from the end when fully extended.
b) Set up a means to operate the brake by applying a force via the force measuring device specified in 5.8 and aligned as shown in Figure A.1.
c) Fully apply the brake via the force measuring device and record to the nearest newton, the maximum operating force.
d) Perform c) three times, rotating the relevant wheel between applications, and calculate to the nearest newton, the arithmetic mean value of the forces measured (ISO, 2002).

Source: ISO /FDIS 7176-3 Wheelchairs — Part 3: Determination of effectiveness of brakes, 2002.

Bibliography: The bibliography provides references to research literature that was used in the standard.

5.3.3 Relationship between National and International Efforts

Many countries have organizations that develop standards for national usage. National standards bodies provide important sources of information and can often move more rapidly to address emerging issues related to standardization. However, national bodies have more limited resources than when multiple countries work together. Also, international standards bodies bring a much wider range of experiences and may identify unique issues that some nations face. Essentially by definition, the scope of national standards is limited to locally produced or distributed products. Imported products may be disadvantaged when national standards are not harmonized with international standards.

International standards development activities bring together the resources of multiple nations, global manufacturers, and numbers of regulatory bodies. This has the advantage of gathering more data, broadly applicable data, and addressing potential international barriers in advance. A drawback of international standards is the potential that the standard is driven towards the lowest acceptable level of performance. An international standard may have been reached through broad consensus, but it may not be as stringent as a national standard in some countries. For example, some countries may wish to enter a particular market by driving down costs, while other countries with established markets may wish to reach higher standards. However, international standards create access to markets in multiple countries and benefit consumers by expanding choice. The ISO provides a rigorous and well-proven framework for the development of standards. The integrity of the ISO process makes their standards respected throughout the world and helps to provide at least a minimal level of assurance that the device is safe and effective.

5.3.4 Organization of Standards Development in the United States

In the United States, standards for most assistive devices are developed under the umbrella of RESNA. RESNA is recognized as a standards development body by ANSI. The process of developing a RESNA standard essentially mirrors that of ISO and most other countries. RESNA sponsors the Technical Standards Board (TSB). The TSB is made up of committee chairs of the various standards subcommittees and some at-large members. The TSB is responsible for ensuring that the ANSI standards development process is properly followed, and for establishing priorities for new work items. The TSB creates and dissolves subcommittees as needed. In reality, standards take years to develop, and therefore subcommittees typically have extended lives. The RESNA TSB has subcommittees on wheelchairs, wheelchair seating, recreational technology, devices for people with low vision or who are blind, and wheelchair transportation. The subcommittees may be responsible for multiple standards. In order to have concentrated expertise on specific standards, the subcommittees establish working groups. The working groups do much of the actual work in producing a standard. The working groups develop the initial test methods, organizing validation testing, and draft the standard. Most RESNA working groups actively participate in equivalent ISO working groups in an effort to achieve as great as harmonization as possible. Once an ISO standard is reached, the FDIS stage it is typically considered by TSB for adoption and publishing as the ANSI/RESNA national standard. This process typically results in only minor differences between the United States and ISO standards.

5.3.5 The User-Responsive Development Model

Standards are developed for the benefit of manufacturers, regulatory bodies, purchasers, people with disabilities, and clinical professionals. The ISO process is designed with the intent that these constituents are ensured participation in the development of standards for AT. Each nation indicates its official participants (experts) on the appropriate subcommittee. However, participation in the working groups is open to anyone who can contribute to the process. This places the ISO subcommittee in the critical position of ensuring that the standard is not biased toward one constituency group. Minutes are required to be filed for all working group and subcommittee meetings. The minutes must include the names of all participants and their affiliations. When voting comments are compiled, each country votes with a unified voice. However, those individuals voting within a country as well as their affiliations must be recorded. All voting comments whether incorporated or rejected must be discussed and, if rejected, justification must be provided. This helps to ensure fair representation and transparency of the working-group process. However, consumers rarely have the financial means of, for example, manufacturers. Hence, the consumer voice is often underrepresented. Another problem that has plagued AT standards development is the lack of human and financial resources to support development of the standards. In the over-25-year history of the development of wheelchair standards, only about 50 volunteers have carried most of the workload. Increased government support is required in order to ensure that standards are responsive to the needs of all the constituents they strive to serve.

5.3.6 Summary of Progress over the Past Several Decades

In general, RESNA adopts the ISO standards with only minor modifications once they have been published as a completed ISO standard. At the time of this writing, there are approximately 30 industry standards related to wheelchairs, wheelchair seating, and wheelchair transport technology. Both standards development bodies require that all standards be reviewed on a 5-year cycle, providing opportunities for continuous review and revision as may be necessary.

5.3.6.1 Impact Benefits for Users, Clinicians, Industry, and Healthcare-Funding Agencies

Unquestionably, standards have improved wheelchair technology, and as the development of standards progress, all of the constituents benefit. In general, people who use ATs gain access to safer, more reliable, and higher-performing products. Standards help users of AT and clinicians by providing means of attaining reliable information that is suitable for product comparison. The standards have helped all constituents by removing trade barriers, opening wider markets, and increasing access to products.

Some may argue that standards increase the cost of developing a new assistive device, but this argument gains little traction. Without the existence of standards, manufacturers must develop in-house test methods and performance criteria without the benefit of broad independent expertise and validation.

Clinical professionals have consistently played critical roles in the development of AT standards. A notable challenge for standards development is seeking agreement on common terminology. This helps to improve the precision of the language of the standards and assists with clarifying product literature. Standards have also helped to improve communication between professionals from different disciplines. Payers benefit in much the same way, plus the standards help to save costs through improvements in assistive device quality, safety, and reliability.

Manufacturers receive numerous benefits from standards. As noted previously, standards facilitate cross-border trade, reduce product liability, assist with regulatory approval, and provide design guidelines. Manufacturers actively participate in the development of national and international standards. Some do so from a defensive posture and others out of a sincere interest in ensuring device safety, reliability, and appropriate performance.

5.4 THE ROLE AND CONTRIBUTION OF REHABILITATION ENGINEERING IN STANDARDS DEVELOPMENT

Before discussing the specific role and contribution of rehabilitation engineers in the standards development and application process, the main activities that require science and engineering involvement will first be briefly discussed.

5.4.1 THE DEVELOPMENT AND VALIDATION OF TEST METHODS

Successful and timely development of standards must rely on an established body of knowledge and the use of engineering principles and methodologies to ensure reproducibility of test results across products and between test laboratories. Experience has taught us that, before the clock is started, by applying for an NWI, test methods need to have been piloted, ideally published in peer-reviewed scientific literature and received endorsement by knowledgeable technical personnel in leading companies associated with the products to be tested. It should be remembered that most leading companies invest extensively in the development of test methods for their products for product safety, performance, and quality assurance purposes. Although some of their more advanced techniques may be proprietary, they are often willing to share information about those most likely to be of interest to international standards developers.

5.4.2 DEVELOPMENT OF TEST-METHOD INSTRUMENTATION AND EQUIPMENT

There are key tenets that should be followed in the development of test equipment to be used to support the test methods prescribed in a standard. They include the following:

Test equipment should provide, whenever possible, measurements that can be calibrated against established international benchmarks.

Test methods should be developed with empathy for the cost to the product developer. The cost of the equipment, particularly if it has to be fabricated specially to conduct the test method, is a significant factor in determining the final cost of the test. Keeping the cost of test methods and equipment to a minimum ensures that start-up companies and those in developing countries are not placed at a commercial disadvantage by the standards that are published.

Where test equipment is available from more than one commercial source, it is important for the standard to embrace as many of the commercial sources as possible. Standards should not provide commercial advantage to particular test-equipment companies. Work may need to be undertaken to demonstrate that existing test equipment yields comparable results when performing the test method.

Test equipment that is protected by patents or other intellectual property (IP) protection measures may be specified. However, ISO regulations require assurance that individuals would not excessively exploit the use of their IP for standards testing purposes (Ferguson-Pell, 2005).

5.4.3 Organizing and Operating a Standards-Testing Facility

Creating and operating an AT test laboratory can be rather expensive. Most of the investment is in the specialized equipment, precision instrumentation, and trained personnel. Each standard requires a specific set of test equipment in order to perform the tests and record the results. For example, for testing a wheelchair's suitability for use as a seat in a motor vehicle, a sled test facility is required, as well as a high-speed motion-capture system. Even for basic wheelchair testing, machines such as the double-drum tester and curb-drop tester are $2 \times 2 \times 2$ m^3 in size. To test for electromagnetic compatibility, nearly $1 million worth of test equipment is required. Because of the costs of equipment and specialized personnel, there is no single laboratory in the world capable of performing the complete battery of wheelchair standards, let alone all assistive device standards. In addition to the test and measurement equipment, access to machine shop facilities is required to construct test fixtures. The test and measurement equipment also needs to be calibrated to accepted standards. This is typically done on at least an annual basis. Several test engineers are required to perform the various AT standards. Knowledge of measurement, electronics, materials, and advanced dynamics is required. This has, in part, caused test facilities to specialize.

5.4.4 Role and Contribution of Rehabilitation Engineering to Standards

In general, standards test methods involve the use of engineering principles and methodologies to ensure reproducibility of test results across products. The purpose of the test data, in addition to confirming that a specific product performs to an expected level of reliability or safety, is to provide data that can be used for clinical decision making regarding the selection of one product vs. another. However, the test data that is produced may have little meaning to clinicians or wheelchair users who have no understanding of the laboratory process by which the data was determined. Therefore, the clinical rehabilitation engineer is in an excellent position to interpret what test data is relevant to a specific situation and then assist his nontechnical clinical colleagues, in conjunction with the consumer, to take advantage of the test data in their decision making related to the selection, for example, of a particular wheelchair and/or wheelchair seating product.

The various standards development committees working at both the national and international levels are always in need of participants who can bring the technical–clinical perspective to the standards development table. Once again, this is an exciting contribution that the clinical rehabilitation engineer can make to the advancement of product improvement on a worldwide scale. This is also an excellent forum for contributions by research rehabilitation engineers who are qualified to conduct research projects designed to answer questions needed to advance the standards development process. The clinical rehabilitation engineer is also often able to make a contribution to this laboratory research process, especially those working within an academically-based clinical setting.

5.4.5 How to Get Involved

In most cases, it is very simple to get involved. Simply contact the committee chair (convener) responsible for leading an area of interest, arrange to attend a meeting, decide if the activity and membership is appealing, and arrange to the get officially signed on as a working group member. In the United States, there is an ANSI requirement that there be a multidisciplinary distribution involving; researchers, clinicians, users, and manufacturers as voting members on a committee. However, even if one is not immediately recognized as a voting member, attendance at meetings is always welcomed and remote electronic contributions are valued.

5.5 HIDDEN RATIONALE FOR PARTICIPATION IN INDUSTRY STANDARDS DEVELOPMENT

5.5.1 A Strange and Amazing Multidisciplinary Model

The model that has evolved for voluntary standards development defies all logic. What other successful development model demands that the participants self-finance their participation for 3 to 5 years, meet rigorous deadlines at the threat of expulsion, travel far and wide to participate, openly volunteer their talents and knowledge, and then relinquish ownership of the final product to the organizers so they can sell it at a profit? There must be hidden benefits that drive so many intelligent and talented people to act so illogically. This section attempts to explore these hidden benefits.

5.5.2 The Magnitude of the Goal and Potential Impact of the Final Result

It is proposed that those drawn to this activity (researchers, engineers, practitioners, and consumers) are stimulated by the prospect of working as part of a collegial group that has set a goal for itself larger than could be accomplished by an individual. The prospect that the realization of that goal, within a stipulated period of time, can positively impact the quality and safety of the technology on a worldwide scale is highly appealing.

5.5.3 Consolidation of Worldwide Knowledge

The research process, in theory, is a process in which new knowledge derived from rigorous scientific methodology will be systematically added to prior knowledge, mainly through archived peer-reviewed publications. The fundamental assumption is that the initial questions or hypotheses being investigated are, in fact, relevant to the void in knowledge being pursued. In reality this process has been slow to advance the field of rehabilitation technology.

One problem has been the randomness with which investigators independently select their research questions and then acquire the limited research funding to pursue the answers, the usual end result being an archived publication. In the standards forum, researchers, clinicians, industry representatives, and users combine their collective experiences to define the goals (scope) of a standard. This in turn defines (focuses) the research that must be carried out to provide the answers needed to advance the standard. This focus then tends to consolidate international leading minds on the problem, who then agree to conduct collaborative investigations and subject the findings to critique by other researchers, clinicians, and manufacturers' technical personnel. After validation, positive findings are then applied directly to the standards, usually in the form of a design or performance requirement and a related test method. The intellectual property issues get sorted out, and high-quality peer-reviewed publications result. It is proposed that this is a very cost-effective model for focusing limited research resources on a worldwide scale, and that those participating in the process find it a professionally rewarding.

In the absence of standards activities, the opportunity for this extensive collaboration between leading researchers, clinicians, and commercial developers would not naturally exist. Certainly, the in-depth discussions that result from a working group developing new test methods cannot be matched in the context of traditional dissemination mechanisms offered by research publications or conferences. A standards working group is an ideal environment to synthesize existing knowledge, focus future initiatives, and achieve outcomes in a time-limited manner. For an emerging researcher or AT clinician, participating in standards development can provide access to a powerful network of opinion leaders and new technical concepts.

5.5.4 Development of Collegial Networks

In the scheme of medical technology, rehabilitation AT is a small enterprise. Annual conferences provide opportunities for information sharing and meeting new and old colleagues. However, only rarely does this interchange translate into the building of any significant collegial relationships unless the parties are jointly engaged in a group activity that stimulates more extensive communication between annual meetings. Also, if one's research interests are narrowly focused, it may be difficult to find like-minded colleagues with whom one can openly discuss ideas and concepts of common interest. It is proposed that the standards forum, especially the international arena, enhances the opportunity for these professional interchanges to take place as part of the collegial networks that seem to form naturally.

5.5.5 Clinical/User Application of Industry Standards

Another strong motivator, especially for clinicians and product users, is the expectation that the application of the information generated by the test methods in the standards will improve the quality of clinical decision making and client outcomes. Many leading wheelchair and wheelchair seating clinicians (therapists, CREs, and RTSs) have contributed countless volunteer hours, preparing guidelines and giving presentations to their colleagues on the application of the standards in clinical practice. It is proposed that their knowledge and understanding of this largely technical information places them in a leadership position in the eyes of their clinical colleagues, and therefore they see this and their obvious contribution to their chosen specialty a rewarding professional experience.

5.5.6 The Direct Rewards to Industry

Perhaps the greatest rewards are derived by design engineers and quality control personnel from the manufacturers that produce and market AT products. Most industry personal have little background training in research methodology or the clinical application of their products. To be partners in a working group comprised of top-level clinicians and research engineers, who openly share information of direct importance to the safety and usability testing of their products, is immensely valuable. To gain insights into how test equipment can be built and validated and to influence the direction and cost of such equipment are equally important. Finally, the standards development forum is one of the few places where manufacturers can openly meet and exchange information to the extent possible within the limits proprietary concerns, without the threat of infringement of laws that govern industry collusion.

5.6 FUTURE OPPORTUNITIES

5.6.1 Standards Development in Other Areas of Assistive Technology

Mainly due to the traditional dominance of wheelchairs and wheelchair seating products in terms of costs to government insurance providers, this area of AT received the initial focus for standards development. However, now that the model has been successfully demonstrated to be so effective, other areas of AT are ripe for similar standards development: full-body support surfaces (mattresses), aids for the blind and hearing impaired, computer access technologies, augmented and alternative (AAC) communication technologies, and recreation technologies, only to name the obvious. In some cases, ISO technical committees or subcommittees already exist under which these activities can be incubated and allowed to flourish. Also, the rapid evolution of Internet communication resources will greatly facilitate the integration of scarce talents and resources on an international scale to facilitate involvement in these new areas of AT standards development.

5.6.2 Sustaining Research and Development Resources

The majority of the research in support of AT standards has come from government funds. Most of the countries actively involved in the development of AT standards have directly or indirectly provided support for the research necessary to create validated test methods. In some cases, the support has come through competitive grants or contracts, whereas in others, it has been through funds directly allocated by a government subsidized laboratory. It is a constant struggle to find funding for developing valid and meaningful standards. A challenge is that research in support of standards does not fit into a traditional research model, and few funding agencies have interest in supporting long-term work, especially the revision of an existing standard. However, standards development can have far-reaching positive effects for governments, consumers, and the industry. Ideally, a private–public partnership should be developed to support the cost of developing standards. This is done in other industries where industrial consortia are helpful in supporting standards work and often partner with government agencies to share costs.

5.6.3 Resolving Barriers to Clinical and User Application of Standards

There are two significant barriers to the application of AT standards among clinicians and AT users. The first barrier is the lack of support to participate in the development of the standards themselves. In order to fully participate in the development of the standards, clinicians and consumers need to attend the working group meetings and participate in electronic discussion groups. Greater participation by consumers and clinicians would help to improve the relevance of standards to meet their needs and to make the language of the standards more accessible to clinicians and consumers. This is especially important for standards that focus on terms, definitions, and performance values.

The second barrier relates to dissemination. ISO and national standards are typically sold. Because the market for copies of the standards has been small, the cost for a single set of standards (e.g., wheelchair standards) may cost in excess of $1000. The standards are also highly technical and require technical knowledge in order to be able to use them. Most clinicians and consumers do not actually need to read the standards; they need to be able to interpret and implement the results for their use. This can be accomplished by incorporating information about the standards into publications and workshops and by preparing clinical guidance documents related to the use and interpretation of various AT standards. Participating clinicians can slip into "high gear" for this need by preparing these materials and offering educational opportunities for their clinical colleagues. Participating researchers can prepare application materials on test methods that will facilitate adoption by manufacturers, especially those who may not have been able to participate directly in the development process, which are often smaller companies with limited research and development budgets. All these activities can greatly facilitate the widespread dissemination and adoption of the standard and, therefore, the ultimate impact intended by the initiators and dedicated working group that made it happen.

5.7 SUMMARY

In this chapter, we have discussed the standards development process and provided the basic structure of AT standards. We have not attempted to describe each of the standards applicable to AT. This would cause the materials to go out of date rather quickly. Standards play a critical role in ensuring minimal product quality and for objectively comparing performance. More specifically, this chapter has provided the reader with information in the following:

- Purpose and rationale for industry standards
- Historical overview of U.S. involvement in national and international standards development
- Organization of national and international industry standards
- Summary of progress over the past several decades
- Impact and benefits for users, clinicians, industry, and healthcare funding agencies
- The role and contribution of rehabilitation engineering in standards development
- The hidden rationale for participation in industry standards development
- Future needs and opportunities for continued progress

5.8 STUDY QUESTIONS

- What is the rationale behind having voluntary industrial standards for AT?
- What are the benefits for having harmonized international standards as opposed to each country having its own standard?
- What percentage of the participating member countries must vote to approve an ISO-DIS for it to move onto the next stage?
- What is the purpose of the "terms and definitions" section of a standard?
- What is the difference between a performance standard and a test method?
- What are the benefits of AT standards to manufacturers?
- What are the benefits of AT standards to clinicians and AT users?
- Describe several barriers to the greater application and usage of AT standards?
- What is the purpose of the "disclosure section" of a standard?
- Who may participate in a working group developing an AT standard?
- Briefly explain the five levels that a standard must pass through before becoming a published international standard.

Under current operating procedures of ISO, how long does it typically take to develop and publish and international standard?
List the key parts of a typical ISO standard.
List five ways that engineers can participate in the standards development process.
List three key tenets that should be followed in the development of test equipment to be used to support the test methods prescribed in a standard.
Discuss the process of getting involved in the standards development process as a rehabilitation engineer.
List three new areas of AT that are in need of industry standards.
List five indirect rewards that appear to motivate people to contribute their time and talents to the standards development process.
List three barriers that need to be overcome in order to facilitate effective standards development.

ACKNOWLEDGEMENTS

AT standards development in the US has been supported by many organizations and agencies since its inception in the early 1980s. Early supporters were the Paralyzed Veterans of America (PVA) and the Department of Veterans Affairs (VA). Indirect support via Government programs such as NIDRR/RERC on Wheeled Mobility, followed by the RERC on Wheelchair Transportation and the VA-HERL, has provided indirect research and travel support for researchers to attend both RESNA and ISO working group meetings since the very beginning. These research funds have also supported much of the US and ISO leadership efforts. Beneficial Designs, a small at research and development company, has contributed countless volunteer leadership and standards management services, especially to US programs. ANSI/RESNA Part 19 — Wheelchairs for Use in Motor Vehicles, a consortium of national school transportation entities, secured the funding to initiate this work in the US. These were: the National Congress on School Transportation (NCST), the National Association of State Directors of Pupil Transportation Services (NASDPTS), National Association for Pupil Transportation (NAPT), and National School Transportation Association (NSTA). Of course, the wheelchair and wheelchair tie-down industries have supported this effort by sending technical personnel to meetings, reviewing, and commenting on drafts of standards, as well conducting laboratory validation testing. This work could not have progressed to this extent without the aforementioned contributions, in addition to those of others too numerous to include in this acknowledgement.

BIBLIOGRAPHY

ANSI/RESNA Wheelchairs, Volume 1, 2000, Requirements and test methods for wheelchairs (including scooters). Section 19: Wheelchairs used as seats in motor vehicles, RESNA, 1700 N, Moore St, Suite 1540, Arlington, VA, 22209–1903.

Durable Medical Equipment Regional Carriers (DMERC), December 3, 2001, Wheelchair seating, Draft Medical Review Policy No. WCS.

Ferguson-Pell, M. et al., The role of wheelchair seating standards in determining clinical practices and funding policies, *Assistive Technology*, 17, 1–6, 2005.

International Standards Organization (ISO), 1999, General requirements for the competence of testing and calibration laboratories, ISO/IEC 17025, Geneva, Switzerland.

International Standards Organization (ISO), 2000, Technical systems and aids for disabled or handicapped persons — Wheelchair tiedown and occupant restraint systems — Part1: Requirements and test methods for all systems, ISO FDIS 10542-1, Geneva, Switzerland.

International Standards Organization (ISO), 2001–1, Technical systems and aids for disabled or handicapped persons — Wheelchairs — Part 19: Wheeled Mobility Devices for Use in Motor Vehicles, ISO 7176-19, Geneva, Switzerland.

International Standards Organization (ISO), 2001–2, Technical systems and aids for disabled or handicapped persons — Wheelchairs — Part 3: Determination of effectiveness of brakes, ISO 7176-3, Geneva, Switzerland.

International Standards Organization (ISO), 2004, Technical systems and aids for disabled or handicapped persons — Wheelchairs — Part 26: Vocabulary, ISO DIS7176-26, Geneva, Switzerland.

Raflo, B., 2002, A code of one's own, May 1st Homecare.

6 Seating Biomechanics and Systems

Mark Schmeler, Bengt Engstrom, Barbara Crane, and Rosemarie Cooper

CONTENTS

6.1 LEARNING OBJECTIVES OF THIS CHAPTER

Upon completion of this chapter, the reader will be able to:

Have an understanding of seating biomechanics
Know the current principles behind seating systems
Understand the practical applications of seating biomechanics
Understand how to apply seating systems
Have an understanding of the principles of postural support
Have some knowledge of future seating technologies

6.2 INTRODUCTION

The concept of the human body in a seated position seems simple, yet it is complex. A reason for this is that the human body is not designed to sit, at least not for prolonged periods of time, in the same position. People alter their positions frequently from various lying, sitting, and standing positions, depending on the task being performed, be it an active or a resting task. For people with disabilities who are unable to walk and require the use of wheeled mobility and seating devices, sitting is necessary, and identifying the most effective seating system for a such a specific need can be challenging. The purpose of this chapter is to review the principles of seating biomechanics and the components of seating systems designed to accommodate specific needs based on the person's preferences, functional capabilities, and environmental demands (both physical and social).

The human body is a dynamic structure that is designed to perform and engage in a number of tasks in a variety of positions. Even in a lying position, the body frequently (and subconsciously) changes position. The same applies to a standing or sitting position. In a standing position, the body is a constantly dynamic structure, able to change its base of support frequently, depending on movement and placement of the feet (base of support) in relation to the torso. The same applies to the body in a seated position, where the pelvis and thighs act as the base of support and the hip joints are placed in a flexed position, resulting in a potentially unstable base of support. In a seated posture, the pelvis is free to tilt and rotate in multiple directions, ultimately affecting the position and posture of the rest of the upper body that includes the entire spine, trunk, upper extremities, and head. The seated body, therefore, can be dynamic but at the same time unstable (Figure 6.1).

The human body also has an inherent need to move and change position in order to find comfort (relieve pressure or strain) or find a better point of stability to function

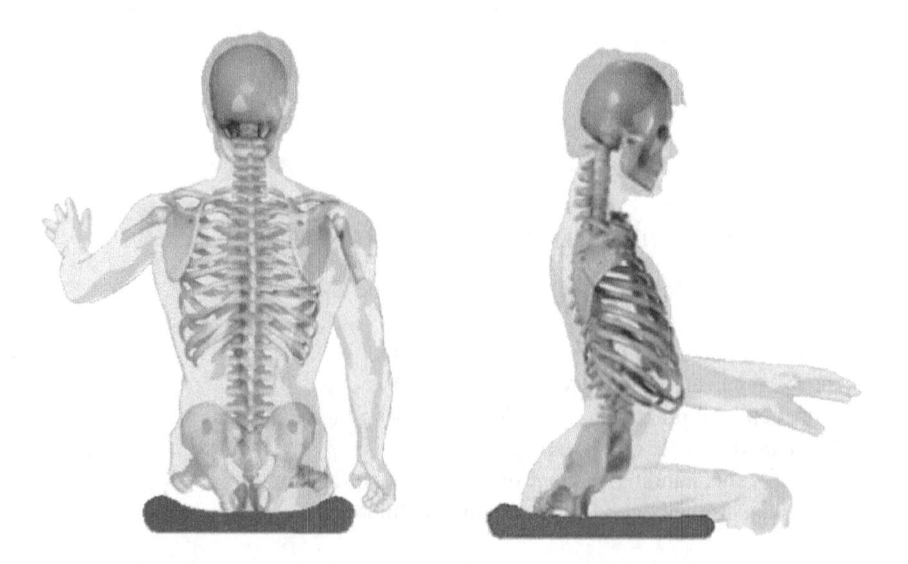

FIGURE 6.1 Posterior and lateral view of stable seated postures.

in a particular task. A simple way to grasp the concepts of seating and positioning is to pay attention to various postures one might assume throughout the day or the postural behaviors engaged in when sitting for prolonged periods, such as at computer workstations, behind the wheel of a vehicle, or on a long flight.

6.3 SEATING AND COMMON PATHOLOGIES

The human body maintains balance through its complex neuromuscular system as well as through bones, joints, tendons, and ligaments that hold joints in place. A compromise to these body structures, including, for example, loss of muscle activity due to spinal cord injury or multiple sclerosis, limitation in joint range of motion as a result of arthritis, increased muscle tone from cerebral palsy or brain injury, and amputation of a limb, can impair balance of the seated body structure and lead to other complications. For people with these types of conditions, long-term sitting, especially in a poorly fitted seating system, can become pathological. Sitting has been associated with secondary complications such as pressure sores, back pain, joint contractures, postural deformities, and edema in the lower extremities. Therefore, the design of the seating system becomes an even more important consideration. The wheeled mobility and seating device or wheelchair is the user's primary means of mobility. Therefore, the person uses the same seating system for prolonged periods across an entire range of daily activities.

Given the inherent structure and function of the human body systems, there are common postures associated with sitting. Improper posture can result in long-term formation of contractures and deformities, which need to be managed or avoided altogether. The most common is the pelvis rotating in a posterior direction, resulting in sacral sitting, flexion of the lumbar spine, and kyphotic posture. For some people with decreased trunk control, sacral sitting is a necessary posture to maintain trunk stabilization against a back support. However, there are concerns with pressure over the coccyx and sacral areas and further collapse of the spine and trunk, which may eventually compromise internal vital organ capacity such as breathing and digestion (Figure 6.2 and Figure 6.3). The pelvis can also tilt to the left or right, whereby the spine and trunk will compensate in the opposite direction, causing a scoliosis or lateral flexion deformity of the spine (Figure 6.4). In some cases, the pelvis rotates in an anterior direction, resulting in increased flexion of the lumbar spine, known as *lordosis* (Figure 6.5). This can be due to the mass of the abdominal area pulling the body forward in the case of morbid obesity or the need to find a sense of balance due to proximal trunk weakness in people with muscular dystrophy.

The pelvis can also rotate to the left or right in combination with any of the other mentioned movements, causing what is termed a "windswept deformity," whereby the thighs migrate to the left or right (Figure 6.6).

6.4 SEATING ASSESSMENT

Prior to the provision of a seating system, a physical motor assessment, including muscle strength, joint range of motion, coordination, balance, posture, tone,

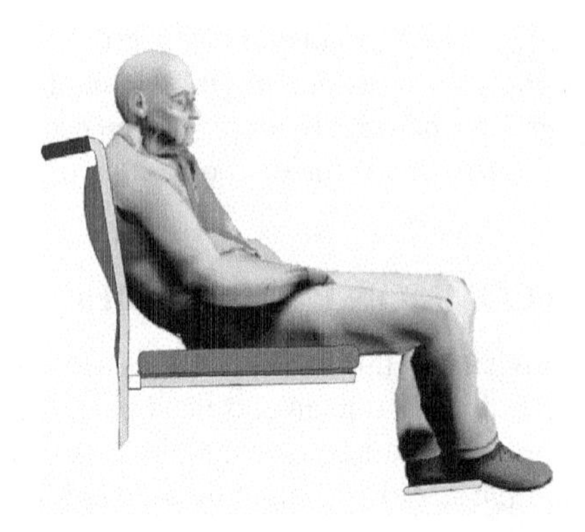

FIGURE 6.2 Slouched posture with sacral sitting.

endurance, sitting posture, and assessment of cognition and perception, is necessary to obtain a basic understanding of the capacity and needs of the person. These assessments should be performed by a qualified professional with specialty training and certification as well as knowledge of wheelchair seating and mobility applications. Physical therapists are well suited to assess physical capacities and limitations, especially as these affect mobility. Occupational therapists are well suited to assess functional capacity and deficits in basic and instrumental activity of daily living (ADL). It is important to know how people perform tasks, where the deficits are, and how wheelchairs and seating systems can compensate for the deficits in order to augment task performance. The rehabilitation engineer has an important role in understanding the capabilities and application of various technologies to assist in the selection process and product design. A qualified equipment supplier is also an important team member to include early in the process, because the supplier will be familiar with available devices and how they can be applied to solve problems. The end user of the equipment, family, or caregiver is the most important member of the team. A proper assessment will begin with listening to their needs, concerns, and goals of using a device. Realistic goal setting fosters discussion related to what the technology is and is not capable of.

The physical motor assessment begins by observing and noting the person's posture in their existing seating system or in a seated position. This provides a baseline to the seated postures that a person may assume. It is important to note that a person may sit in a particular position due to preference, limitations, or the design of the seating system. It is, therefore, necessary to then remove the person from the existing system and place him or her in an unsupported seated position on a flat surface, such as a therapy mat table, to observe their posture without postural supports. If necessary, the examiner may need to provide support if the person has limited unsupported sitting balance. This procedure provides a means of determining the person's postural capabilities and needs. Careful attention should be paid to pelvic and spinal alignment. While the person is sitting on the mat table, the examiner should apply appropriate forces through the use of their hands to determine if any observed deformities of the pelvis and spine are fixed or flexible. A fixed deformity cannot be corrected by a

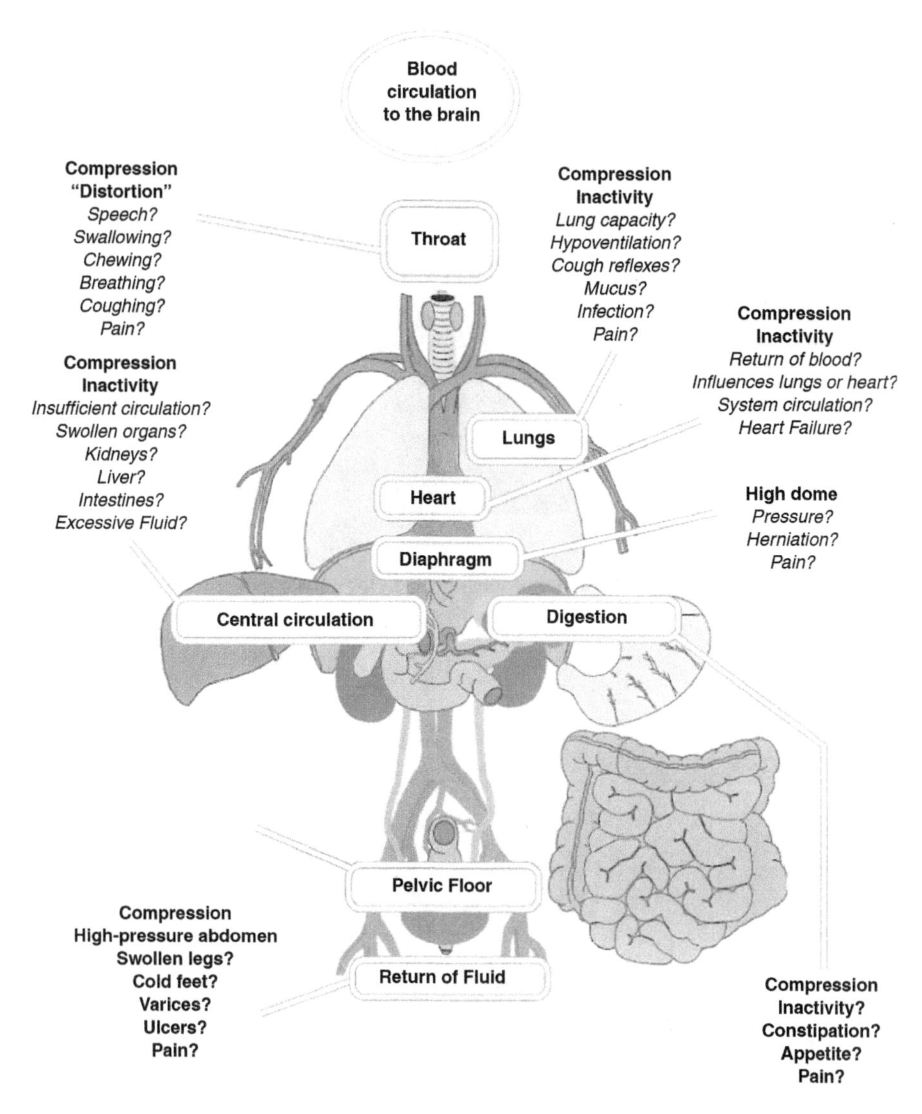

FIGURE 6.3 Possible influences on organs in prolonged slouched sitting posture.

seating system and has to be accommodated by the system. A flexible deformity can be corrected in varying degrees depending on complexity and has to be supported by the seating system.

The person should then be asked to transition or be assisted to a supine position, where pelvic and spinal alignment can be reassessed with the pull of gravity working in a different plane on the body. In this position, the range of motion at the hips and knees should be assessed. Limitations in hip flexion will affect the seat-to-back angle configuration of the seating system. With the hip flexed, the knee's range of motion (especially extension) needs to be assessed, considering the hamstrings cross the hip

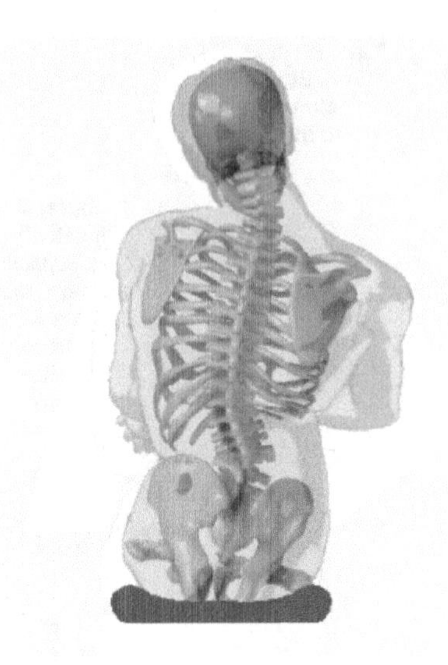

FIGURE 6.4 Scoliosis of the spine.

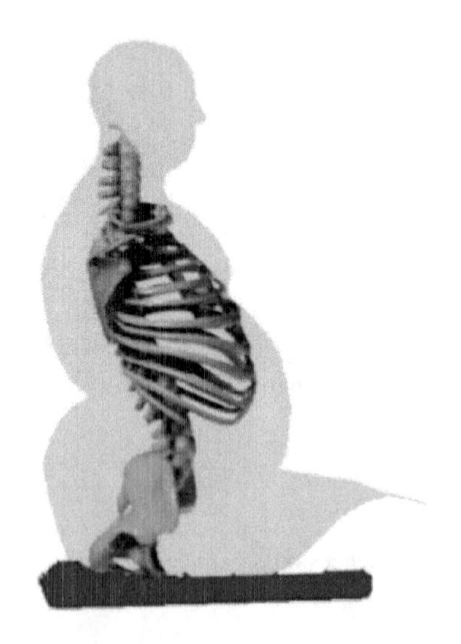

FIGURE 6.5 Lordosis of the spine.

and knee joints, and this will determine optimal angle of the leg rests and position of the feet. In either a seated or supine position, strength and range of motion of the upper extremities need to be examined using commonly accepted measures and procedures. During this assessment, it is also necessary to note quality of movements related to coordination because this is affected by tone, spasticity, tremors, and primitive reflexes.

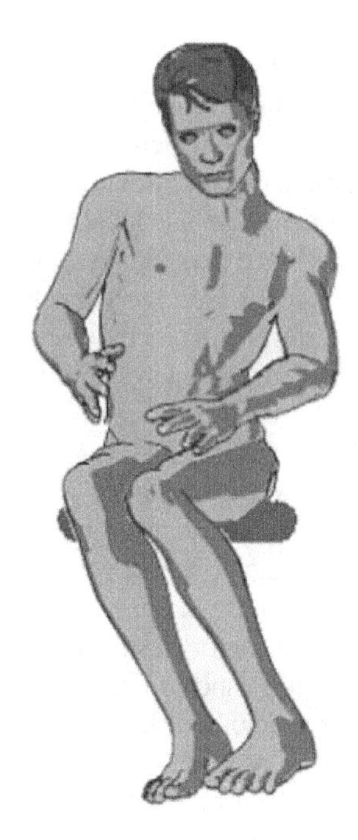

FIGURE 6.6 Windswept deformity.

6.5 INTERVENTIONS: SEATING SYSTEMS

Seating systems can be classified into three general categories: prefabricated, modular or adjustable, and custom-contoured systems. These categories have products that overlap two categories. Every person should be given a seating system designed and provided specific to the user's medical, functional, and personal preferences or needs. Medically, a system should address issues of soft tissue management, comfort, reducing the potential for or accommodation of orthopedic deformities, and maintain vital organ capacity. Functionally, the system should address the movements and supports that the user may need to perform such activities, such as to reach or access objects, transfers, sit at tables, and other activities of daily living. This requires careful matching of critical chair dimensions to body dimensions, user ability, and intended use. The user's goals and priorities must be the primary considerations. For example, a user may forgo pressure relief and comfort for a firmer system that provides greater stability and allows the user to slide off the seat for transfers.

The simplest form of prefabricated or premanufactured seating systems is a linear seating system. A linear seating system refers to a planar seat and back with fixed angles and orientations. These types of systems may not provide a great deal contour or support but may be better suited for users who need more freedom of movement or who will experience rapid changes in size such as growing children. This is followed by generically contoured systems that are meant for people who have minimum to moderate postural support needs and who have body shapes and dimensions that fit within these generic contours. Modular systems are comprised of components that can be adjusted, added, or removed to address a specific postural need such as lateral trunk supports and thigh guides. These systems are ideal for people with progressive conditions, such as multiple sclerosis, in which postural needs may increase over time. They are also intended for people with conditions that might improve, such as a stroke or brain injury. Modular components are also ideal when needs change across a given day or set of activities. For example, a person may fatigue towards the end of the day and need more lateral trunk support added to the backrest, but may want these supports removed because they interfere with reaching activities or transferring out of the wheelchair. Custom-molded and contoured systems are more complicated to design and apply clinically. These systems are for people with moderate to severe fixed or semifixed postural deformities, including curvatures of the spine, pelvic obliquities, and windswept postures of the lower extremities. A disadvantage of custom seating is that it may limit function and movement because the system is very specific to the shape of the body. There are also concerns for people who might grow, change, gain or lose weight, or have seasonal variations in the thickness of their clothing. Custom contoured seating systems can be designed using several techniques such as liquid foam in place, molding bags, and/or computer-aided design/computer-aided manufacturing (CAD/CAM) technologies. It is also important to note that seating systems should not attempt to overcorrect postural deformities but should accommodate them as sitting should be comfortable; they should also accommodate many other needs. Reduction or correction of postural deformities and joint contractures is best addressed through other interventions that include surgery, controlled passive stretching, or orthotic devices.

Seating systems are comprised of several components including at least a seat and back support. Other components also often include supports for the arms, legs, and the head. Adjustable features may include reclining backrests, tilt-in-space, seat elevators, and standing systems. Each of these components and their applications will be reviewed in the following subsections.

6.5.1 Seat Supports

The seat interface must provide a balance of pelvic stabilization and pressure distribution and yet allow for some degree of movement and be functional to the individual's needs. The seat is the base of support for the pelvis, which supports the upper body. A seat should have an underlying rigid base of support. A sling seat, commonly found on a standard manual wheelchair, provides limited postural support and is intended only to easily allow the wheelchair to fold. Sling upholstery tends to stretch and bow over time, which may contribute to adduction and internal rotation of the hips as well as a posterior pelvic tilt and kyphotic posture of the spine. The angle as well as the shape of the seat contributes to pelvic stabilization. Without stabilization, the pelvis tends to rotate in a posterior orientation, causing the buttocks to slide forward in the seat. This can cause shearing forces in the buttocks and may lead the spine to further collapse into a slouched posture. Providing a posterior slope to the seat angle, wedging the seat, or designing an ischial shelf in the shape of the cushion can provide greater pelvic stabilization and a better sense of balance; however, these may not always be ideal. These could make it more difficult to slide forward out of the seat for transfers, increase pressure in the ischial tuberosity and coccyx regions, or promote more of a posterior pelvic tilt if the back support is not adequate. A posterior slope may also not be ideal for people who self-propel with their feet. As stated previously, it is also important to understand that people sometimes want to slide forward in their seats and slouch to relax and change position at various times throughout the day, as sitting upright continuously is difficult and causes fatigue.

6.5.2 Back Supports

As stated previously, wheelchair users are vulnerable to long-term collapsing deformities of the spine that include kyphosis, scoliosis, or a combination of both. These can lead to compromised breathing and other vital organ functions, whereas properly applied back supports can slow the onset of these problems. Back supports can be categorized by their shape, height, and stiffness. Ideally, a person looks for a backrest that provides posterior and lateral support but does not inhibit freedom of movement of the trunk and upper extremities, which may be needed for the propulsion of a manual wheelchair or other activities such as reaching. A contoured or curved back should accommodate the width of the user and follow the natural curves of the spine so as to provide enough trunk support without compromising movement or function. Standard sling-back upholstery (as with sling-seat upholstery) also tends to stretch over time, which can result in a posterior pelvic tilt and contribute to a kyphotic posture. Flat or planar backs allow for more movement but might be less comfortable and stable as they neither provide lateral stabilization nor conform to the natural curves of the spine. A high back will provide greater support to the spine but may interfere with trunk rotation and shoulder movements. Back supports below the scapulae and thoracic spine allow for greater freedom of movement but may result in long-term spinal deformities. Fabric backs with adjustable tension straps are now more common, especially in manual wheelchairs, and tend to be lightweight as compared to rigid-shell back supports and can be adjusted for varying needs and shapes. However, they do stretch and wear out over time and do not provide a sense of rigid stabilization.

Current trends in backrest design are focused on taller backs with lateral curves around the thoracic–lumbar regions and cutouts in the scapular region to provide spinal support and yet allow for movements at the shoulders. Attention is also being focused on the area of using adjustable yet durable carbon fiber materials in the construction of back supports to address previous design shortcomings.

6.5.3 Arm Supports

Arm supports may be necessary to rest the arms, provide a greater sense of lateral trunk stability, decrease pressure loads in the buttocks by bearing some weight through the upper extremities, to hold onto with one hand while reaching with the other, or to push from for weight shifts or transfers in and out of the wheelchair. Armrest assemblies can also serve to contain the pelvis and thighs to keep the thighs in alignment as well as to protect clothing and skin from rubbing on the tires of a manual wheelchair. They do, however, add weight to a manual wheelchair and can get in the way of accessing the pushrims for effective propulsion. Therefore, many active manual wheelchair users choose to forgo the use of arm supports. If arm supports are to be used, it is ideal for them to be removable or capable of swinging out of the way for accessing pushrims, sitting at tables, or lateral transfers out of the wheelchair. They should also be set at an optimal height for the user or be adjustable.

6.5.4 Foot and Leg Supports

Foot and leg supports can be classified by the angle to which they position the knee and location of the feet. Traditional swing-away footrests are designed to keep the user's feet out in front with about 60° to 70° of knee flexion to avoid interference with caster swivel and rotation. This is not necessarily a natural seated position as it promotes a slouched posture by pulling the hamstrings and the pelvis into a posterior pelvic tilt.

Elevating leg rests poses an even greater problem. A perceived purpose of elevating leg rests is to assist with edema (swelling) management in the lower extremities. However, being seated in an upright position with the hips flexed and the knees extended will do little to assist with edema unless the feet and legs can get at least to or above the heart level, which can only be accomplished by tilting the seat back and reclining the backrest. Elevating leg rests without these features will also possibly pull further on the hamstrings, promote more of a slouched posture, and cause significant problems for people with limited knee extension.

Some people (such as active manual wheelchair users) prefer to sit with their knees at a more natural 90° flexed position or flexed further with their feet tucked under the seat to provide greater postural stability as well as shorten the length of the wheelchair for maneuverability. This position also warrants tapering the footrests inward to keep the feet from interfering with front caster swivel. In a power wheelchair, a front- or mid-wheel drive configuration allows for position of the feet with a natural 90° flexion of the knee as the need for front casters is eliminated. Some active wheelchair users also prefer to have the foot supports fixed in a one-piece structure to add more rigidity and durability. However, removable or swing-away foot supports are essential for people who stand for transfers out of the wheelchair or propel a manual wheelchair with their feet.

6.5.5 Head Supports

The head can be complicated to position due to its size and the movements in the cervical spine or neck. For a person with poor head control, head support is warranted. Prior to positioning the head, the pelvic and spinal support and balance should first be addressed because the head is more distal to these other structures and will tend to position itself based on their position. A head support is also almost always required for use with a wheelchair that includes tilt and/or reclining backrest because the head needs to be supported in these positions. A head support may also be warranted if the user has the need to rest in the wheelchair. Head supports are also recommended if the wheelchair is to be used as a seat in a vehicle in order to provide support in the event of a collision.

As with all other seating components, head supports come in various shapes and sizes to address different needs and preferences. Some people prefer no head support as they interfere with head movements and can reduce the field of vision. Flat head supports merely provide a surface to rest against, whereas curved supports provide some degree of lateral stability. Wide head supports allow for greater space for placing the head but tend to interfere with field of vision, and therefore narrower ones are sometimes preferred. By adding lateral head and suboccipital supports, head supports can also be very helpful for people with very poor head control (such as people with amyotrophic lateral sclerosis). Head supports should be mounted to the seating system using adjustable hardware so that the support can be moved, adjusted, and removed as needed.

6.5.6 Tilt Frames and Reclining Backrests

As it is unnatural to sit in one position for prolonged periods of time, tilt frames and reclining backrest systems are options that benefit people who are unable to move or reposition themselves effectively. These options redistribute pressure, manage posture and tone, provide comfort, and help with personal care activities. These options are available for both manual and power wheelchairs, and can be operated by the wheelchair user as a powered feature or manually by an attendant (Figure 6.7).

Tilt allows the entire seating system to pivot posteriorly while keeping the hip and knee joints in the same position. This assists with pressure redistribution away from the buttocks region and into the back support. It is also useful in repositioning, as people tend to slouch and slide forward when sitting upright for prolonged periods of time. In a tilted position, the trunk is also able to extend, thus countering the kyphotic posture and potentially reducing or slowing the development of trunk and spinal deformities.

A reclining backrest allows for stretch of the hip flexors and opens the hip angle to assist with attending to catheters, toileting, dressing, and dependent transfers. Reclining the backrest does, however, create shear forces in the seat and back as the user often tends to slide down in the seating system. The addition of tilt-in-space

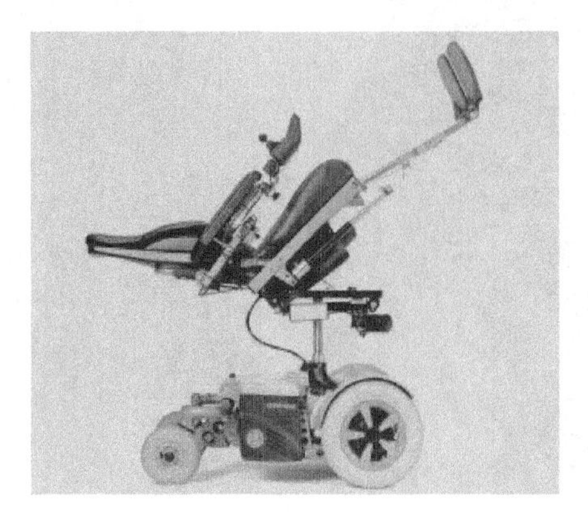

FIGURE 6.7 Adjustable power tilt, recline, elevating leg rests, and seat elevation.

will counter this tendency of sliding down, and should therefore be considered when reclining is warranted. Both systems also allow a person to rest in the wheelchair without having to be transferred to a bed. Research supports use of combined tilt and reclining backrest interventions to provide optimal pressure distribution.

6.6 STANDING SYSTEMS

Wheelchairs that allow a person to passively stand benefit individuals who would typically be unable to do so otherwise (Figure 6.8). The benefits of passive standing may include decreased bladder infections, reduced osteoporosis, and decreased lower extremity spasticity. In addition, there are likely psychological benefits due to the feeling of an upright posture and the ability to interact at eye level, as well as functional benefits associated with being able to reach objects in higher locations or higher

FIGURE 6.8 Passive-standing wheelchair.

surfaces. It is critical to carefully assess a person's posture and range of motion prior to the consideration of a standing device, because certain people with limitations in range of motion and postural deformities may not be able to stand upright. A candidate for a standing wheelchair should be carefully assessed by a physician or other qualified practitioner prior to standing because there are also concerns with orthostatic hypotension in cases where people have not stood for a prolonged period of time.

6.7 SEAT ELEVATION SYSTEMS

Some of the benefits obtained from a standing system, such as reaching objects in higher locations, accessing high surfaces, and being at eye level with standing people, can be accomplished through the use of a seat elevating system (Figure 6.9). This is a feature typically available only on power wheelchairs. Raising the seat is also sometimes critical to facilitating safer and more efficient transfers. For a person using a sliding board or performing a lateral transfer, the use of a seat elevator allows for transfers in a downhill direction, which has been shown to require less strain on the upper extremities as compared to transferring to a level or higher surface. In other cases, the seat elevator can facilitate transfers for people who stand to transfer but have difficulty rising from a low seat.

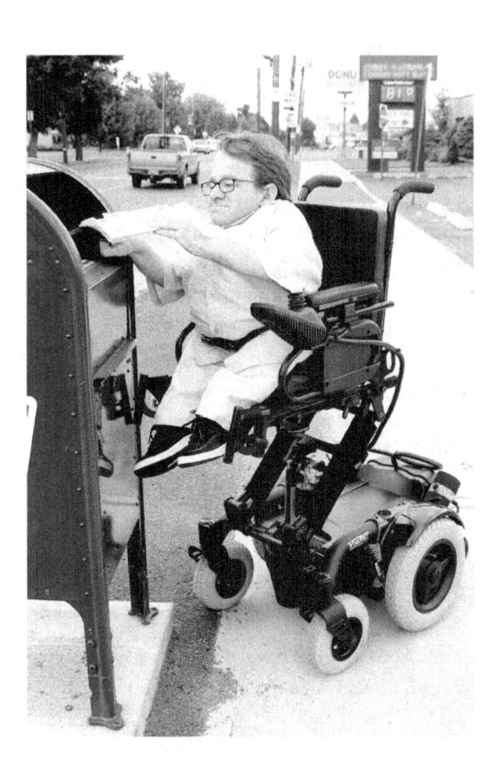

FIGURE 6.9 Seat elevator.

6.8 SUMMARY

The past decade has seen vast improvement in innovative seating technologies. A successful seating recommendation requires a systematic assessment approach, appropriate interpretation of assessments findings, and a keen understanding of how the findings relate to the individual's positioning needs. It also requires current knowledge of available seating systems, their intended application, and how they relate to the individual's seating and positioning parameters. It is important to consider the person's seating and mobility goals; medical, physical, and functional variables; environments in which the device and its seating system will be used; past experience with wheelchair or seating technology; and reimbursement. Adequate fitting, training, and follow-up are vital in ensuring comfort, satisfaction, and a successful match between an individual and a seating system.

The interdisciplinary assessment team should consist of members who are qualified professionals with specialty training and certification. Training for the team members can be obtained by attending meetings such as RESNA (www.resna.org), the International Seating Symposium (www.iss.pitt.edu), or Medtrade (www.medtrade.com). The Rehabilitation Engineering and Assistive Technology Society of North America (RESNA) offers certifications for assistive technology practitioner (ATP), assistive technology supplier (ATS), and rehabilitation engineering technologist (RET).

6.9 STUDY QUESTIONS

Name and describe the following:

1. Three common postural deformities of the human body
2. Three types of body system impairments that influence a person's ability to sit without appropriate postural supports
3. The three common goals of postural supports
4. Three movements of the pelvis against the spine
5. Two indications to use a generic seating system
6. Two indications to use a modular seating system
7. One reason to use a custom-contoured seating system
8. An indication to use a tilt-in-space seating system
9. A functional benefit for the use of low-height backrest
10. A disadvantage of a low-height back support

BIBLIOGRAPHY

American Physical Therapy Association (2001) *Physical Therapy*, **81**, 9–744.

Cook, A.M. and Hussey, S.M. (2002) *Assistive Technologies: Principles and Practice*, Mosby, Inc., St Louis, MO.

Cooper, R.A., Schmeler, M.R., Cooper, R. and Boninger, M.L. (2000a) *Rehab Management: The Interdisciplinary Journal of Rehabilitation*, **13**, 58.

Cooper, R.A., Schmeler, M.R., Cooper, R. and Boninger, M.L. (2000b) *Rehab Management: The Interdisciplinary Journal of Rehabilitation*, **13**, 60.

Dewey, A., Rice-Oxley, M. and Dean, T. (2004) *British Journal of Occupational Therapy*, **67**, 65–74.

Dunn, R.B., Walter, J.S., Lucero, Y., Weaver, F., Langbein, E., Fehr, L., Johnson, P. and Riedy, L. (1998) *Assistive Technology*, **10**, 84–93.

Edlich, R.F., Nelson, K.P., Foley, M.L., Buschbacher, R.M., Long, W.B. and Ma, E.K. (2004) *Journal of Long Term Effects of Medical Implants*, **14**, 107–130.

Engstrom, B. (2002) *Ergonomic Seating; A True Challenge When Using Wheelchairs*, Posturalis Books, Stockholm.

Geyer, M.J., Brienza, D.M., Karg, P., Trefler, E. and Kelsey, S. (2001) *Adv Skin Wound Care*, **14**, 120–129; quiz 131–132.

Goemaere, S., Van Laere, M., De Neve, P. and Kaufman, J.M. (1994) *Osteoporos Int*, **4**, 138–43.

Harms, M. (1990) *Physiotherapy*, **76**, 266–271.

Hobson, D.A. (1992) *J Rehabil Res Dev*, **29**, 21–31.

Janssen, W.G.M., Bussmann, H.B.J. and Stam, H.J. (2002) *Phys Ther*, **82**, 866–79.

Kaplan, P.E., Roden, W., Gilbert, E., Richards, L. and Goldschmidt, J.W. (1981) *Paraplegia*, **19**, 289–293.

Lacoste, M., Weiss-Lambrou, R., Allard, M. and Dansereau, J. (2003) *Assistive Technology*, **15**, 58–68.

Lange, M.L. (2000) *OT Practice*, **5**, 21–22.

Pellow, T.R. (1999) *Canadian Journal of Occupational Therapy*, **66**, 140–149.

Schmeler, M.R., Boninger, M.L., Cooper, R.A. and Cooper, R. (2004) *Rehab Management: The Interdisciplinary Journal of Rehabilitation*, **17**, 38.

Sprigle, S. and Sposato, B. (1997) *Orthopaedic Physical Therapy Clinics of North America*, **6**, 99–122.

Vachon, T., Weiss-Lambrou, R., Lacoste, M. and Dansereau, J. (1999) In *RESNA '99 Annual Conference: Spotlight on Technology* (Ed. Sprigle, S.) RESNA Press, Long Beach, CA, pp. 221–223.

Vaisbuch, N., Meyer, S. and Weiss, P.L. (2000) *Disability and Rehabilitation*, **22**, 749–755.

Wang, Y.T., Kim, C.K., Ford, H.T., 3rd and Ford, H.T., Jr. (1994) *Percept Mot Skills*, **79**, 763–766.

7 Tissue Integrity Management

David Brienza, Yihkuen Jan, and Jeanne Zanca

CONTENTS

7.1 LEARNING OBJECTIVES OF THIS CHAPTER

Upon completion of this chapter, the reader will be able to:

Have an understanding of tissue integrity management
Know the current principles behind maintaining tissue integrity
Understand the practical applications of tissue integrity management
Understand how to apply various cushion principles
Have an understanding of the principles of cushion technologies
Have some knowledge of the biological aspects of tissue health

7.2 INTRODUCTION

Because the body weight is transmitted through the bony prominences such as the scapula, sacrum, greater trochanters, ischial tuberosities, and heels, it can cause significant concentrations of pressure at the skin's surface and in the underlying soft tissue. The pressure peaks and pressure gradients surrounding these peaks can put the soft tissue at risk of breakdown. Wheelchair users, especially those who have only limited ability to reposition them and/or have a loss of sensation in the areas where weight is being supported, are at high risk for developing pressure ulcers. However, high pressure alone usually is not a sufficient condition for the development of a pressure ulcer. Research has clearly demonstrated that the damaging effects of pressure are related to both its magnitude and duration. Simply stated, tissues can withstand higher loads for shorter periods of time.

This chapter will focus on the possible effects of support-surface characteristics — pressure distribution, shear, temperature control, and moisture control — on pressure ulcer prevention. We will examine these relationships relative to elastic, viscoelastic, fluid-filled, low-air-loss, air-fluidized, and alternating-pressure support surfaces; and explain support-surface characteristics that are related to tissue.

7.3 PRESSURE ULCERS

7.3.1 ETIOLOGY

Prolonged external pressure over bony prominences has long been identified as the primary etiology in pressure ulcer development. Other related causes include the magnitude of shear and friction forces and the additive effects of temperature and moisture. Each of these factors can be affected by (and is related to) the characteristics of the support surface chosen for a given person.

However, it is clear that while extrinsic factors such as temperature, moisture, and mechanical characteristics are critical to the development of pressure ulcers, factors intrinsic to the patient's skin and its supporting structures, vasculature, or lymphatics also play a significant role in the patient's susceptibility to these factors. The relationships between support-surface characteristics and their effects on pressure ulcer prevention. These characteristics are related to the following classifications of support surfaces: elastic, viscoelastic, fluid-filled, low-air-loss, air-fluidized, and alternating-pressure. It should be noted that these classifications and terms are being examined by National and International standards committees and are likely to change, however the information presented is valid relative to the definitions used here. The physiological responses and the associated clinical measures related to support-surface characteristics are also discussed.

7.4 PRESSURE ULCER STAGING SYSTEM

7.4.1 RISK ASSESSMENT

Although interface pressure is the most common parameter used for comparing support-surface performance, the relationship between interface pressure and pressure ulcer (PU) incidence has not been adequately studied. Excessive pressure applied to skin over a bony prominence for prolonged periods of time is the most important extrinsic factor leading to the development of pressure ulcers. Thus, measurement of pressure on the skin — skin interface pressure — is a logical method for assessing this important factor. However, the use and interpretation of interface pressure measurements for comparing surfaces is confounded by several factors. Interface pressure measurements are highly dependent on the material properties of the pressure transducer, soft tissue, and support surface. Pressure measurements tend to have a relatively high degree of variability. In addition, the relationship between interface pressure and the pressure in subcutaneous tissues is dependent on the composition and properties of the underlying tissues. Despite these problems, many clinicians, researchers, and product developers use interface pressure measurements to assess support-surface performance.

The wide variance in tissue tolerance between people means that researchers cannot pinpoint a general pressure threshold below which pressure ulcers will not develop. The often-quoted safe pressure threshold of 32 mmHg is unsubstantiated. This value was derived from research dating back to the 1930s, when Eugene Landis measured the pressures within a capillary loop of fingernail beds and found average pressures to be 12 mmHg near venules and 32 mmHg near arterioles. There is no proven relationship to pressure ulcer susceptibility.

Researchers have used interface pressure measurement to try to determine the relative pressure ulcer risk of individuals. For example, we studied the relationship between buttock–wheelchair seat cushion interface pressure measurements and pressure ulcer incidence in a skilled-nursing facility. Interface pressure measured on wheelchair seat cushions was higher for patients who developed sitting-induced pressure ulcers, compared to patients who did not. The mean peak pressure was 114 mmHg with a standard deviation of 46 mmHg for subjects with pressure ulcers and 77 mmHg with a standard deviation of 22 mmHg for subjects remaining ulcer-free. But this was a relatively small study and is not generalizable.

Another group of researchers found significantly more pressure ulcers among elderly wheelchair users who experienced peak interface pressure recorded at 60 mmHg or higher. Conine et al. (1993) studied the effectiveness of Jay® cushions in preventing pressure ulcers in a group of 163 elderly wheelchair users. They found that "the incidence of pressure ulcers was significantly higher among those patients who experienced peak interface pressure recorded at 60 mmHg or higher … ." Of the subjects, 40% who developed pressure ulcers had peak pressure higher than 60 mmHg, compared to 19% of the subjects who did not develop pressure ulcers. The Scimedics Evaluator (Next Generation Co., Temecula, CA) was used to measure pressure under five bony prominences: the ischial tuberosities, trochantors, and in the sacrococcygeal area.

The results of these two studies are not comparable, however, because the pressure measurement devices used in these two studies were different. Thus while high pressure is associated with increased risk of developing pressure ulcers, we cannot determine a safe threshold below which an individual will not develop pressure ulcer.

Many experts suggest that interface pressure measurement is better for identifying inappropriate support surfaces than for determining appropriate ones. In other words, clinicians should use pressure measurement to exclude certain interventions due to high loading, but should not use it as a singular assessment measure. To choose an appropriate support surface, you must consider more than just the normal pressure. When choosing a wheelchair cushion, for example, also consider its effect on transfers, posture, propulsion, comfort, and stability.

In general, pressure maps are a visual tool for evaluating pressure distribution. Special attention needs to be given to areas of highest risk. For example, when evaluating a person sitting on a wheelchair seat cushion, look at the areas under the ischial tuberocities and greater trochantors and ask: "Is this peak pressure too high? Is the pressure gradient too great?" If the answer is "yes," then try another cushion. If you answer "no," then consider the symmetry of weight distribution, and other characteristics of the cushion such as temperature control, and heat and water vapor transferability.

Pressure mapping can also help patients become more involved in their own care. If they can see how tilting back 45° reduces pressure on their buttocks, for example, they are more likely to change their seating habits. Along the same lines, pressure mapping can help convince third-party payers that a particular intervention is necessary.

7.4.2 Alternative Assessment Techniques and Confounding Factors

Dodd and Gross examined how pressure is transferred to the subcutaneous tissue. An animal model, weanling white-haired pigs, which match human skin anatomy and physiology, was used for their study and was subjected to a near-uniaxial compression load over the wings of the ilea. They found a variation by anatomical site of the amount of interface pressure that was transferred to the interstitial fluid. They measured a 28 and 43% transfer to the interstitial fluid over the wing of the ileum and dorsal aspect of a spinous process, respectively. They believe that the specific underlying bony geometry is key to determining how much load is transferred. Reddy et al. (1981) concluded that nearly the entire external load applied is transferred to the interstitial tissue. They found that 65 to 75% of the interface pressure was transferred to the interstitial fluid. In their study a uniform compressive load was applied with a pressure cuff over the bone of a limb. Later, they demonstrated that 100% of the interface pressure should be transferred to the interstitial pressure measured in the posterior thigh of a seated person.

Sangeorzan et al. studied the effect of interface pressure on subcutaneous pressure and tissue oxygenation via transcutaneous partial pressure of oxygen ($TcPO_2$). $TcPO_2$ quantifies the tissue oxygen tension as oxygen diffuses from the dermal capillaries through the epidermis to the skin surface. $TcPO_2$ has been used extensively as a measure of the perfusion of the skin in response to external loading. Sangeozan used simultaneous measurement of $TcPO_2$, interface pressure, subcutaneous pressure, and skin deformation to investigate the relative effect of skin interface pressure over bone vs. skin interface pressure over muscle on the lower leg. The results showed that less interface pressure was required to reduce $TcPO_2$ to zero for the skin over bone as compared to skin over muscle. In contrast, at the points of zero $TcPO_2$, the subcutaneous pressures were not significantly different for skin over bone and skin over muscle. The subcutaneous pressure and $TcPO_2$ relationship was consistent for the over-bone and over-muscle sites, while the interface pressure and $TcPO_2$ relationship varied according to the site tested. The implication of this result is that the relationship between skin interface pressure measurements and cutaneous perfusion, as measured by $TcPO_2$, is dependent on the mechanical properties of the underlying tissue. Hence, the interpretation of a skin interface pressure measurement must account for the properties of the underlying tissues. That is, high skin interface pressure over bone is possibly more harmful than high interface pressure over softer tissues.

Dodd and Gross's, Reddy and coworkers', and Sangeorzan's results all suggest that the relationships between skin interface pressure, subcutaneous pressure, and skin oxygenation vary from person to person and from measurement site to measurement site, based on the tissue thickness and composition and, thus, how it deforms. In our study, we recognize that these factors exist, but hypothesize that pressure ulcer incidence is associated with high skin interface pressure.

7.5 SUPPORT-SURFACE CLASSIFICATION

7.5.1 ELASTIC FOAM

An elastic material deforms in proportion to the applied load: greater loads result in predictably greater alterations in the shape of the material and vice versa. Support surfaces that are made from resilient foam exhibit this type of elastic response. (If time is also a factor in the load vs. deformation characteristic, then the response is considered to be viscoelastic; this will be discussed later.)

Foam support-surface products are made from two basic types of foam — open cell or closed cell. Foam is said to have "memory" because of its tendency to return to its normal shape or thickness. The minimum density (weight per cubic foot) of bed support-surface material should be 1.3 to 1.6 lb/ft^3. Convoluted foam should have a minimum depth of 4 in. from the bottom of the foam to the lowest point of the convolution to achieve the optimal pressure-reducing effects of the material.

Foam products typically consist of either foam layers of varying densities, combinations of gel and foam, or fluid-filled bladders and foam. The advantage of support surfaces with a combination of fluid-filled bladders and resilient foam would be to provide a degree of postural stability with a resilient shell and improved immersion and envelopment with a fluid- or viscous-fluid-filled layer at the skin interface.

An elastic foam support surface should have a resistance to pressure that is high enough to fully support the load (prevent bottoming out) without providing a reactive force (memory) that is too high to keep the interface pressure low. Over time and with extended use, foam degrades and loses its stiffness. This results in higher interface pressures. Mattresses typically wear out in 3 years, and the pressure is then transferred to the underlying supporting structure used to support the foam. In other words, the mattress bottoms out.

Foam is limited in its ability to immerse and envelop by its stiffness and thickness. Soft foams will envelop better than stiffer foams, but will necessarily be thicker to avoid bottoming out. Foam seat cushions are frequently contoured to improve their performance. Precontouring the seat cushion to provide a better match between the buttocks and the cushion increases the contact area, thus reducing the average pressure. Precontouring also increases immersion and envelopment properties, thus decreasing the pressure peaks.

Foam tends to increase skin temperature because foam materials and the air they entrap are generally poor conductors of heat. The heat-transfer characteristics of foam mattresses are less than normal physiological resting heat losses. Heat-transfer rates for mattresses with nonstretch and two-way stretch covers are less (by nearly half) than heat-transfer rates for mattresses without covers.

Moisture does not increase as much on foam products with porous covers as on most foams because the open-cell structure of the covers provides a pathway through which moisture can diffuse. Water vapor transmission rates can be reduced by more than half when foam mattresses are covered with nonstretch and two-way stretch covers.

7.5.2 Viscoelastic Foam

Viscoelastic foam products consist of viscoelastic, open-cell foam that is temperature sensitive. The foam becomes softer at operating temperatures near body temperature, the effect of which is that the layer of foam nearest the body provides improved pressure distribution through envelopment as compared to high-resilient foam. The viscoelastic foam acts like a self-contouring surface because the elastic response diminishes over time, even after the foam is compressed. The disadvantage of the temperature and time-sensitive response is that the desirable effects may not be realized when the ambient temperature is too low. The properties of viscoelastic foams vary widely and must be chosen according to the specific needs of the patient for seat and mattress applications. Solid gel products also are viscoelastic in nature and are included in this category.

Mean temperature increases of 2.8°C have been reported for viscoelastic foam. Gel products, on the other hand, tend to maintain a constant skin contact temperature or may even reduce the contact temperature for some time. Gel pads have higher heat flux than foam due to the high specific heat of the gel material. However, in one study, the heat transfer rate decreased after 2 h, indicating that the heat reservoir was filling. This suggests that temperature may increase during longer periods of unrelieved sitting (>2 h). In addition, this study found that moisture increased 22.8% during a 1-h period. The relative humidity of the skin surface increases considerably because of the nonporous nature of the gel pads.

7.5.3 Fluid-Filled Products

Fluid-filled products may consist of small or large chambers filled with air, water, or other viscous fluid materials such as silicon elastomer, silicon, or polyvinyl. The fluid flow from chamber to chamber or within a single chamber is passive in response to movement and requires no supplemental power. The term *air flotation* is sometimes used to describe interconnected multichamber surfaces, such as those manufactured by ROHO, Inc. For air cushions, care must be taken to maintain the correct level of inflation to achieve optimal pressure reduction. Underinflation causes bottoming out, and overinflation increases the interface pressure. For viscous-fluid-filled seating surfaces, such as the Jay® seat cushion (Sunrise Medical, Inc.), distribution of viscous material must be carefully monitored, and the material must be manually moved back to the areas under bony prominences if it has migrated.

Most fluid-filled products permit a high degree of immersion, allowing the body to sink into the surface as the surface conforms to bony prominences. This effectively increases the surface pressure distribution area and lowers the interface pressure by transferring the pressure to adjacent areas. These products are capable of achieving small-to-modest deformations without large restoring forces. In a direct comparison of interface pressures with air-fluidized (Clinitron, Hill-Rom, Inc.) and low-air-loss beds (Therapulse, Kinetic Concepts International), the RIK mattress (a fluid-filled product) was shown to relieve pressure as effectively as the aforementioned technology.

Skin temperature is affected by the specific heat of the fluid material contained in the support surface. Air has low specific heat (limited ability to conduct heat), and water has high specific heat (greater capacity to increase heat flux). The viscous material used in the RIK mattress also has a high specific heat; skin temperature decreases have been demonstrated with this product.

Given the large variety of materials used as covers for products in the fluid-filled category, it is difficult to generalize on moisture control characteristics. However, the insulating effects of rubber and plastic used in some fluid-filled products have been shown to increase the relative humidity due to perspiration.

7.5.4 Air-Fluidized Beds

Air-fluidized beds have been available since the late 1960s and were originally developed for use with burn patients. These products consist of solid particles, usually glass (75 to 150 mm), encased in a cover sheet. The solid particles take on the characteristics of a fluid when pressurized air is forced up through them. Feces and other body fluids flow freely through the sheet. In order to prevent bacterial contamination, the bed must be pressurized at all times, and the sheet must be properly disinfected after use by each patient and at least once a week with long-term use by a single patient.

Air-fluidized beds use fluid technology to decrease pressure through the principle of immersion while simultaneously reducing shear. Air-fluidized products permit the highest degree of immersion currently available, allowing the surface to conform to bony prominences. Almost two-thirds of the body may be immersed into the surface, effectively lowering the interface pressure by increasing the surface pressure distribution area. The high degree of immersion possible with this technology enables the transfer of pressure to adjacent body areas and other bony prominences. Shear force is minimized by the use of a loose (reduced surface tension) cover sheet.

The pressurized air in these products is generally warmed to a temperature of 28 to 35°C. This warming feature could be beneficial or harmful, depending on specific patient characteristics. For example, the heat may be harmful to patients with multiple sclerosis or beneficial for patients in pain. In any case, the beneficial effects must be balanced against the increasing metabolic demands of tissue.

The high degree of moisture vapor permeability of the system is very effective in managing body fluids. In cases of severe burns, air-fluidized beds have been known to cause dehydration.

7.5.5 Low-Air-Loss Systems

Low-air-loss describes a support-surface feature in which air passes through the pores of the cover material. The covers are usually made of a special nylon or polytetrafluoroethylene fabric with high moisture vapor permeability. Many support surfaces employing this feature use a series of connected, air-filled cushions or compartments. These cushions are inflated to specific pressures to provide loading resistance based on the patient's height, weight, and distribution of body weight. An air pump circulates a continuous flow of air through the device, replacing any air that is lost through the surface's pores. The inflation pressures of the cushions vary with patient weight distribution; some systems have individually adjustable sections for the head, trunk, pelvic, or foot areas. One manufacturer offers the ability to individualize each of the compartments rather than just the sections. Support surfaces are available that combine low-air-loss with alternating- and pulsating-pressure features.

In low-air-loss systems, the patient lies on a loose-fitting, waterproof cover that is placed over the cushions. The waterproof covers are designed to allow air to pass through the pores of the fabric; they are usually made of a special nylon or polytetrafluoroethylene fabric with high moisture vapor permeability. Manufacturers have addressed the problem of dehydration of the skin by altering the number, size, and configuration of the pores in the covers. The material is very smooth, has low coefficient of friction, is bacteria-impermeable, and easy to clean.

Low-air-loss beds use fluid (air) technology to distribute pressure through the principle of immersion. Deeper body-tissue immersion results in greater surface area for pressure distribution. Most surfaces employing the low-air-loss feature allow the air inflation level to be adjusted so that the immersion and pressure redistribution can be increased. These devices are capable of achieving moderate-to-large deformations without large restoring or shear forces.

The volume of air may be adjusted to provide more or less immersion for the entire body, for specific sections, or even for individual chambers or cells. The loose-fitting covers envelop and decrease friction. In fact, care must be taken to avoid sliding patients off the mattress during bed transfers.

As with other fluid-filled surfaces, the temperature of the skin is affected by the specific heat of the fluid material; air does not have a high specific heat. In addition, the circulating air is warmed. However, the constant air circulation and evaporation tend to keep the skin from overheating.

In systems with low air loss, the patient's skin is in contact with the cover. The local tissue environment is a function of the moisture vapor permeability of the cover and cushion materials, the airflow and porosity of the cover and cushion materials, and the thermal insulation of the cover. The ideal combination of these factors would be a material with high thermal insulation to prevent excessive loss of body heat, high moisture vapor permeability to prevent accumulation of excess moisture on the skin, and a moderate airflow to keep the skin from overheating. In one study, low-air-loss cover materials were rated based on a normalized comparison of these parameters. The combination of a cover and a cushion made from nylon/high-air-loss Gore-Tex laminate material had the highest scores and was most likely to promote a favorable local climate at the skin–cover interface. Devices with low airloss have been shown to prevent build-up of moisture and subsequent skin maceration.

7.5.6 Alternating-Pressure Technology

Alternating pressure describes a support-surface feature in which the pressure distribution is periodically altered. Most surfaces employing this feature contain air-filled chambers or cylinders arranged lengthwise, interdigitated, or in various other patterns. Air is pumped into the chambers at periodic intervals to inflate and deflate the chambers in opposite phases, thereby changing the location of the contact pressure. Pulsating pressure differs from alternating therapy in that the duration of peak inflation is shorter and the cycling time is more frequent. The latter appears to have a dramatic effect on increasing lymphatic flow.

The concept of alternating pressure for prevention of tissue ischemia is not new. Kosiak concluded in 1961 that "since it is impossible to completely eliminate all pressure for a long period of time, it becomes imperative that the pressure be completely eliminated at frequent intervals in order to allow circulation to the ischemic tissue." Houle's conclusion (1969) that a dynamic device that alternately shifts the pressure from one area to another would be "the choice to provide adequate protection against the development of ischemic ulcers" has been supported over the years by many others. Rather than increasing the surface area for distribution through immersion and envelopment, alternating-pressure devices distribute the pressure by shifting the body weight to a different surface contact area. This may increase the interface pressure of that area during the inflation phase.

The lack of sufficient study on tissue responses to alternating pressure leaves many questions unanswered regarding this type of support surface. For example, what are the ideal characteristics of the support surface (geometry of the surface [size/shape of cells and space between cells], material, depth, composition, and shape of the supporting structure)? Also, what are the ideal characteristics of the alternating cycle (rise time, hold time, duration of total cycle, and pattern of relief)?

Alternating-pressure technology has the same potential as any other fluid-filled support surface to influence temperature at the interface and care must be taken to maintain the correct levels of inflation. The skin moisture control and temperature control characteristics of an alternating-pressure surface will also depend on the characteristics of the cover and supporting material.

7.6 SUMMARY

Although support-surface technologies have been designed to reduce mechanical (pressure, shear, and friction) and additive (moisture and temperature) factors implicated in pressure ulcer formation, most support-surface comparisons have relied solely on interface pressure measurement. In more recent years, research has gone beyond assuming that pressure ulcers are the result of external pressure alone. Current studies are focusing on the physiological, biochemical, and biomechanical characteristics of tissue and their interactions. The results of these investigations reflect the limitation of using interface pressure as a sole indicator of threshold for pressure ulcer formation. Although based on an individual's relative responses, interface pressure measurements may effectively aid in selecting the best support surface for that individual. However, interface pressure alone is not sufficient to evaluate the efficacy of a particular device or class of devices. Many factors make the results of support-surface studies difficult to compare. For now, the best evidence regarding the effectiveness of support surfaces appears to be the outcome of a decrease in the incidence of pressure ulcers coupled with multiple measures of tissue response.

7.7 STUDY QUESTIONS

1. Name and describe three common reasons for developing pressure ulcers.
2. Name and describe three types seating cushions that influence a person's ability to sit.
3. Name and describe the three common goals of tissue integrity.
4. Name and describe three different tissue integrity principles for cushions.
5. Name and describe two methods for controlling temperature.
6. Name and describe two indications to use a viscoelastic fluid cushion.
7. Name and describe one reason to use a custom-contoured cushion.
8. Name and describe an indication to use an active cushion system.
9. Name and describe a functional benefit of ensuring cushion stability.
10. Name and describe one disadvantage of a fluid-filled cushion.

BIBLIOGRAPHY

Bennett L., Kavner D., Lee B.Y., Trainor F.S., Lewis J.M. Skin stress and blood flow in sitting paraplegic patients. *Arch Phys Med Rehabil* 1984;65:189–90.

Brienza D.M., Iñigo R.M., Chung K-C., Brubaker C.E. Seat support surface optimization using force feedback. *IEEE Trans Biomed Eng* 1993;40:95–104.

Brienza D.M, Karg P.E., Brubaker C.E. Seat cushion design for elderly wheelchair users based on minimization of soft tissue deformation using stiffness and pressure measurements. *IEEE Trans Rehabil Eng* 1996;4:320–8.

Brienza D.M., Karg P.E., Geyer M.J., Kelsey S., Trefler E. The relationship between pressure ulcer incidence and buttock-seat cushion interface pressure in AT-risk elderly wheelchair users. *Arch Phys Med Rehabil* 2001;82:529–33.

Brienza D.M., Karg P.E. Seat cushion optimization: a comparison of interface pressure and tissue stiffness characteristics for spinal cord injured and elderly patients. *Arch Phys Med Rehabil* 1998;79:388–94.

Brienza D.M., Pratt S., Sprigle S. Measurment of interface pressure — research versus clinical applications. *Proceedings of the 21st International Seating Symposium*, Orlando, FL, January 20–22, 2005, pp. 65–6.

Bryant R. *Acute and Chronic Wounds Nursing Management*. St Louis: Mosby-Year Book; 1992.

Chow W.W., Odell E.I. Deformations and stresses in soft body tissues of a sitting person. *J Biomed Eng* 1978;100:79–87.

Clark M., Rowland L. Preventing pressure sores: matching patient and mattress using interface pressure measurements. *Decubitus* 1989;2:34–9.

Clark M. Measuring the pressure. *Nurs Times* 1988;84:72–5.

Conine T.A., Daechsel D., Hershler C. Pressure sore prophylaxis in elderly patients using slab foam or customized contoured foam wheelchair cushions. *The Occupational Therapy Journal of Research.* 1993;13:101–17.

Daniel R., Priest D., Wheatly D. Etiologic factors in pressure sores: an experimental model. *Arch Phys Med Rehabil* 1981;62:492–8.

Dodd K.T., Gross D.R. Three-dimensional tissue deformation in subcutaneous tissues overlying bony prominences may help to explain external load transfer to the interstitium. *J Biomech* 1991;24:11–19.

Exton-Smith A.N., Overstall P.W., Wedgwood J., Wallace G. Use of the "air wave system" to prevent pressure sores in hospital. *Lancet* 1982;5:1288–90.

Fisher S., Szymke T., Apte S., Kosiak M. Wheelchair cushion effect on skin temperature. *Arch Phys Med Rehabil* 1978;59:68–72.

Flam E. A new risk factor analysis: a comparison of cutaneous interface environments of low air-loss beds. *Ostomy Wound Manage* 1991;33:28–34.

Gunther R., Brofeldt B. Increased lymphatic flow: effect of a pulsating air suspension bed system. *Wounds* 1996;8:134–40.

Holzapfel S. Support surfaces and their use in the prevention and treatment of pressure ulcers. *J ET Nurs* 1993;20:251–60.

Houle R. Evaluation of seat devices designed to prevent ischemic ulcers in paraplegic patients. *Arch Phys Med Rehabil* 1969;50(10):587–94.

Kokate J., Leland K., Held A., et al. Temperature-modulated pressure ulcers: a porcine model. *Arch Phys Med Rehabil* 1995;76:666–73.

Kosiak M. A mechanical resting surface: its effect on pressure distribution. *Arch Phys Med Rehabil* 1976;57:481–5.

Kosiak M. Etiology of decubitus ulcers. *Arch Phys Med Rehabil* 1961;42:19–28.

Krouskop T., Randall C., Davis J., Garber S., Williams S., Callaghan R. Evaluating the long-term performance of a foam-core hospital replacement mattress. *J Wound Ostomy Continence Nurs* 1994;21:241–6.

Landis E. Micro-injection studies of capillary blood pressure in human skin. *Heart* 1930;15: 209–28.

Levine S.P., Kett R.L., Brooks S.V., Cederna P.S. Electrical muscle stimulation for pressure sore prevention: tissue shape variation. *Arch Phys Med Rehabil* 1990;71:210–15.

Mak A.F., Huang L., Wang Q. A biphasic poroelastic analysis of the flow dependent subcutaneous tissue pressure and compaction due to epidermal loadings: issues in pressure sores. *J Biomech Eng* 1994;116:421–9.

Nicholson G.P. Scales J.T., Clark R.P., de Calcina-Goff M.L. A method for determining the heat transfer and water vapour permeability of patient support systems. *Med Eng Phys* 1999;21:701–12.

Noble P., Goode B., Krouskop T., Crisp B. The influence of environmental aging upon the load-bearing properties of polyurethane foams. *J Rehabil Res Dev* 1984;21:31–8.

Patel S., Knapp C., Donofrio J.C., Salcido R. Temperature effects on surface pressure-induced changes in rat skin perfusion: implications in pressure ulcer development. *J Rehabil Res Dev* 1999;36:189–201.

Peltier G., Poppe S., Twomey J. Controlled air suspension: an advantage in burn care. *J Burn Care Rehabil* 1987;8:558–60.

Reddy N., Cochran G., Krouskop T. Interstitial fluid flow as a factor in decubitus ulcer formation. *J Biomech* 1981;14:879–81.

Reswick J., Rogers J. Experiences at Rancho Los Amigos Hospital with devices and techniques to prevent pressure sores. In: Kennedi R., Cowden J., Eds. *Bedsore Biomechanics*, London: University Park Press;1976, pp. 301–10.

Reuler J., Cooney T. The pressure sore: pathophysiology and principles of management. *Ann Intern Med* 1981;94:661–6.

Sangeorzan B., et al. Circulatory and mechanical responses of skin to loading. *J Orthop Res* 1989;7:425–31.

Seymour R., Lacefield W. Wheelchair cushion effect on pressure and skin temperature. *Arch Phys Med Rehabil* 1985;66:103–8.

Souther S., Carr S., Vistnes L. Wheelchair cushions to reduce pressure under bony prominences. *Arch Phys Med Rehabil* 1974;55:460–4.

Sprigle S.H., Chung K.-C., Brubaker C.E. Reduction of sitting pressures with custom contoured cushions. *J Rehabil Res Dev* 1990;27:135–40.

Stewart S., Eng M., Palmieri V., Van G., Cochran B. Wheelchair cushion effect on skin temperature, heat flux and relative humidity. *Arch Phys Med Rehabil* 1980;61:229–33.

Sulzberger M., Cortese T., Fishman L., et al. Studies on blisters produced by friction. I. Results of linear rubbing and twisting techniques. *J Invest Dermatol* 1966;47:456–65.

Weaver J., Jester J. A clinical tool: updated readings on tissue interface pressure. *Ostomy Wound Manage* 1994;40:34–43.

Wells J., Karr D. Interface pressure, wound healing and satisfaction in the evaluation of a non-powered fluid mattress. *Ostomy Wound Manage* 1998;44:38–54.

West J., Hopf H., Suski M., Hunt T. The effects of a unique alternating-pressure mattress on tissue perfusion and temperature. Paper presented at the European Tissue Repair Society, Padova, Italy, 1995.

Yarkony G. Pressure ulcers: a review. *Arch Phys Med Rehabil* 1994;75:908–17.

8 Wheelchairs

Alicia M. Koontz, Jonathan L. Pearlman, Brad G. Impink, Rory A. Cooper, and Matthew Wilkinson

CONTENTS

8.1 LEARNING OBJECTIVES OF THIS CHAPTER

Upon completion of this chapter, the reader will be able to:

- Understand basic manual wheelchair components, fit, and propulsion techniques
- Understand basic power wheelchair components, drive, and control systems
- Understand the basic control system of a hybrid wheelchair
- Learn how wheelchair standards tests are used to ensure wheelchair durability, reliability, and safety, and estimate cost-effectiveness

8.2 INTRODUCTION

Mobility devices, in particular wheelchairs and scooters, make up a significant portion of assistive devices in use today. There is anticipated continued growth in the wheelchair market due to increasing aging populations, longevity, and increased incidence of traumatic combat-related injuries, and more effective therapies for chronic diseases. More manual wheelchair users are also acquiring electric power wheelchairs (EPWs) when they start to loose function. Thus, there is an overwhelming need for continued advancements in wheelchair technology and for skilled rehabilitation engineers, suppliers, and practitioners to ensure that individuals receive the most functionally appropriate mobility device that meets their needs.

8.3 OVERVIEW

The first part of this chapter discusses manual wheelchairs, beginning with a brief history, followed by typical user profiles, basic structural components, and wheelchair setup and propulsion techniques. Next, a brief history of EPWs is presented, followed by user profiles, basic structural components, power and drive systems, and control algorithms. Hybrid wheelchairs (cross between a manual and power wheelchair) and recent advances in power wheelchair technology that enable users to traverse virtually any mobility barrier (e.g., curbs and stairs) and terrain (e.g., uneven surfaces and slopes) are also reviewed. Wheelchair safety, durability, and reliability are major issues in wheelchair design. Standardized tests are employed to ensure that wheelchairs and scooters comply with minimum standards of durability, reliability, and stability. The current wheelchair standards are presented later in this chapter. The chapter concludes with a discussion of the role of the rehabilitation engineering in wheelchair development, testing, and provision, and the future direction of wheelchair technology.

8.4 MANUAL WHEELCHAIRS

8.4.1 Brief History

Two of man's earliest inventions, the chair and the wheel, were created as early as 4000 BC. The earliest recorded combination of the chair and wheel dates back to spoke wheeled chariots in 1300 BC in China. In the third century, the Chinese developed the wheelbarrow to transport the sick or disabled. The earliest recorded wheelchair was created by the Chinese in 525 AD. The first appearance of wheelchairs as we know them dates back to the late 1500s. King Philip II of Spain was disabled and needed a form of mobility. A special wheeled chair, called an invalid chair, was designed to improve the king mobility. The chair was made of very heavy wood and had footrests similar to today's wheelchairs. However, King Philip II had no way of propelling the chair and needed someone to push his chair. This led to the development of the first self-propelled wheelchair in 1655 by a watchmaker with paraplegia named Stephen Farfler. His wheelchair design involved the use of hand-crank-powered metal cog wheels. By 1677, many chairs used the cog wheel device as well as other accessories such as leg supports and casters. By the 18th century, it was common for a wheelchair to contain two large rear wheels and one small front wheel. Further evolvement of the wheelchair led to considerations of occupant comfort. Wheelchairs began to include reclining backs and adjustable footrests. The 19th century saw the inclusion of several significant modifications such as the conversion from wooden to iron wheels, followed by the addition of hollow rubber tires. In the late 19th century, wire spoke wheels were adopted on most wheelchairs and pushrims were added to allow users to propel themselves. Attempts at producing a lighter chair were made but were unsuccessful. For example, chairs were made of Indian reed rather than wood, but these could not withstand the daily wear and tear. The first power wheelchairs appeared in the early 1900s when a 1¾ hp motor was added to an invalid's tricycle in 1912, and in 1916 London produced the first motorized wheelchairs. In 1932, the first lighter weight, metal, folding wheelchair, which most resembles wheelchairs today, was produced by Harry Jennings and Herbert Everest, who went on to form the Everest and Jennings Company. The wheelchair was made of tubular steel and was capable of folding so that it could fit in a car for transportation. This wheelchair revolutionized the industry and greatly influenced the evolution of wheelchairs. This basic design has been used to develop many of today's wheelchairs by applying design modifications to produce lighter, more durable wheelchairs that provide more independence for people who use them on a daily basis.

8.4.2 Manual Wheelchair User Profiles

Manual wheelchairs are mainly used by individuals who have the necessary upper-body strength, function, and stamina for everyday propulsion. Manual wheelchairs are often used by individuals with (1) spinal cord injury at the first thoracic level or below and some persons with cervical six- or seven-level injuries, (2) spina bifida, (3) early stages of multiple sclerosis, (4) lower limb amputations, (5) postpolio affecting only the lower extremity, (6) arthritis, (7) stroke, (8) older age, or (9) cardiopulmonary disease. Comorbid conditions such as excessive body weight, overuse of the upper limbs, long-time living with a disability, and poor health and nutrition can impair the ability to independently propel a manual wheelchair, and for these individuals it may be more functional to use an EPW. For individuals who cannot use a power wheelchair because, for example, they are unsafe or are physically unable to, a manual wheelchair is prescribed and mobility is facilitated by an attendant or caregiver.

8.4.3 Basic Structural Components

Design is a critical factor in the mobility of manual wheelchair users, because individuals need upper-body strength to bear the load of the wheelchair in addition to the weight of their body. Standard manual wheelchairs, typically used in hospitals and nursing homes, tend to be bulky, heavy, and of a one-size-fits-all nature. These wheelchairs have the benefit of folding, which allows for simple storage and transportation. The seat and backrests are sling upholstery, which provides very little comfort and lacks the ability to provide pressure relief. Owing to the weight and comfort restraints, these wheelchairs are not intended for long-term use, but for temporary use of less than a few hours a day. For individuals requiring long-term use, high strength, ultralight wheelchairs have been designed. These wheelchairs can be customized to fit the user, allowing for better comfort and mobility. They are typically made of materials such as aluminum and titanium rather than the heavy steel used in standard wheelchairs.

In some instances, none of the previously mentioned wheelchairs meet the needs of the user. For example, obese or overweight individuals need a larger, stronger wheelchair to support their weight. This class of wheelchairs is called heavy duty or bariatric wheelchairs. They can support individuals ranging from 300 to 1000 lbs. Another population that needs a wheelchair designed more specifically is children. Pediatric manual wheelchairs are designed to fit children better by reducing the overall seat dimensions and wheelchair weight so that the child can move more easily. They also have adjustable or "add-on" hardware to accommodate for growth. Other types of wheelchairs include those designed for sports. These wheelchairs are built to be very lightweight and maneuverable. Their high maneuverability is accomplished by adjusting the axle position and camber angles of the wheelchair.

The following section will present all the basic features of manual wheelchairs and their importance for user comfort and mobility. These features include seat height, width, and depth, backrest height, armrest height, seat and back angles, rear wheel camber, and rear axle position, as illustrated in Figure 8.1. Many of these features can be adjusted for ultralight wheelchairs; whereas lightweight wheelchairs have limited adjustability and standard wheelchairs have no adjustability.

Frame: In the past, the only material available for wheelchair frames was stainless steel. Now, frames are made of chrome, aluminum, steel tubing, titanium, and other lightweight composite materials. The choice of materials has a great effect on the overall weight of the wheelchair in addition to the strength and durability. Also, the frame structure will affect durability, transportability, and storage. For example, rigid-frame chairs are one piece with removable wheels and foldable back supports for transport and storage, whereas cross-brace frames allow for the wheelchair to collapse in the middle for storage. Some frames have built in or

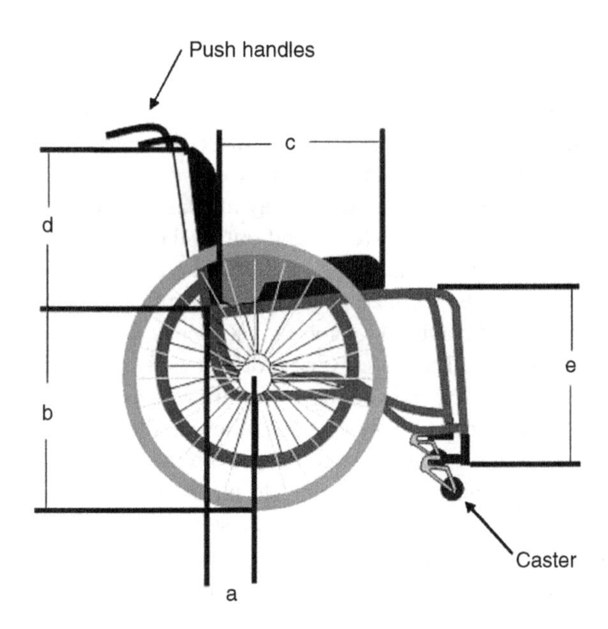

FIGURE 8.1 Ultralight wheelchair with some of its critical dimensions: (a) axle position; (b) seat height; (c) seat depth; (d) backrest height; (e) leg-rest length.

"add-on" suspension elements to decrease shock and vibration and make for a smoother overall ride.

Seat height: Seats are positioned so that leg length is accommodated while leaving sufficient space under the footrest for obstacle clearance. Ideally, seat height is placed at a level where the individual will have the necessary knee clearance to fit under tables, counters, sinks, etc.

Seat depth: The seat depth will support the individual's thighs. Improper depths cause increased pressures on the legs. Shallow seats increase sitting pressure because of less contact between the thighs and seat, which decreases the overall pressure distribution. Deeper seats can put added pressure behind the knees and on the calves.

Seat width: The width of the wheelchair seat should be slightly larger than individual's buttocks. If the seat is too narrow, then there will be added pressure on the sides of the hips, which may result in pressure sores. If the seat is two wide, the user will be forced to use a more abducted arm position for propulsion, which is more difficult.

Back height: The back height varies from person to person, depending on the amount of postural support needed and comfort. Ideally, the backrest should be low enough that it does not impede the range of motion of the arms during propulsion. If the backrest is too high, then individuals will not be able to contact as much of the pushrim, whereas, if it is too low, individuals will not have the necessary postural support.

Seat and back angles: Seat angle is defined as the angle between the seat and the horizontal, and back angle is the angle between the backrest and the vertical. Both of these can be adjusted to provide the best comfort and support for a particular user. Increasing the seat dump (sloping seat down toward the rear) can help stabilize the pelvis and spine in individuals with decreased trunk control, which improves their ability to propel. However, too much dump may increase the sitting pressure and can also make it more difficult to transfer in and out of the wheelchair. Increasing the back angle is another way to ensure user comfort. If an individual has difficulty with hip flexion, the back can be reclined to prevent discomfort. However, reclining too far can shift the center of gravity (CG) too far backward, making the chair unstable.

Armrests and footrests: There are a variety of armrest lengths and styles to accommodate the specific user. The positioning of the armrest should be at a comfortable height so that, when the arm is on the armrest, it lies parallel to the ground. Armrests that are too high or too low may cause discomfort in the user's neck and shoulders. Many ultralight wheelchair users prefer not to have armrests because they tend to increase the overall width of the wheelchair, cause the arms to abduct more during propulsion, and get in the way of accessing certain desks and tables. These individuals are also safe and comfortable with using other parts of the wheelchair (e.g., frame or wheels) to perform transfers and pressure-relief. Cross-brace-frame chairs typically have footrests that can flip up or down, swivel, are angle-adjustable, or can be removed, whereas rigid-frame chairs have the footrest built into the frame. They normally are not removable, but can be slightly adjusted to accommodate various leg lengths.

Wheels and tires: A typical wheelchair has four wheels, two large rear wheels and two small front wheels called casters. These caster wheels are typically solid rubber, but can be plastic, pneumatic, or a combination of them. The caster wheel size varies from 2 to 8 in. in diameter, depending on the wheelchair type and user preference. The standard rear wheel tire is a pneumatic 24 in. (0.6096 m) spoked wheel. Different tire types are available, such as solid insert tires, semipneumatic, or radial tires. Additionally, different wheels such as mag wheels and off-road wheels can be purchased for some chairs. A majority of wheelchairs have quick release wheels that can be easily removed to change the wheels or for storage and transport.

Rear-wheel camber: The camber angle is the amount of rear-wheel tilt. At 0° camber, the wheels are vertical, lining up directly with the sides of the wheelchair. To increase the camber, the top of the wheel is tilted inward, thus moving the bottom of the wheel further outward. Everyday wheelchairs typically have up to 8° of camber. Increasing camber provides more stability at the cost of increased wheelchair width. More than 8° is possible and seen on many sport wheelchairs, but can make the wheelchair too wide to fit through doorways and navigate everyday environments.

Rear-axle position: The rear axle can be adjusted both horizontally and vertically. Horizontal adjustments allow the rear wheels to be placed in an optimal position with respect to the arms. This positioning can have a great effect on propulsion ability and technique. Moving the axle position forward will allow the user to grasp more of the wheel during propulsion, whereas moving the position further back limits the amount of pushrim that can be reached by the individual. Although moving the axle forward allows users to reach more of the wheel, it also increases the likelihood of the wheelchair to tip backwards. To a degree, this can be useful as it will allow users to perform wheelies and traverse obstacles more easily. To prevent the danger of tipping backwards, anti-tippers can be added to the rear of the chair so that users can have the benefit of reaching more of the pushrim without the risk of tipping over unexpectedly. Additional benefits of a forward axle position are a decrease in the number of strokes used to propel as well as less effort for propulsion. Vertical adjustments of the axle can provide similar propulsion advantages. Moving the axle position up with respect to the seat will also in effect lower the seat height of the chair. Lowering the seat height improves stability, but if it is too low, it can induce poor propulsion patterns due to abducted arms. An appropriately placed axle can help to decrease the risk of secondary injury as a result of everyday wheelchair propulsion.

8.5 WHEELCHAIR PROPULSION

As previously mentioned, the various features of a wheelchair can affect the overall propulsion capabilities of the user. Some of the major factors are the wheelchair weight and seat positioning. Many researchers are concerned with investigating wheelchair propulsion patterns in order to determine the most optimal wheelchair configurations and implement new design criteria for future wheelchair development. The study of wheelchair propulsion falls into two main categories: kinematics and kinetics.

Kinematics is the study of the motion of a body without considering its mass or forces acting upon the body. In this case, the bodies of interest are the arms and trunk. Kinematic variables consist of angular displacements, velocities, and accelerations. In order to obtain these variables, some knowledge of the position of the bodies of interest is required. Position data are typically gathered from anatomical landmarks or centers of mass of limbs. Three-dimensional position data are commonly obtained using motion-capture systems. Markers specific to the motion-capture system are placed on the upper extremities. As a person propels the wheelchair, the location of these markers is recorded by the system and saved for later use. This data is then used to calculate the joint angles, velocities, and accelerations of the upper extremities.

Kinetics is the study of the effects of forces on the motion of a body or a system of bodies. For wheelchair propulsion, the interest is focused on the forces and moments applied to the pushrims by the arms. In order to obtain accurate force information during propulsion, a specially designed wheel instrumented with strain gauges or a load cell is required. The instrumented wheel provides the forces and moments at the pushrim only, which can be very useful, but of greater interest are the individual joint forces and moments. Using an inverse dynamics model in combination with the kinematic data, the forces at the pushrim can be translated to the upper extremity. The end result will yield the forces and moments imposed at each of the joints of the arm (wrist, elbow, or shoulder). Propulsion analyses are generally conducted to understand propulsion inefficiencies, mechanisms of upper-limb pain and injuries, develop interventions (e.g., modifications to wheelchair setup, design, and techniques) that minimize effort and protect the joints from developing pain or injury.

8.6 ELECTRIC POWER WHEELCHAIRS

8.6.1 Brief History

The first electric power wheelchairs were developed in the early 1900s, but it was not until the 1940s that they were patented. The earliest models relied on automobile batteries and starter motors, making them inefficient and unreliable. Advances in the electronics led to major progress in electric power and drive systems. While EPW reliability and efficiency improved over the decades, it was not until the late 1970s and 1980s that EPWs became a highly reliable form of mobility for people with disabilities. The late 1970s brought solid state components, improving reliability and allowing the use of a true proportional controller for the EPW. The major change was in the use of the metal-oxide-semiconductor field-effect transistor (MOSFET) in place of relays for switching on the drive motors. Prior to this advance, EPWs operated on bang–bang controllers utilizing relays to switch the motors, making them unreliable, difficult to maneuver, and thus limited to the outdoors. In the 1980s, microprocessors became widespread, and were used as the brain behind the EPW controller. Microprocessor-based controllers allow for a wide range of control algorithms and programmability in the EPW. Furthermore, they can accommodate a wide range of control devices used by wheelchair users (joystick, sip-and-puff switches, etc.), and can also control secondary systems, such as an augmentative communication device or environmental control (doors and/or lights in a house). Because of these advances, along with drive-train advances such as independent suspension, the current EPWs are an efficient and effective mobility device, affording people with disabilities a great deal of independence. The state-of-the-art EPWs on the market and in development phases are considered mobile robots, integrating complex control features that can balance the EPW on two wheels, track their own motion, and use environmental cues to restrict or direct the EPW driving paths.

8.6.2 Electric Power Wheelchair User Profiles

Electric power wheelchairs are indicated for a wide range of individuals with disabilities. For some individuals with severe sensory-motor impairments, EPW are the only functional mode of mobility. Examples of this are individuals who have high-level tetraplegia, advanced multiple sclerosis, or severe cerebral palsy. Sensory-motor impairments due to these diseases can limit functional motor control to individual parts of the user's body. In these cases, EPW input devices into the controller must be customized to take advantage of the intact motor function of the individual. Microswitching mechanisms that can be actuated with the mouth (sip-and-puff) or other parts of the body (feet, head, etc.) must be used as controller input. EPW are also indicated for people at risk of falling or are in pain when ambulating and who cannot safely or functionally propel a manual wheelchair. For these individuals the EPW represents a safe and pain-free means of mobility. For all users, the goal of the EPW is to provide greater independence and quality of life for the user. As EPWs have become more technologically advanced, they have been able to achieve this goal for a broader range of people with disabilities. Because people with disabilities have diverse

FIGURE 8.2 Example conventional (left) and power-base (right) electric power wheelchairs.

and complex needs to achieve independence, a more customizable EPW can reach a larger proportion of the population with disabilities. The technological advances of controller programmability, special seating systems, and integrated control units (with augmentative communication devices, environmental control units, etc.) have allowed a wider population of people with disabilities to become independent.

8.6.3 Basic Structural Components

The frame design of an EPW falls into two types: either a traditional style (based on a manual wheelchair design) or a power-base style (Figure 8.2). Because the manual wheelchair design predates the power wheelchair, first designs for power wheelchairs were simply add-on power devices to the manual wheelchairs. These devices still exist, especially for portable-style EPWs that can be easily disassembled, but the majority of EPWs are power-base styles. The conventional frame design was an easy first step in the EPW development, but adding the special seating features (tilt and recline, etc.) made the wheelchair heavy and unstable. To accommodate these additional features, the EPW was redesigned as a two-component power-base style. The bottom "power-base" component houses the batteries and motors and, in most cases, the controller. The seat component simply attaches to the power base, and thus can be swapped for whatever style of seat best meets the needs of the user. A secondary benefit of separating the seating and the power and control system was that the power-base design could be optimized nearly independent of the seating requirements of the user. Consequently, power-base designs have progressed quite rapidly to include complex suspension systems, a low CG, and excellent maneuverability. Concomitantly, seating technology has advanced to meet the needs of a wide range of users.

An EPW seating system can be as low tech as a "captain's chair" style seat or as complex as a power-actuated seat, which can have any combination of the following motions: tilt (changing the whole seat angle relative to the horizontal), recline (changing the seat-back to seat-bottom angle), elevating leg rests (leg rest to seat-bottom angle), seat elevator (changing the height of the seat), and a standing mechanism (raising the person into a standing posture). Depending on the user's pressure management needs and state of their musculoskeletal system, any combination of the aforementioned motions may be prescribed. Tilt of the seat is one of the most common features, because it allows for pressure distribution to be shifted from the buttocks to the back; this can help prevent dangerous pressures ulcers that are common for wheelchair users with sensory loss.

8.6.4 Power and Drive Systems

Most EPWs currently on the market utilize 24-V DC power and drive systems. Two 12-V wet- or gell-cell batteries are linked in series to provide 24 V to the drive motors; typically, the controllers draw from only one of these 12-V batteries. The ampere-hour rating of a battery indicates the number of amperes the battery can provide in a given period of time (typically 20 h). Thus, a 100 A h battery can provide 5 A/h over a 20-h time span. The theoretical range (R km) of an EPW can be calculated by knowing the battery capacity (C ampere-hours), and the amount of depletion of the battery (E ampere-hours) while traveling a known distance (D meters) with the following formula:

$$R = \frac{C * D}{E * 1000} \tag{8.1}$$

Clearly, improving both the efficiency of the drive train (lower E) or using higher capacity batteries (higher C) will extend the range of the EPWs. EPWs in the 1980s had a short range, limiting the users mobility (Woods and Watson, 2003). Research has shown that users typically travel a mean (±SD) of 8.35 (7.07) km over a 5-day period, and wheelchair theoretical ranges have been shown to be significantly longer than this (R between 32 and 25 km). Thus, the range of current EPWs allow for most users to ride their wheelchair all day without a recharge. This holds true for new batteries, but over time their capacity is reduced based on the number times they are recharged: typically, life cycles (number of recharges) of an EPW battery will range from 50 to 300 cycles. Thus, a user who recharges the battery nightly can typically expect to replace their batteries once a year.

Two DC permanent-magnet electric motors are typically used on EPWs. Each motor is attached to a drive wheel through a gear or belt reduction system. The relays used in the 1970s were replaced with MOSFET bridges to switch on/off the drive motors. Bidirectional switching of the DC motor is achieved using a H-bridge with four MOSFETs (Figure 8.3). With M_1 and M_4 switched on, the motor rotates forward, and with M_2 and M_3 switched on, the motor rotates in reverse. Switching both M_1 and M_4 would allow full power to be supplied to the motor. Instead, controllers typically switch either M_2 or M_4 on to select direction, and then pulse M_1 and M_3 to modulate the current going to the motors (and thus controlling the speed). The output of the controller is thus to pulse these MOSFETs based on various inputs.

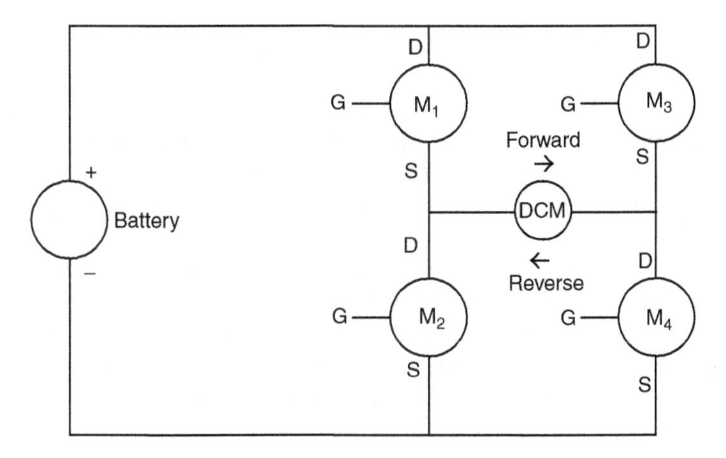

FIGURE 8.3 H-bridge MOSFET circuit for bidirection motor control (S = supply, D = drain, G = gate. M_1–M_4 = MOSFETS, DCM = DC drive motor).

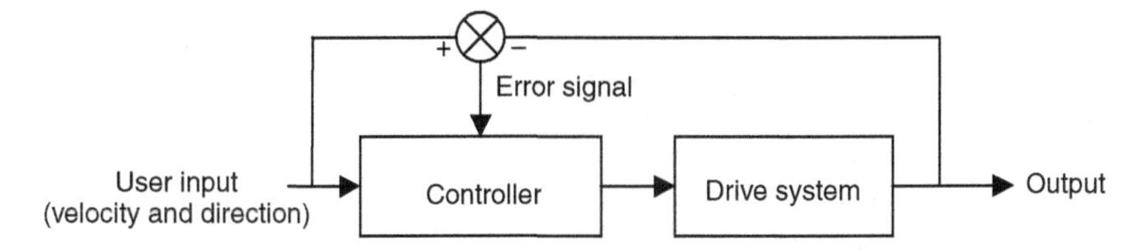

FIGURE 8.4 Basic block diagram of an electric-powered wheelchair control system.

8.6.5 Control System

Microprocessor-based controllers are the standard in current EPW because of their programmability and the variety of input and output channels that can be used to control the EPW and, in some cases, other devices. The primary task of the wheelchair controller is to integrate the user input (typically through a joystick) with the error signal (difference between desired and achieved velocity and direction) to produce an output that results in the desired motion of the EPW. User input, such as joystick displacement, is interpreted as a "desired" velocity input. Acceleration magnitude can typically be programmed into the controller, and feedback based on optical encoders or back electromotive force (EMF) of the motors is used for error correction to achieve the desired velocity. A generic block diagram for this control system is shown in Figure 8.4.

For the controller to effectively integrate the user input and the error signal to produce the appropriate output, it must understand the drive system and wheelchair dynamics. A complete set of dynamical equations of the wheelchair can be found elsewhere — see bibliography. Progress is being made on more sophisticated controllers that consider more input channels for guidance control and can control other devices. Power seating functions on current EPW are commonly controlled with the main EPW controller. By using a common controller, one can program safety features, such as limiting the velocity of the wheelchair when the seating system is positioned such that the wheelchair is statically or dynamically unstable. These unstable positions are common when the user is tilted back or in a standing posture. Additionally, EPW controllers can be used to interface with other assistive technology systems that the user may have for environmental control (door openers or lighting control) or augmentative communication devices.

The most common input device for speed and direction control of the EPW is the proportional control joystick. As described earlier, joystick displacement represents a direction and velocity goal. For some people with disabilities, sensory-motor loss or motor spasticity makes it impossible for the user to operate a conventional joystick. In these cases, several alternatives are available. For motor spasticity, it is possible to use a low-pass filtering scheme to extract the desired signal (direction and velocity) from the input device. In cases in which this is not successful, and for users with low sensory-motor function, direct-selection switches can be used (user pushes a separate switch for each desired direction) or single-switch scanning method is used. The scanning method uses a set of cursors (representing a particular direction) that are sequentially highlighted after a certain dwell period. A user with limited sensory-motor function or spasticity can thus control a wheelchair using a single switch to select the desired direction when it is highlighted. This control scheme mimics that of a bang-bang controller, although acceleration profiles can be programmed into the controller to minimize jerkiness of the motion. These users, and those with limited cognitive or visual function, are the candidates for the new generation of EPW controllers, which use environmental input (such as position and obstacle sensors) to help guide the EPW.

8.7 POWER-ASSISTED WHEELCHAIRS

Power-assist devices include stand-alone powered units that are external to the wheelchair, and the wheelchair user holds onto it, power add-on devices that attach to the wheelchair and have a steering mechanism or input device for controlling the wheelchair, or a pushrim-activated system (PAS) with motors in the wheelchair hubs (Figure 8.5). PASs offer an alternative between manual wheelchair mobility and EPW driving. A PAS operates much like a manual wheelchair but with less effort. This may make PAS suitable for people with or at risk for upper-extremity joint degeneration, reduced exercise capacity, and low upper-extremity strength or endurance.

One type of PAS measures pushrim forces using a linear compression spring and simple potentiometer that senses the relative motion between the pushrim and the hub. A microprocessor uses these signals to control a permanent-magnet DC motor attached to each rear wheel. Each motor is connected to a ring gear with a resulting 27:1 gear reduction. The PAS controller provides the feel of a traditional manual wheelchair by simulating inertia and compensating for discrepancies between the two wheels (e.g., differences in friction), and increases user safety with an automatic speed limiting and braking system through the use of regenerative braking. Power is supplied by either a single custom-designed nickel–cadmium battery (NiCd) or a nickel–metal hydride battery (NiMH).

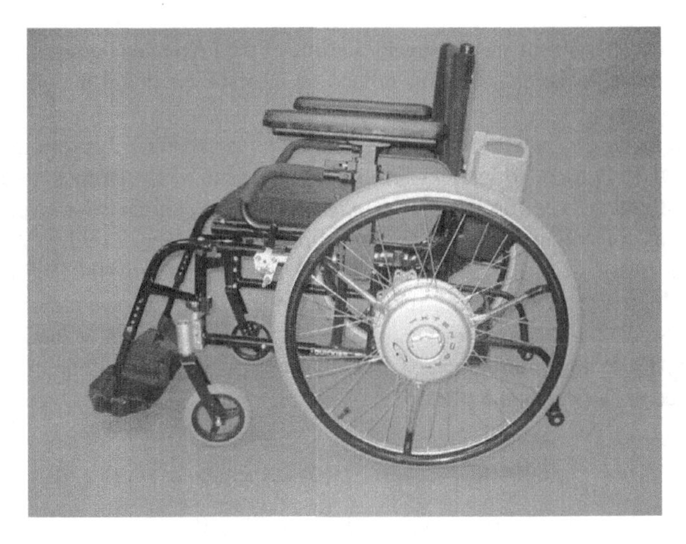

FIGURE 8.5 A pushrim-activated system.

8.7.1 PAS Control Algorithm

Manual wheelchair propulsion is a bilateral discontinuous activity powered by the user applying torque to each pushrim. The hand-to-pushrim coupling and its accompanying torque may differ depending on the speed at which the hand contacts the pushrim. When optimal hand coupling/decoupling occurs, the pushrim torque of the PAS should be similar in form to that of a skilled manual wheelchair user but lower in amplitude. To accomplish this, the PAS control system begins with a simple scaling approach as described by.

$$k_w k_g k_m i(t) = k_w k_g \tau_m(t) = u = k_w k \tau_p(t) \tag{8.2}$$

The motor torque is represented by $\tau_m(t)$ and is assumed to be proportional to the motor current, $i(t)$, times the motor torque constant (k_m); k_w is the wheel gear ratio, and k_g is the transmission gear ratio. The control signal, u, during the propulsion phase is the pushrim torque, $\tau_p(t)$, times a constant. The output torque to the wheel is proportional to the motor torque times the transmission gear ratio (k_g) and the wheel gear ratio, k_w. For a proportional controller to achieve the desired effect, certain criteria must be satisfied. For example, by specifying a dead zone (ε), defined in Equation 8.3, unintentional movements of the wheelchair due to sensors noise are avoided.

$$\begin{aligned} u &= 0, \quad |\tau_p(t)| < \varepsilon \\ u &= k_w k \tau_p(t), \quad |\tau_p(t)| \geq \varepsilon \end{aligned} \tag{8.3}$$

The coupling between the pushrim and the torque sensor must have similar stiffness as that of a standard wheelchair pushrim so as not to alter the coupling and decoupling of the hand and the pushrim. The response time of the PAS must be faster than that of the user's musculoskeletal system; otherwise, a response delay may cause torque oscillations.

The velocity of the wheelchair experiences exponential decay from the moment the pushrim is released, which plays an important role in it effectiveness as a mobility device. Without momentum simulation when the pushrim torque is removed, a simple proportional controller would drive the motor torque to zero rapidly. With a gear transmission ratio of about 27:1 (i.e., $k_g = 27$), this would cause the PAS to come to an abrupt halt. Therefore, during the recovery phase, the PAS switches to a control law that models the pushrim torque as decaying linearly from its value just prior to roll-off, $\tau_p(t_r)$. While simulating coasting, the control law given in Equation 8.2 must be modified as in Equation 8.4.

$$\begin{aligned} &u = \tau_p(t_r)(t_c - mt), \quad |\tau_p(t)| < \varepsilon, \quad \text{and} \quad (t_c - t) > 0 \\ &u = 0, \quad |\tau_p(t)| < \varepsilon, \quad \text{and} \quad (t_c - t) \leq 0 \\ &u = k_w k \tau_p(t), \quad |\tau_p(t)| \geq \varepsilon \end{aligned} \tag{8.4}$$

The time just prior to release, t_r, is defined as when the pushrim torque starts its decent towards the threshold torque of the dead zone. The cost time, t_c, determines the length of time to zero torque (hence, zero velocity), and the rate of decay is defined by m. EPW controls commonly limit the forward and rearward speed and acceleration as well as the yaw speed and acceleration. The limiting values for speed and acceleration are variable, depending on the device and the user. Speed limiting, as described in Equation 8.5, with a PAS is accomplished by applying regenerative braking after the maximum speed threshold is achieved.

$$\begin{aligned} &u = \tau_p(t_r)(t_c - mt), \quad |\tau_p(t)| < \varepsilon \quad \text{and} \quad (t_c - t) > 0, \quad \text{and} \quad v(t) \leq v_{\text{thresh}} \\ &u = 0, \quad |\tau_p(t)| < \varepsilon \quad \text{and} \quad (t_c - t) \leq 0, \quad \text{and} \quad v(t) \leq v_{\text{thresh}} \\ &u = k_w k \tau_p(t), \quad |\tau_p(t)| \geq \varepsilon, \quad \text{and} \quad v(t) \leq v_{\text{thresh}} \\ &u = (F_f + \tau_b), \quad v(t) > v_{\text{thresh}} \end{aligned} \tag{8.5}$$

Here, gearbox and wheel friction are represented by F_f, and the braking torque given by τ_b is assumed proportional to k_b, the braking constant, and v, the speed of the wheelchair. The maximum speed threshold, v_{thresh}, is typically set at 6 km/h.

$$\tau_b = k_b v(t), \quad v(t) > v_{\text{thresh}} \tag{8.6}$$

The PAS has the command, control, and feedback signals all interfaced to a single microprocessor and control algorithm in order to balance the loads due to terrain, user ability, and component variations (e.g., motors). Equation 8.7 shows the load compensation algorithm that is used to drive the torque of the two wheels to the same trajectory given that the desired trajectories (i.e., input values from the pushrim) are within a predefined threshold of one another.

$$
\begin{aligned}
&u = \tau_{\mathrm{p}}(t_{\mathrm{r}})(t_{\mathrm{c}} - mt), \quad |\tau_{\mathrm{p}}(t)| < \varepsilon, \quad \text{and} \quad (t_{\mathrm{c}} - t) > 0 \quad \text{and} \quad v(t) \leq v_{\mathrm{thresh}} \\
&u = 0, \quad |\tau_{\mathrm{p}}(t)| < \varepsilon, \quad \text{and} \quad (t_{\mathrm{c}} - t) \leq 0, \quad \text{and} \quad v(t) \leq v_{\mathrm{thresh}} \\
&u = k_{\mathrm{w}} k \tau_{\mathrm{p}}(t), \quad |\tau_{\mathrm{p}}(t)| \geq \varepsilon, \quad \text{and} \quad v(t) \leq v_{\mathrm{thresh}} \\
&u = (F_{\mathrm{f}} + \tau_{\mathrm{b}}), \quad v(t) > v_{\mathrm{thresh}}
\end{aligned}
\tag{8.7}
$$

where

$$
\begin{cases}
\tau_{\mathrm{p}}(t) = \dfrac{(\tau_{\mathrm{pl}}(t) + \tau_{\mathrm{pr}}(t))}{2} & |\tau_{\mathrm{pl}}(t) - \tau_{\mathrm{pr}}(t)| < \Delta\tau_{\mathrm{p}} \\
\tau_{\mathrm{p}}(t) = \tau_{\mathrm{p}}(t) & |\tau_{\mathrm{pl}}(t) - \tau_{\mathrm{pr}}(t)| \geq \Delta\tau_{\mathrm{p}}
\end{cases}
$$

In this equation, $\Delta\tau_{\mathrm{p}}$ is the threshold difference between the left and right pushrim torques (τ_{pl} and τ_{pr}, respectively). This variable is used to compensate for side-to-side differences in the user's ability to provide torque as well to compensate for sensor variations. Cooper et al. (2001) found that using a pushrim-activated wheelchair resulted in significant reductions in the physiological demand of propulsion when compared to using a manual wheelchair. Using a pushrim-activated power-assist wheelchair also reduces upper-extremity range of motion at the wrist and shoulder, which may help reduce the incidence of soft tissue injuries. Power-assist devices may increase the width or length of the manual wheelchair base a few inches. They are more difficult to transport than manual wheelchairs mainly because the equipment adds about 40 to 50 lb to the wheelchair; however, this is still considerably lighter than a fully powered wheelchair.

8.8 MULTIFUNCTIONAL WHEELCHAIRS

Multifunctional wheelchairs are devices that are capable of changing drive configurations to adapt to environmental changes or challenges. An example of this is the iBOT® 4000 transporter, which utilizes a two-wheel cluster design (Figure 8.6). With this design, the iBOT® can assume four different drive configurations: standard, four-wheel, balance, and stair. In the standard function, the transporter resembles a rear-wheel-drive wheelchair (Figure 8.6a) by raising the foremost drive wheels and lowering the front casters. In the four-wheel configuration (Figure 8.6b), all four drive wheels are in contact with the ground and the front casters are elevated. This configuration enables the user to climb curbs (up to 127 mm high) and easily traverse uneven and sloped terrain. The stair-climbing ability (Figure 8.6c) is achieved by controlling the cluster rotation on the basis of the position of the CG. The device strives to keep the system's CG above the ground-contacting wheels and between the front and rear wheels at all times, regardless of disturbances and forces operating on the system. Pitch and roll motion recorded with gyroscopes are used by a closed-loop servo-control algorithm (Figure 8.7). The joystick is deactivated during

FIGURE 8.6 (a)–(d): iBOT® shown in each drive configuration.

stair climbing to prevent unintentional deflection. Users can initiate the function on their own and maintain stability by holding the stair handrails, or assistants can control the rate of climbing through the assist handle. When a user leans either forward or back, or an assistant leans on the device, shifting the CG, the device rotates the wheel cluster, which results in the device climbing down or up one stair. The control system requires the device to dwell on each step for a few seconds before allowing another cluster operation. This hiatus helps keep the device from running away on stairs. The stair configuration performs well in descending and ascending stairs that are standard height, width, and depth in private homes, in public buildings, and outdoors.

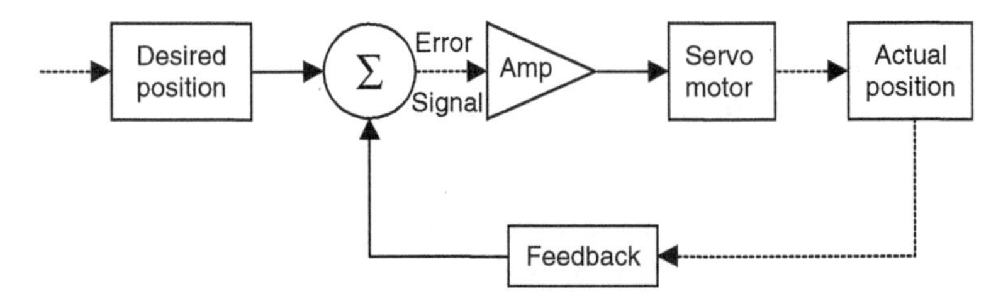

FIGURE 8.7 Simple block diagram of a servo closed-feedback control system. The Greek summation character is the summation of the input signal and the response signal. Initially, the system will compare the desired position with the actual position. If these are not equivalent, the feedback circuit will produce an error signal or difference, which will in turn be corrected by the motor. The loop continues until the actual position is reached and a new desired position is requested.

In balance configuration (Figure 8.6d), the clusters are rotated so that one set of drive wheels is positioned directly over the other set, leaving only two wheels in contact with the ground. The gyroscopes and closed-loop control system enable the user to remain balanced even during perturbations; for instance, if the person were to shift their weight forward as if to reach something off the shelf, the transporter responds by bringing the wheels forward to compensate for the change in occupant CG and keep the occupant balanced. While driving is possible in this configuration, the main purpose of the balance mode is to enable the person to interact at eye level with their able-bodied peers and to extend their reaching height. The iBOT® is the first stair-climbing device to appear on the market that does not use a track-based design. The major advantage of tracked-based stair-climbing wheelchairs is simple control and robustness in operation on irregular stairs. However, a disadvantage is the high pressure exerted on the stair edges and the difficulties with using a track-based design as an everyday wheelchair (e.g., tends to be bulkier, heavier, and less maneuverable).

8.9 WHEELCHAIR STANDARDS

Wheelchair standards were developed to provide a means to objectively compare the durability, strength, stability, and cost-effectiveness of commercial products. They are designed and implemented to improve the quality of available wheelchairs and to assist persons in making informed selections of wheelchairs. Wheelchair standards are comprised of test methods that apply to all types of wheelchairs and scooters and were developed through years of laboratory validation and arbitration. In the United States, wheelchair standards are voluntary; however, the test results are accepted by the United States Food and Drug Administration, which approves commercial marketability of the device. Most countries have adopted the International Organization for Standards (ISO), which acts to continually develop and refine wheelchair standards. The American National Standards Institute (ANSI) and RESNA are member organizations of ISO for the United States. The ANSI/RESNA Wheelchair Standards are made up of 22 distinct tests that

TABLE 8.1
Current ISO 7176 Standards for Wheelchairs

ISO number	Document title
00	Nomenclature, Terms, and Definitions
01	Determination of Static Stability
02	Determination of Dynamic Stability of Electric Wheelchairs
03	Determination of Effectiveness of Brakes
04	Determination of Estimated Range of Electric Wheelchairs
05	Determination of Overall Dimensions, Weight, and Turning Space
06	Determination of Maximum Speed, Acceleration, and Retardation for Electric Wheelchairs
07	Determination of Seating and Dimensions
08	Static, Impact, and Fatigue Strength Testing
09	Climatic Tests for Electric Wheelchairs
10	Determination of Obstacle Climbing Ability of Electric Wheelchairs
11	Wheelchair Test Dummies
13	Determination of the Coefficient of Friction Test Surfaces
14	Testing of Power and Control Systems for Electric Wheelchairs
15	Guidelines for Information Disclosure
16	Flammability Characteristics
17	Serial Interface Compatibility (Multiple Master Multiple Slave)
18	Stair Traversing Wheelchair Testing
19	Wheelchair Tie-Downs and Occupant Restraints
20	Stand-up Wheelchair Testing
21	Electromagnetic Compatibility for Electric Wheelchairs
22	Wheelchair Setup Procedures for Testing
23	Battery Testing

address safety, performance, and measurement uniformity. The number of tests needed depends on the type of wheelchair being evaluated. Power wheelchairs undergo all the tests that the manual wheelchairs do in addition to a series of tests on the power and control systems and batteries. These include testing the battery life and distance traveled on a full charge, electric circuit connections, automatic braking system operation, and climatic effects on the controllers. All the standards tests are listed in Table 8.1. A few of the tests are described in the following paragraphs.

8.9.1 Fatigue-Strength Tests

The most destructive test evaluates the long-term performance of the device. Because this test is the most destructive, it is the last mechanical test performed on the wheelchair. The test involves two parts; double-drum testing (Figure 8.8a) and curb-drop testing (Figure 8.8b), with double-drum testing administered first. The double-drum tester consists of two rollers with 1-cm-high slats attached to each to simulate wheeling over door thresholds, cracks in the road, and small obstacles.

(a)

(b)

FIGURE 8.8 (a) Double-drum machine; (b) curb-drop machine.

A motor in the tester drives the rear rollers at an equivalent speed of 1 m/sec for manual wheelchairs, whereas power wheelchairs are set to drive at the same speed by securing the joystick in a forward drive position. The front rollers turn at rate 5 to 7% faster than the rear, which imparts uneven loads on the wheelchair to ensure that one side is not favored over the other. The wheelchair is balanced over the rollers with a swing arm attached to the wheelchair's rear axles for stability. The purpose of double-drum testing is to expose the wheelchair to a large number of low-level stresses, similar to stresses experienced during daily wheelchair use. The curb-drop machine is designed to simulate repeated loads experienced by wheelchairs when going down small curbs. This test machine lifts the wheelchair 5 cm and allows it to free-fall to a hard surface. As described by ISO 7176-08, each wheelchair is to be tested for 200,000 cycles on the double-drum tester (1 cycle = 1 revolution of the rear roller). For both tests, the wheelchair is loaded with a standardized 25, 50, 75, and 100 kg test dummy, depending on the appropriate body size for the wheelchair. Recently, a 250 kg test dummy was developed to evaluate bariatric type wheelchairs.

If the wheelchair completes this part of the test without sustaining a *high-risk failure*, it is then transferred to the curb-drop tester for 6666 drops. A *high-risk failure* is defined as permanent damage, deformation, or failure that significantly affects the operability of the wheelchair or compromises the individual's safety. A *low-risk failure* is one that requires attention and can be easily repaired by the user or a technician but does not compromise the safety of the individual. Examples are loose bolts and screws, flat tires, and loose spokes. A device is allowed to sustain two low-risk failures over the course of testing, but beyond that the device fails the test. If a high-risk failure occurs, the test is over. When the initial cost of the wheelchair is considered, a measure of cost-effectiveness can be ascertained. The life-cycle cost of the wheelchair is computed by taking the initial purchase price and dividing it by the total number of equivalent cycles until a high-risk failure occurred. The total number of cycles (also known as *equivalent cycles*) is computed based on the ratio of the total number of double-drum cycles and curb-drop drops before failure:

$$\text{Total cycles} = (\text{double-drum tester cycles}) + 30 \times (\text{curb-drop tester drops})$$

8.9.1.1 Example

Wheelchair 1 (WC1), a depot style wheelchair, encountered a high-risk failure at 114,000 cycles on the double-drum. Because it did not meet the required 200,000 cycles, it was not tested on the curb-drop machine.

Wheelchair 2 (WC2), an ultralight style wheelchair, encountered one low-risk failure at 157,000 cycles but completed the 200,000 cycles. The wheelchair moved on to the curb-drop machine, where it completed the required number of drops with additional failure. For a more accurate cost analysis, the testing was continued, alternating between the double-drum test for 200,000 cycles and proceeding to the curb-drop tester for 6666 cycles. The wheelchair completed three rounds of testing on the double-drum and curb-drop machines. On the fourth round on the double-drum, the wheelchair encountered a high-risk failure at 122,300 cycles.

Determine the life-cycle cost of both wheelchairs provided that the depot-style wheelchair was \$350 and the ultralight wheelchair initial cost was \$3200.

Life-cycle cost for

$$
\begin{aligned}
\text{WC1} &= \frac{350}{114{,}000 + (30 \times 0)} = \$0.00307/\text{equivalent cycle} \\
\text{WC2} &= \frac{3200}{722{,}300 + (30 \times 19{,}998)} \\
&= \$0.00242/\text{equivalent cycle}
\end{aligned}
$$

So while the initial cost of the ultralight wheelchair was higher, it withstood a higher number of equivalent cycles before failure resulting in a better investment.

8.9.2 Static-Strength Tests

Another constituent of the standards is static-strength testing of the armrests, anti-tip levers, push handles, hand grips, and footrests. Static strength is tested using a constant force or weight as prescribed in the standards. The amount of force applied to the component depends on the projected amount of force. In some cases a factor of safety is introduced to ensure that the part will withstand more force than expected. An example of this is the static-strength testing of armrests. In a pressure-relief exercise, the individual's mass is applied to each armrest almost vertically. When transferring out of the wheelchair the force is exerted at an angle. The equation for determining the amount of force to apply to each armrest for testing is:

$$F = \frac{mgS}{2\cos 15^\circ} \tag{8.8}$$

where F is the applied force, $mg/2$ the amount of "user" weight supported by each armrest, S the factor of safety, and $\cos 15^\circ$ the estimated angle of force application.

For tests in which safety is of greater significance, the loads are increased by a factor of 1.5. The weight of the "user" is set equivalent to the maximum dummy size/weight assigned to the wheelchair. The standard also includes impact strength testing for the seat, backrest, wheels, casters, pushrims, and armrests. In this test, a pendulum is used to strike parts of the wheelchair that are subjected to impacts typically applied, for example, when a user falls against the wheelchair backrest or when the footrest, casters, etc., collide with obstructions.

8.9.3 Static and Dynamic Stability

Static and dynamic stability testing is another measure to evaluate wheelchair user safety. Tips and falls account for a majority of serious wheelchair-related injuries, and so it is important that wheelchair manufacturers consider stability as an important criterion for their designs. In this section of the standards, the wheelchair is evaluated for static stability in the rearward and forward directions with and without brakes by placing the wheelchair on a hard, test plane, which is slowly inclined to the point where a wheel looses contact with the surface. This angle of incline is noted and

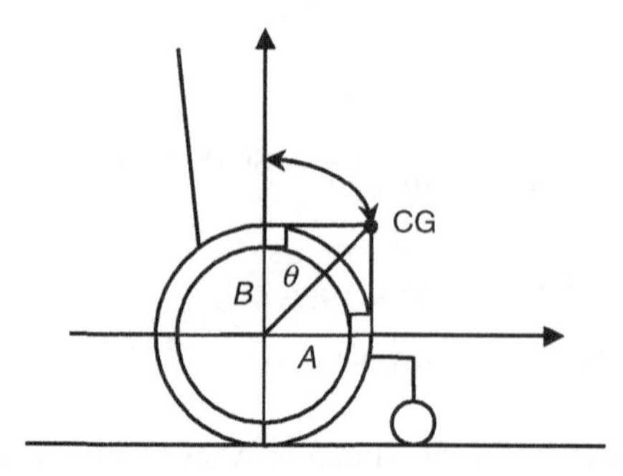

FIGURE 8.9 Geometrical model of the wheelchair–user system. The wheelchair will tip when the center of gravity (CG) rotates through an angle of θ (in degrees), which is calculated using the equation $\theta = \arctan(A/B)$, where A and B are the horizontal and vertical distances from the CG to the axis of rotation (the axle center of the rear wheels when wheel locks are disengaged).

compared to the minimum angle noted in the standards. Test dummies as specified in the standards are used in the tests to simulate realistic wheelchair loading.

Majaess et al. (1993) developed computer models to understand the effect of user positioning in the wheelchair, and Kirby et al. (1996) studied adding loads to various parts of the wheelchair on static stability of the wheelchair–user system. A simple geometrical model of the wheelchair–user system to illustrate how changing the CG of the system (which occurs when repositioning the seat relative to the rear wheels) affects stability is shown in Figure 8.9. Moving the seat forward moves the CG of the wheelchair–user system forward, which increases the rear stability while decreasing the front stability. Adding loads near the front of the wheelchair increases rear stability while decreasing front stability. Dynamic stability tests are conducted on power wheelchairs to evaluate their rearward, forward, and lateral stability on specified slopes. The test includes observing stability when accelerating uphill, sudden braking when traveling uphill and downhill, traveling from a sloped surface to a level surface, turning in a circle on an uphill/downhill slope, traveling up/down a step transition, and traveling with one side of wheelchair driving down a step transition. Testers record their observations on a scale of 1 to 4, ranging from 4 = no tip (at least one uphill wheel remains on the test plane) to 0 = full tip (wheelchair completely tips over).

8.9.4 Manual Wheelchair Performance

Ultralight wheelchairs have been shown to last about 13 times longer than standard wheelchairs and cost about $3^1/_2$ times less to operate. In comparison to lightweight wheelchairs, the ultralight wheelchairs last about five times longer and cost approximately two times less to operate than lightweight wheelchairs. When tested to failure, ultralight wheelchairs had the longest survival rate and fewer catastrophic failures in comparison to standard and lightweight wheelchairs. Premature failures of wheelchairs can place the user at risk for injury. Ultralight wheelchairs, while durable and cost-effective, are also safer for individuals. Wheelchair standard test results of

TABLE 8.2
Three Classifications of Power Wheelchairs Evaluated Using Wheelchair Test Standards

Class	Description
A	Conventional power wheelchair: low-cost power wheelchair; basic seating; nonprogrammable
B	Indoor/outdoor power wheelchair with a programmable controller for adjusting the wheelchair's acceleration, deceleration, turning speed, braking, and tremor dampening according to the individual's function and needs
C	Heavy-duty power wheelchair with "specialized" features for individuals who either weigh over 250 lb, need seat functions (e.g., tilt, recline, elevation), are active outdoors, or need a ventilator tray

ultralight models with suspension elements incorporated into the frame and/or casters show that they perform similarly to lightweight wheelchairs with respect to wheelchair durability and cost effectiveness.

8.9.5 Electric Power Wheelchair Performance

Component failures and engineering factors are responsible for 40 to 60% of the injuries to power wheelchair users. As with manual wheelchairs, power wheelchairs are separated into classes, which are listed in Table 8.2. While differences between wheelchair make and models within the classifications have been reported, overall class C power wheelchairs outperform the other classes in terms of durability (total equivalent cycles on the fatigue test machines) and class A power wheelchairs perform the worst in several areas (poorer durability and reliability, dynamically unstable, and prone to wiring failures). In comparison to manual wheelchairs, power wheelchairs perform similar to ultralight wheelchairs in terms of their durability. If power wheelchairs experienced failures during testing, they typically did not interfere with the wheelchair's operability during the required number of cycles on fatigue-test machines. In contrast, manual wheelchairs experienced failures that would affect the user's operation or safety, such as broken casters, cracked seat and frame, and other faults difficult to repair. Regarding value, power wheelchairs are more costly than ultralight and lightweight manual wheelchairs, but are more economical than standard wheelchairs considering how long they last, number of repairs and cost of repairs, and the initial cost of the device. Very little test data have been reported on scooter performance. Three-wheel scooters are reportedly less stable than power and manual wheelchairs in the lateral direction.

8.10 SUMMARY

Rehabilitation engineers contribute to the design, development, safety, evaluation, and procurement of wheelchair technology. Many are employed by the wheelchair manufacturers directly to assist with (1) identifying design specifications based on user preferences, (2) design, development, and fabrication of new wheelchair designs, (3) prototype testing, (4) refining prototypes based on test results, and (5) preparing the end product for commercialization. Rehabilitation engineers also play an important role in the wheelchair clinic by assisting the assistive technology team with conducting wheelchair assessments, identifying potential devices, making adjustments to the wheelchair to improve performance and fit, educating the user on maintenance, safety precautions and operation, and providing ongoing follow-up and technical support.

The EPW is poised to undergo revolutionary design changes in the future. While innovative devices such as the PAS represent important advances for people whose abilities balance between using a manual wheelchair and an EPW, there are many more people who could benefit from advances in EPWs. People with disabilities and people who are elderly are becoming more empowered to insist upon maintaining or increasing independence and mobility. This has prompted the investigation of technologies that will negotiate uneven terrain, traverse stairs, and detect obstacles in the environment. Scooters and EPWs are becoming more similar. The demand for electric powered mobility devices that do not look like wheelchairs and that can provide both indoor and outdoor mobility is creating innovation in the marketplace. Midwheel drive scooters that provide good indoor mobility and yet have the lightweight and intuitive use of a scooter will emerge, and light, more transportable power products are being introduced. In the future, modular type designs may evolve that allow wheeled mobility systems to be configured (e.g., wheelbase, track width, and steering interface) for the user and the activity.

8.11 STUDY QUESTIONS

1. Standards reports show that in recent testing, ultralight wheelchairs lasted 894,500 equivalent cycles, lightweight wheelchairs lasted 187,370 equivalent cycles, and depot wheelchairs lasted 56,139 equivalent cycles.
   (a) Explain why this stratification possibly occurred.
   (b) If the ultralight wheelchairs were purchased at $3500 each, lightweight wheelchairs at $1500 each, and depot wheelchairs at $500 each. Which type of wheelchair type provides the best value for the patient?
2. If you needed to specify manual wheelchair for long-term use, which type of wheelchair would you choose, and why?
3. A patient has diabetes and uses a manual wheelchair for mobility. After a couple years of using the wheelchair, the diabetes worsens, resulting in reduced circulation and eventually a double leg amputation above the knee. If we do not make adjustments to his wheelchair, what might we expect to see when he tries to use it? Explain what adjustments would be necessary to ensure safe mobility.
4. A power wheelchair with a mass of 113 kg is on a tilting platform. The wheelchair center of mass is located at 13 cm horizontal and 20 cm vertical relative to the center of rotation of the drive wheel. A test dummy of mass 72 kg is seated in the wheelchair. The dummy center of mass is located at

23 cm horizontal and 50 cm vertical relative to the center of rotation of the drive wheel axle. The platform is slowly raised until slope of 20°. Is the wheelchair still stable at this angle?

5. Describe how the control system of a pushrim-activated wheelchair emulates wheelchair/user inertia and momentum.
6. (a) Determine at $t = 4$ sec, the pushrim-activated wheelchair controller output (u) if:

   $t_c = 5$ sec
   t_r is initialized to 0
   Threshold torque is 1 Nm
   Pushrim torque at release is 10 Nm
   The wheelchair is traveling at a speed that is less than the threshold velocity

   (b) What is the controller output when $t = 0.5$ and $t = 6$ sec?
7. Describe the consequence of:
   A seat depth that is too short
   A seat width that is too wide
   A seat that has too much dump (seat angle)
   A back height that is too high
   Arm rests that are too high
   Large camber angles
   Having the rear wheels in the most rearward (horizontal) axle position
8. Describe the benefit of having a seating system that is separate from a powered base.
9. What is the minimal rate of motor torque for a PAS given an ε of 13, constant of 3, and a gear ratio of 7:1?
10. If a power wheelchair uses 2.3 A h after traveling 2 km on a flat surface and uses a 100 A h battery, what is the theoretical range of the wheelchair? How will this range differ from the actual range, and will it increase or decrease as the battery is recharged and why?
11. Discuss and describe an approach to modifying the input device for a power wheelchair (joystick) to accommodate a person with motor spasticity of the hand. Assume the spasticity occurs constantly at 2 Hz in the right (dominant) hand.

BIBLIOGRAPHY

Arva J, Fitzgerald SG, Cooper RA, Boninger ML. 2001. Mechanical efficiency and user power requirement with a pushrim-activated power assisted wheelchair. *Med Eng Phys* 23:699–705.

Boninger ML, Cooper RA, Robertson RN, Shimada SD. 1997. 3-D pushrim forces during two speeds of wheelchair propulsion. *Am J Phys Med Rehabil* 76:420–426.

Boninger ML, Cooper RA, Baldwin MA, Shimada SD, Koontz AM. 1999. Wheelchair pushrim kinetics: body weight and median nerve function. *Arch Phys Med Rehabil* 80:910–915.

Boninger ML, Baldwin MA, Cooper RA, Koontz AM, Chan L. 2000. Manual wheelchair pushrim biomechanics and axle position. *Arch Phys Med Rehabil* 81:608–613.

Cooper RA. 1995. *Rehabilitation Engineering Applied to Mobility and Manipulation*. Bristol, UK and Philadelphia, PA: Institute of Physics Publishing.

Cooper RA, Robertson RN, Lawrence B, Heil T, Albright SJ, VanSickle DP, Gonzalez J. 1996. Life-cycle analysis of depot versus rehabilitation manual wheelchairs. *J Rehabil Res Dev* 33:45–55.

Cooper RA, Gonzalez J, Lawrence B, Renschler A, Boninger ML, VanSickle DP. 1997. Performance of selected lightweight wheelchairs on ANSI/RESNA tests. American National Standards Institute-Rehabilitation Engineering and Assistive Technology Society of North America. *Arch Phys Med Rehabil* 78:1138–1144.

Cooper RA. 1998. *Wheelchair Selection and Configuration*. Demos Medical Publishing, Inc., New York, NY.

Cooper RA, Fitzgerald SG, Boninger ML, Prins K, Rentschler AJ, Arva J, O'Connor TJ. 2001. Evaluation of a pushrim-activated, power-assisted wheelchair. *Arch Phys Med Rehabil* 82:702–708.

Cooper RA, Thorman TT, Cooper R, Dvorznak MJ, Fitzgerald SG, Ammer W, Song-Feng G, Boninger ML. 2002. Driving characteristics of electric powered wheelchair users: how far, fast, and often do people drive. *Arch Phys Med Rehabil* 83:250–255.

Cooper RA, Boninger ML, Cooper R, Dobson AR, Kessler J, Schmeler M, Fitzgerald SG. 2003. Use of the Independence (TM) 3000 IBOT (TM) transporter at home and in the community. *J Spinal Cord Med* 26:79–85.

Corfman TA, Cooper RA, Boninger ML, Koontz AM, Fitzgerald SG. 2003. Range of motion and stroke frequency differences between manual wheelchair propulsion and pushrim-activated power-assisted wheelchair propulsion. *J Spinal Cord Med* 26:135–140.

Fass MV, Cooper RA, Fitzgerald SG, Schmeler M, Boninger ML, Algood SD, Ammer WA, Rentschler AJ, Duncan J. 2004. Durability, value, and reliability of selected electric powered wheelchairs. *Arch Phys Med Rehabil* 85:805–814.

Fitzgerald SG, Cooper RA, Rentschler AJ, Boninger ML. 1999. Comparison of fatigue life for three types of manual wheelchairs. *Arch Phys Med Rehabil* 82:1484–1488.

Gaal RP, Rebholtz N, Hotchkiss RD, Pfaelzer PF. 1997. Wheelchair rider injuries: causes and consequences for wheelchair design and selection. *J Rehabil Res Dev* 34:58–71.

Kirby RL, Ackroyd-Stolarz SA. 1995. Wheelchair safety — adverse reports to the United States Food and Drug Administration. *Am J Phys Med Rehabil* 74:308–312.

Kirby RL, Ashton BD, Ackroyd-Stolarz SA, MacLeod DA. 1996. Adding loads to occupied wheelchairs: effect on static rear and forward stability. *Arch Phys Med Rehabil* 77:183–186.

Kwarciak AM, Cooper RA, Ammer WA, Fitzgerald SG, Boninger ML, Cooper R. 2004. Fatigue testing of selected suspension manual wheelchairs using ANSI/RESNA standards. *Arch Phys Med Rehabil* 86:123–129.

LaPlante MP, Carlson D. 1996. Disability in the United States: Prevalence and Causes. Washington, D.C., National Institute on Disability and Rehabilitation Research.

Majaess GG, Kirby RL, Ackroyd-Stolarz SA, Charlebois PB. 1993. Influence of seat position on the static and dynamic forward and rear stability of occupied wheelchairs. *Arch Phys Med Rehabil* 74:977–982.

Masse LC, Lamontagne M, O'Riain MD. 1992. Biomechanical analysis of wheelchair propulsion for various seating positions. *J Rehabil Res Dev* 29:12–28.

Pearlman J, Cooper RA, Karnawat J, Cooper R, Boninger ML. 2005. Evaluation of the safety and durability of low-cost nonprogrammable electric powered wheelchairs. *Arch Phys Med Rehabil.* 86:2361–2370.

Rentschler AJ, Cooper RA. 1999. A comparison of the dynamic and static stability of power wheelchairs versus scooters. *Proceedings of the First Joint BMES/EMBS Conference*, Atlanta, GA, October 13–16, p. 613.

Simpson R, LoPresti E, Hayashi S, Guo S, Ding D, Ammer WA, Sharma V, Cooper RA. 2005. A prototype power assist wheelchair that provides for obstacle detection and avoidance for those with visual impairments. *J Neuroeng Rehabil* 2.

Ummat S, Kirby RL. 1994. Nonfatal wheelchair-related accidents reported to the National Electronic Injury Surveillance System. *Am J Phys Med Rehabil* 73:163–167.

van der Woude LHV, Veeger DJ, Rozendal RH, Sargeant TJ. 1989. Seat height in handrim wheelchair propulsion. *J Rehabil Res Dev* 26:31–50.

Woods B, Watson N. 2003. A short history of powered wheelchairs. *Assist Technol* 15:164–180.

9 Functional Electrical Stimulation

Songfeng Guo, Karl Brown, Emily Zipfel, Yusheng Yang, Jue Wang, Tong Zhang, and Elizabeth Leister

CONTENTS

9.1 LEARNING OBJECTIVES OF THIS CHAPTER

Upon completion of this chapter, the reader will be able to:

Understand the basic theory of functional electrical stimulation (FES)
Understand basic electrophysiology
Be able to describe the applications of FES
Learn how FES may be used for rehabilitation

9.2 INTRODUCTION

The application of electricity for therapeutic purposes dates back to antiquity. However, the development of new technologies and innovative applications provide new and exciting potential for applications in rehabilitation. Electrical stimulation therapy involves the application of external electrical energy to tissue, which results in the stimulation of a physiological response. The degree and type of response depends on the mode, quantity, and quality of energy and the anatomical location of the application site.

Physiological responses evoked by electrical stimulation can include relaxation of spasticity, controlled muscle contractions, increased production of endorphins, increased muscle fiber recruitment, circulatory stimulation, and enhancement of the reticuloendothelial response. Treatments using electrical stimulation include pain management; control of seizures; the controlled stimulation of muscles to return function in case of paralysis due to stroke, spinal cord injury (SCI) or peripheral nerve damage; cardiac pacemakers; tissue repair; increases in functional activity; and stimulation of de-enervated muscle.

One of the most promising applications of electrical stimulation is for the purposes of rehabilitation and for increasing function in people with neuromuscular disabilities.

9.3 HISTORY OF FES

In ancient Greece, Egypt, and Rome, electric eels were used to treat pain. Around the year 50 AD, Roman writer Scribonius Largus reported success in treating pain, headache, and gout with electric torpedo fish. Other writers from antiquity claimed this technique was useful in treating athralgia, migraine, melancholy, and epilepsy.

In 1745 the Leyden jar came into use for the storage of electrical energy. At this time, Luigi Galvani (1737–1798) performed his famous experiments: around 1780, when he found that charges applied to the spinal cord of a frog resulted in muscle contractions, he first discovered that electricity flows through the body. This discovery led to the important new hypothesis that nerves were electrical conductors. Galvani's work was pivotal in that of his colleague Alessandro Volta (1745–1827), and led to Volta's invention of the first electric battery in 1800. This battery, the voltaic pile, was the first mechanism developed for generating a sustained electric current, and prompted more controlled experiments with electricity.

The discoveries of Galvani and Volta created the mechanisms with which doctors and scientists, such as Guillaume Duchenne (1806–1875), were able to experiment and use to treat disease, thereby developing the field of "electrotherapy." Thus, in the 19th century, electrical therapy began to play a role in medicine and the treatment of both pain and paralysis. After the development of Volta's battery and the induction coil by Michael Faraday (1791–1867) in 1840, a period of clinical discovery and thus a period in the late 19th century known as the "golden age of medical electricity" began.

Duchenne, considered the father of electrotherapy and neurology, began to experiment with applying voltage to muscles of individuals with paralysis and even those recently dead. In doing this, he observed the resulting muscle twitch. In the 1840s, using Faraday's recent invention of the induction coil, Duchenne experimented with the method of connecting the induction coil to the skin near the muscle and concluded that most human movements were caused by groups of muscles working in tandem rather than individual muscles working alone. In addition, Duchenne experimented with using the induction coil to help patients with paralysis regain muscle function.

Paralysis, as well as various complaints summarized as "rheumatism," became the major areas for electrotherapy. In 1836, Guy's Hospital in London established an electrical department under Golding Bird, who became the leading advocate for electrotherapy. Due to his research, electrotherapy became one of the predominant treatments for paralysis. Around this time, electrotherapy for the purpose of pain relief was being explored.

The therapeutic use of electricity was exploded in diversity during the 20th century, when researchers explored its effect on the arrested heart, diaphragm, auditory nerve, muscular system, visual cortex, and central nervous system (CNS). Figure 9.1 depicts the timeline of important firsts in FES during the 20th century.

Technological advances such as the discovery of the diode, triode, transistor, and integrated circuit gave researchers significant freedom to explore electricity's affect in humans. During the development of heart pacemakers, researchers found that surface electrodes induced burns a couple days after operation. Later, electrode wires penetrating the skin resulted in bad infections. Improved biomaterials by 1960 allowed the first heart pacemakers to be fully implanted in 10 patients. The radio frequency inductive method allowed pacing the diaphragm to become routine by the late 1960s and early 1970s. The problem of "drop foot" was first addressed with FES in 1961 by Liberson and coworkers, who stimulated the proneal nerve to induce dorsiflexion, inversion, and eversion at the ankle.

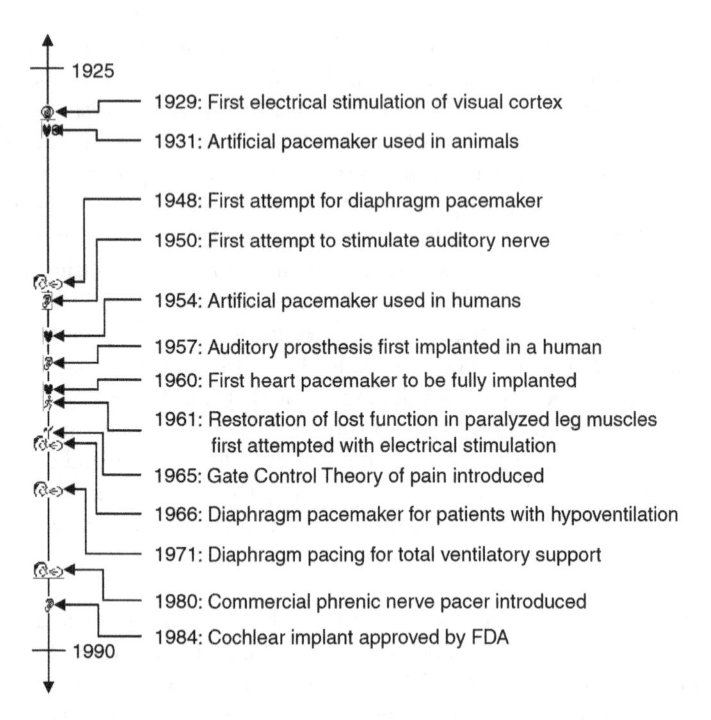

FIGURE 9.1 The timeline of important firsts in FES during the 20th century.

9.4 CLINICAL CONSIDERATIONS OF FES

9.4.1 NeuroMuscular Stimulation (NMS) of Atrophied Muscles

Muscles with impaired innervations suffer from progressive atrophy, decreased capacity to generate force, and increased fatigue. When these muscles are electrically stimulated to contract, the contractions may be few and with rapidly decreasing amplitude. Before these muscles can be used for FES, they must be rebuilt with regular electrical stimulation. Then, the patient must go through physical therapy to increase the endurance of the whole body in addition to the stimulated muscles.

9.4.2 NMS to Modify Patterns of Movement

Individuals who are ambulatory but hemiparetic after a stroke or an SCI may exhibit strong extensor thrust and a weak flexor pattern resulting in a slow gait, and they may require crutches or other walking aids. For these patients, electrical muscle conditioning can be used to improve muscle resistance, and electrical stimulation can be used simultaneously with their voluntarily initiated movements.

9.4.3 Foot Drop and Wrist Drop

Drop-foot and drop-wrist syndromes occur in some patients who have experienced stroke, traumatic brain injury, or SCI. FES is particularly useful in their treatment because most often the deficit is within a single muscle group whereas the surrounding muscle groups are functional. It is relatively simple to stimulate the contraction of one muscle group and coordinate that movement with the volitional movements of the surrounding muscles.

9.4.4 FES for SCI

Although many individuals with paraplegia and tetraplegia and researchers have hoped that FES would provide the functional capability of normal walking, this has not yet occurred. Walking is a complex task, which relies on the rapid recruitment of muscles from many groups. Most FES systems use only six channels: one to each quadriceps, one to each branch of the peroneal nerve, and one to each gluteus. This small number of stimulation points allows a person to stand, but leads to an awkward and fatiguing gait. In addition, the motion and position of the feet must be controlled through the use of braces, and the FES system must be used in conjunction with a walker or other walking aids.

Study results indicate that patients are reluctant to use the FES system because of time taken out of the day to stand or exercise, discomfort when wearing the hardware and embarrassment at wearing the hardware. However, when subjects did use the FES for standing and exercise, it was because they perceived health benefits from the activities, enjoyed being able to stand up purely for the sake of standing, and the ability to reach and perform tasks that required standing.

Several studies have also found that, in the case of ambulatory SCI patients, prolonged use of FES can lead to increased motor function which persists beyond the period electrical stimulation. However, the extent of this regained function may be minimal, and patients should understand the modest functional improvement that this treatment may provide.

Researchers tend to agree about patient selection criteria, the need for muscle conditioning prior to functional use of FES, and the possible psychological benefits to both FES standing and FES exercise for individuals with SCI. In studies, many subjects have been able to stand, ambulate to some degree, and in few cases climb stairs and side-step.

9.5 ELECTRODES

Electrodes are the interface for the electrical connection between the stimulator device and the body's tissue. They are the means by which the electron current emanating from the stimulator is converted to an ionic current in living tissue. The quality and efficacy of the clinical outcome relies on the electrode–electrolyte interface; waveform characteristics of the current; tissue impedance; and electrode material, size and placement.

Two types of electrodes are used in electrical stimulation: those attached to the surface of the skin (transcutaneous) or below the skin (subcutaneous and percutaneous) (Figure 9.2). Transcutaneous electrodes are more common and are placed over the target area where the excitatory response is required. The target area could be an area

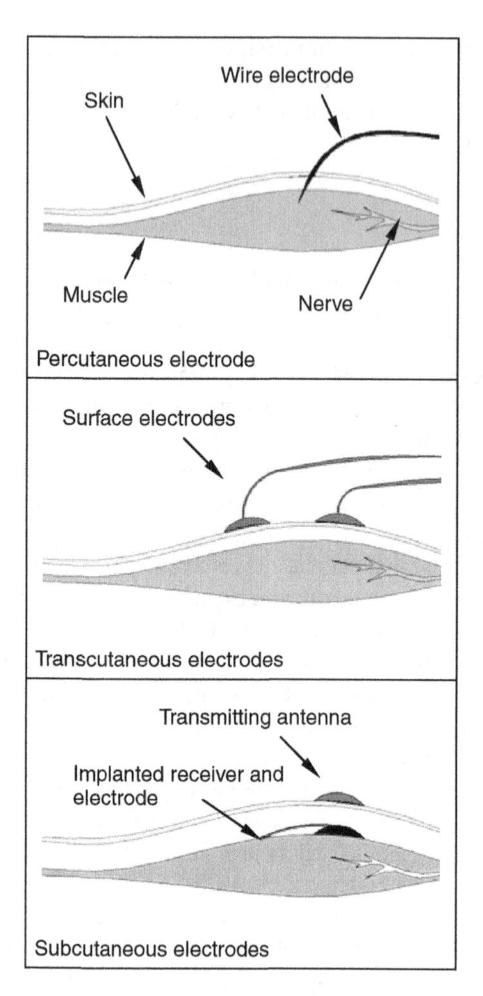

FIGURE 9.2 Two types of electrodes are used in electrical stimulation — those attached to the surface of the skin (transcutaneous) or those placed below the skin (percutaneous or implanted).

of muscle, bursae, joints, wounds or acupuncture points. Subcutaneous and percutaneous electrodes are typically used for walking devices and are inserted directly into muscle tissue.

9.5.1 Electrode–Electrolyte Interface

Electrical stimulation at an atomic level occurs due to an exchange of ions. Living tissue is rich in ions and serves as the electrolyte in the interface with the electrode. At any electrode–electrolyte interface, the electrode tends to discharge ions into the electrolytic solution and ions in the solution combine with the electrode.

When ions cross the cell membrane, this constitutes an electrical current. Current is defined as the flow of positive charges, and it flows from positive to negative regions of potential. In order to create an ionic current in living tissue, it is necessary for the electrodes to have opposite charges. The electrode with the greater concentration of electrons, the cathode or negative electrode, is negatively charged and attracts positive ions from the underlying tissue. The electrode with fewer electrons, the anode or positive electrode, attracts negative ions and free electrons from the underlying tissue.

Positive ions flow in the opposite direction of negative ions and free electrons. The density of the ionic current is greatest at the electrode sites. The resting membrane potential of muscle fibers and nerve axons in the vicinity of the electrodes is affected by the concentration of charge at each electrode site. At the cathode, the membrane potential is reduced or depolarized. At the anode, the membrane potential is increased or hyperpolarized. If the current at the cathode is sufficient to reduce the membrane potential to the threshold value, an action potential is created.

Because the cathode is the site of the lowest threshold for depolarization, it is also know as the "active" electrode and is the site where stimulation is first sensed by the patient. The anode is sometimes referred to as the "indifferent" electrode. In the vicinity of the anode, the concentration of positive ions hyperpolarizes the cell membrane, making it less sensitive to depolarization.

Good mechanical attachment of electrodes to the skin does not necessarily yield a good electrical connection because the electrode material must be in contact with the body's electrolyte. The application of an electrolytic gel, liquid, or paste can facilitate the electron–ion exchange. However, the conductive medium should never be applied so that it spans the area between the electrodes, thereby creating a path of current flow on the surface of the skin.

9.5.2 Electrical Currents

Electrical stimulators can be classified by three primary forms of electrical current delivered through electrodes: direct current (DC), alternating current (AC), and pulsed current (PC). DC occurs when the electrons flow continuously in one direction. Current that flows unidirectional for about 1 sec or longer can be defined as direct current. The basic pattern of DC flow is a square wave characterized by the current flowing continuously on until the circuit is disconnected or the battery is turned off. If one wishes to grade the on and off so that current amplitude will increase and decrease gradually, a modulation termed "ramp" can be added. Ramp-up and ramp-down can be termed surge-up and surge-down, respectively, and the period of time can last from 0.5 to several seconds.

Alternating current is a type of current that periodically changes the direction of flow. The current flow remains uninterrupted but is bidirectional. If the waveform is symmetrically biphasic, the sensation of stimulation is equal at both electrode sites. In the case of unbalanced waveforms, the quantity of charge delivered is unequal for the negative and positive phases of the signal. As a result, with identical electrodes, stimulation will be felt more strongly at one of the electrodes. The typical AC pattern is symmetrical and can have various shapes including sinusoidal, rectangular, trapezoidal, and triangular. AC is commonly used in household electricity and is not designed to excite peripheral nerves. Therefore, AC is not used for any electrical stimulators.

Pulsed current is an electrical current that is conducted as a signal (or signals) of short duration. Each pulse lasts for only a few milliseconds or microseconds (μsec) followed by an interpulse interval. PC flow can be unidirectional (monophasic) or bidirectional (biphasic). A monophasic waveform has one phase to a single pulse with a unidirectional current flow and may possess either negative or positive polarity. Monophasic currents have the potential disadvantages of causing electrode deterioration and skin irritation with prolonged use. Despite the short duration of the pulse, monophasic waveforms are fully capable of altering ionic distributions and causing polarization that lead to skin breakdown and burns. A biphasic waveform consists of two opposing phases within a single pulse. Current flow is bidirectional with the lead phase of the pulse above the baseline and the final phase below the baseline. A pulse can be either a symmetrical biphasic pulse, when the two phases deviate from the baseline in an identical and equal manner, or an asymmetrical biphasic, when the two phases are not identical. In an asymmetric biphasic waveform, one direction of current flow typically is adequate to cause excitable tissue to depolarize while the current flowing in the opposite direction is low in amplitude and long in duration, which allows little or no neural excitation. The advantage of an asymmetric biphasic waveform is to provide a monophasic-like excitability while minimizing ion redistribution and subsequent risk of skin irritation. However, some chemical reactions may still occur while using an asymmetric biphasic waveform. These disadvantages are not as apparent with symmetrical biphasic waveforms.

9.5.3 Tissue Impedance

An important concept to understand when choosing the appropriate electrode is that of impedance, a material's opposition to the flow of electric current. Tissue impedance is related to many factors, resulting in higher impedance in different parts of the body. It is of particular importance when trying to stimulate deeper tissues. The higher the tissue impedance, the less the current penetrates. As a result, a higher voltage from the stimulator must be used to pass an equivalent amount of current through high-impedance, low-conductivity tissue than through low-impedance tissue.

Skin impedance is also affected by the condition of the skin. For example, sweaty skin has lower impedance than dry, scaly skin. Techniques such as mild exfoliation or using a finely grained sandpaper or alcohol wipe to remove dry skin can decrease the impedance and improve the transcutaneous electrode performance. Subcutaneous and percutaneous electrodes are not as influenced by the condition of the skin.

9.5.4 Transcutaneous Electrodes

Transcutaneous electrodes need to be durable enough to be used repeatedly, adhere well to the skin and flexible enough to conform to varying body surfaces. In general, metals are excellent conductors, but they have two drawbacks when used as electrodes. First, they often form toxic metal salts. Second, they may be inflexible and require an electrolytic gel to conform to the body. Silver–silver chloride is relatively nontoxic, and therefore, can be used as an electrode material. Carbon film has some advantages over metals when used as an electrode. It is inert, hypoallergenic, and can be combined with a silicone rubber for flexibility and durability. Carbon has a lower conductance than metal; therefore a carbon film is coated with a thin coat of silver to improve current dispersion while maintaining flexibility. Usually these electrodes contain a self-adhesive. The adhesive type is stuck directly to the skin at the location of desired electrical stimulation.

The other common type of surface electrode consists of a metallic base with a sponge rubber covering. This type is inexpensive and can be fabricated in the laboratory or clinic. The sponge is wetted with water or another electrolytic medium such as water-based gel. The electrode must then be secured to the body with either tape or straps.

Self-adhering electrodes are much easier and faster to use than carbon–silicon and tend to shift around less as the body moves. However, they tend to lose both their adhesiveness and their uniform conductivity, and therefore cannot be reused many times. In addition, they tend to be less conductive than the carbon–silicon types and can be uncomfortable to remove because of the sticky adhesive. Because they cannot be reused by multiple patients, they tend to be more expensive. They are, however, more flexible on the body. Adhesion of these electrodes can be problematic if there is excess sweat or hair on the target area of the body.

An additional consideration regarding this problem is that self-adhesive electrodes lose uniform conductivity more rapidly than the other type of electrodes. This means that the area of the pad that is actually conductive becomes smaller. This causes the current density to increase, which can lead to skin irritation and burns. In order to prevent this, self-adhering electrodes should be inspected for adhesion and conductance frequently. Evidence of improper adhesion and nonuniform conductance can present as either an area of redness that is smaller in size than the area of the electrode pad or as the patient's perception that the stimulation area is smaller than the electrode.

Natural and synthetic polymer hydrocolloids such as karaya gum have been used for application to carbon-based electrodes. These materials act as a sponge by forming a semiliquid, adhesive conductive layer when wet and do not require additional conductive medium to be used. But these materials have been associated with some skin irritation.

The choice of a particular size of electrode will be determined by the size of the area to be stimulated, size of patient's body, skin impedance, root-mean-squared amplitude (RMSa), current density, and the patient's sensation of the stimulation. In general, small electrodes concentrate the current for localized effects, whereas larger electrodes disperse the charge.

In order to maximize the treatment, the clinician must determine the size of the area to be stimulated. Using an electrode that is smaller than the target area creates high electrode–skin interface impedance. This scenario is typically seen in the stimulation of large joints such as the knee or hip and large muscles such as the quadriceps. Using larger electrodes is the best alternative in these cases.

Current density is the amount of current flow per unit area. It is a measure of the quantity of charged ions moving through a specific cross-sectional area of body tissue. The unit of measure is milliamperes per square centimeter. In general, as the current density increases so does the potential for stimulation. Two determinants of current density are size of the electrode and amplitude of the current. Small electrodes concentrate the current. The farther apart the electrodes are placed, the deeper the current penetration.

Root-mean-squared amplitude current is also known as "average current" to many clinicians. RMSa can be defined as the absolute value sum of all phase charges averaged over 1 sec (coulombs/second). Most stimulators used in the clinical setting deliver PC; therefore, the peak current amplitude will be higher than the RMSa value. It is of particular importance for a clinician to be aware of the RMSa current, because excessive RMSa current can be harmful to the tissue. RMSa current is directly proportional to the amount of heat production in the tissue. If the electrode size is known, RMSa current density is determined by dividing the RMSa current by the electrode area. Current density should not exceed safety limits. In order to minimize this, larger electrodes should be used when higher RMSa is used.

A patient may perceive more stimulation under one electrode than another even though the electrodes may be of identical size. Increasing the size of the electrode receiving more stimulation will decrease this problem. In cases in which deep stimulation is required and the stimulator output is insufficient, the RMSa current density can be increased by reducing the size of the electrode; however, skin impedance may thus prevent optimal stimulation from being achieved.

Studies have indicated that larger electrodes produce stronger motor responses without pain, whereas smaller electrodes elicit painful stimulation soon after motor excitation is achieved. A couple of explanations for this have been proposed. First, larger electrodes recruit more motor units simultaneously and exhibit less electrode–skin interface impedance. Therefore, motor unit recruitment can occur with lower phase charges, making the stimulation more comfortable. However, larger electrodes disperse current flow through the tissue and thus make the stimulation less specific. This can cause multiple muscle contractions in addition to the targeted muscle.

9.5.5 Subcutaneous and Percutaneous Electrodes

Limitations of surface electrode stimulation are that small muscles cannot be selectively activated and that the stimulation of deep muscles often inadvertently leads to the stimulation of the superficial overlying muscles. Fine force gradation and repeatable, accurate placements are both difficult. The use of more invasive electrodes helps to remedy some of these problems. Subcutaneous electrodes and percutaneous electrodes can minimize some of these problems.

The benefits of the subcutaneous and percutaneous electrodes vs. the transcutaneous electrode are that they allow for better selectivity, repeatable excitation and more permanent positioning. Percutaneous electrodes consist of a wire inserted through the skin, which is also externally connected to the lead. Potential adverse effects are the possibility of tissue damage, spillover patterns during stimulation, pain, superficial infection, and lead breakage. Infection rates are generally low, and the survival rates for percutaneous electrodes are 90% at 6 months and 82% at 1 year. Percutaneous electrodes have the additional adverse effects that include daily care of the electrode exit site and irritation of the site on the skin.

Subcutaneous electrical stimulation units are the most practical consideration for long-term FES. They consist of an implantable receiver/stimulator, an externally worn transmitter/control unit, and implanted electrodes. These electrodes fall into two categories: epimysial and intramuscular. Epimysial electrodes are attached to the surface of the target muscle, whereas intramuscular electrodes are embedded within the muscle fibers. Because the electrodes are fully implanted, no leads penetrate the skin. The implanted receiver allows the stimulation signals to be transmitted through an antenna placed on the skin over the receiver. The stimulation hardware and power source are external and maintained in proximity to the patient, usually in a backpack attached to the wheelchair. Daily setup requires the patient to don the stimulation antenna and external control system, which are to be attached to the sternum and the shoulder.

Percutaneous electrodes are particularly useful in the research and development phases to allow patients to evaluate different control and coordination paradigms without the expense, risk, and permanency of surgically implanted electrodes. These electrodes are inserted via hypodermic needles. When the needle is removed, the electrodes are left in place. A protecting cover is placed on the skin at the site where the wire emerges from the skin. Although fully implanted electrodes, when properly implanted, have a lower chance of lead breakage and fewer complications, some patients have continued to use percutaneous systems for up to 15 years. Electrodes implanted in upper limbs tend to have a higher survival rate than those in lower limbs because of the lower forces and muscle movements experienced.

Generally, the electrodes used fall into two categories: fine wire and cuff electrodes. Typically, the materials used are stainless steel or the noble metals such as platinum, iridium, and gold due to their noncorrosive nature. However, even these metals can corrode under the right conditions; therefore, the electrodes are sometimes insulated with a thin layer of dielectric. For most applications, intramuscular electrodes are made of stainless steel, and epimyseal electrodes are made of platinum. Fine-wire electrodes may be wound as a single-wire coil or as double helix. The double-helix design tends to be more durable than the single. Intramuscular electrodes may contain fine barbs at the end for enhanced anchoring to the tissue. Epimyseal electrodes typically contain a pad that is sutured to the muscle surface.

The cuff electrode (Figure 9.3) is a type of implantable electrode that is intended for the stimulation of nerves in the peripheral and central nervous systems. The design comprises electrode contacts contained within a springy, curled cuff. The cuff is somewhat stretchy to accommodate the size of the individual nerve. This spring-like quality of the material allows the cuff to stay in place without the use of sutures.

Leads are a critical component of any stimulation system. In the case of subcutaneous and percutaneous electrodes, they must be durable enough to withstand fatigue failure from shear and joint movement. The most common design is a helically wound, single-stranded stainless steel multifilament wire insulated with Teflon. There are intrinsic differences in the application of these two designs. In myelinated nerve axons, under a uniform electrical field, such as that of an externally applied electrical stimulation, the large fibers will fire before the smaller fibers. This is opposite to the natural physiological recruitment pattern. This does not affect applications in which the stimulation is delivered through epimysial or intramuscular electrodes,

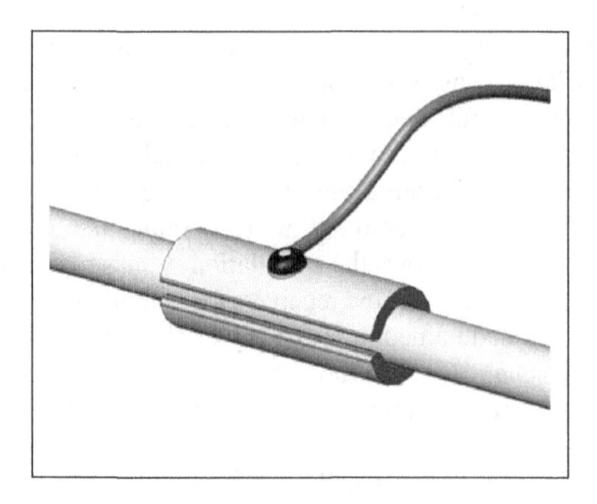

FIGURE 9.3 Drawing of cuff electrode.

but it does have an affect in the nerve cuff application. Therefore, special waveforms must be used that recruit small fibers before large ones.

9.6 CLINICAL APPLICATION OF FES

9.6.1 FES FOR FOOT DROP

The concept of FES was put forward by Liberson in 1960, when he and his team produced the first electrical stimulation device for the correction of dropped foot from an upper motor neuron lesion. The stimulation was applied via electrodes on the skin and was synchronized with the gait phase by a heel switch worn in the shoe. Stimulation was turned on when the heel was lifted at the beginning of the swing phase, which produced dorsiflexion and eversion of the ankle joint. Stimulation was turned off when the heel was on the floor again. His concept was that functional movement could be produced by applying electrical stimulation to paralyzed muscles. Currently, the two-channel implantable stimulator provides separate control of the dorsiflexion and eversion movement by stimulating both deep and superficial peroneal nerves, respectively. In the literature it has been reported that FES can improve walking speed in patients with dropped foot. For example, Waters and colleagues made a comparison between preoperative walking speed without an FES orthosis and walking speed after surgery with an implanted FES orthosis. They found that walking speed was increased significantly (36%) with FES. In a systematic review study of the effect of FES on stroke patients with dropped foot performed by Kottink and coworkers, the pooled analysis of both controlled and uncontrolled trials with FES showed an improvement of 38% in walking speed.

While the FES device has been shown to improve walking by correction of dropped foot, problems often remain with movement of other joints, in particular the knee and hip. Additional stimulation channels might need to be added to smooth movements, such as hip extension during stance phase and knee flexion during terminal stance and initial swing phases. Overall, FES device for dropped foot is an alternative method for managing drop-foot deformity and improving mobility

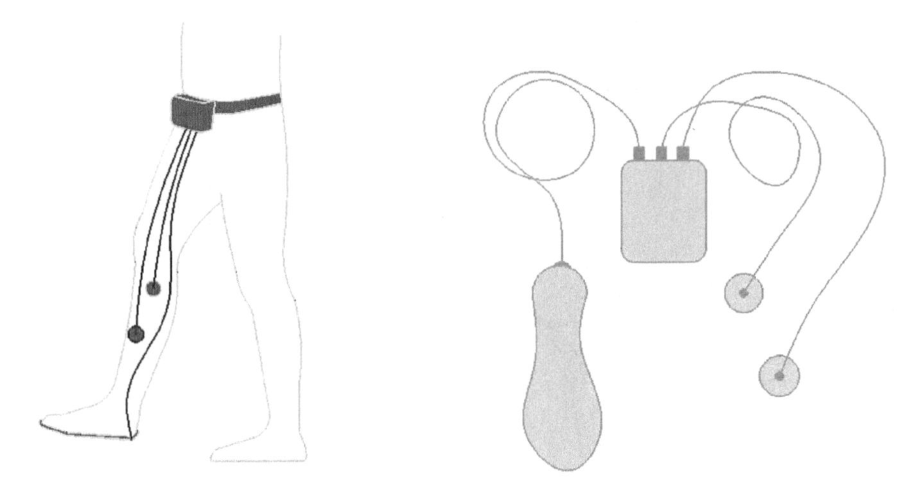

FIGURE 9.4 Drawing of person wearing drop-foot FES orthosis and close up of components.

in people who have dropped feet, such as those with multiple sclerosis (MS), stroke, cerebral palsy, or SCI. See Figure 9.4 for an illustration of drop-foot FES orthosis and its components.

9.6.2 FES FOR WALKING

The FES devices that were designed to support or enable locomotion in individuals with SCI require voluntary control of their upper extremities to maintain stability and balance during walking. In the majority of cases, FES users have to support part of their body weight by using walkers. Therefore, only paraplegic patients who have strong and functional upper extremities can benefit from the FES systems for locomotion (Figure 9.5).

Since Liberson and colleagues proposed the first walking neuroprosthesis in 1961, a number of neuroprostheses for walking have been designed and tested with various patients. The "Parastep" system is the first FES system that received FDA approval and is one of the few FES systems that are commercially available. The Parastep system consists of a belt- or pocket-held microprocessor and stimulator unit, a walker for patient support, 12 reusable self-adhesive skin electrodes, and connecting wires.

The 12 electrodes are placed as follows: one pair over each quadriceps muscle, one pair over each common peroneal nerve, and two pairs over the paraspinals or over the gluteus maximus/minimus muscle. The last two pairs may not be needed by patients with T9–T12 level injuries. The electrodes connect to the Parastep microprocessor unit worn on the patient's belt. The microprocessor serves to generate trains of impulses (stimuli) that grossly imitate the neural triggers that would have passed through the spinal cord to the appropriate peripheral nerves below the spinal cord lesion. These stimuli thus trigger action potentials in the peripheral nerves to which they penetrate (through skin conduction); and the action potentials, in turn, activate muscle contractions in the associated muscle fibers. The stimuli of 150 μsec last at

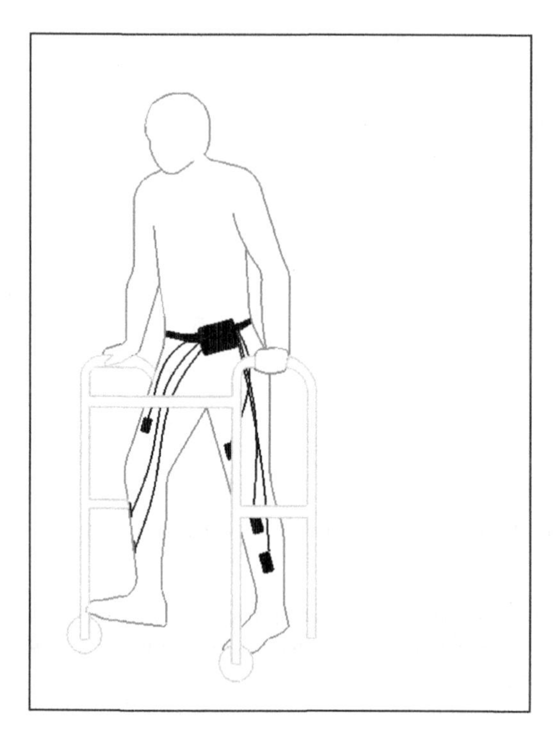

FIGURE 9.5 Drawing of FES walking system.

42 msec intervals. The maximum pulse is 15.0 V. The microprocessor receives commands from manual switches housed in the handlebars of the walker, which serves for balance and also as a safety net to prevent falls. The push-button commands (which require only a single, light-finger touch of the button) serve to select a program menu in the Parastep's microcomputer. The menu itself adjusts shapes and levels of trains of stimuli and distributes them to appropriate electrode pairs. Hence, a single touch-command from the patient allows gradual step-up or step-down adjustment of stimuli levels to change the degree of nerve fiber recruitment at a given stimulation site. The stimuli generator is housed in the same microprocessor unit. The control menus are: stand up, sit down, left step, right step, increase stimulation and decrease stimulation.

Because walking combines the aforementioned gait phases to produce a smooth movement, a feedback sensor is necessary for the FES control system to determine the beginning and end of each phase. The FES walking user may wear shoes fitted with sensors, which detect the force on the user's feet. These forces are then used to determine the location of the leg. As the user walks, the sensors send information to the gait detector, which uses information preset in the software to determine the patient's gait cycle and activate muscle groups that are part of the cascade of walking.

9.6.3 FES for Upper Extremity Function

Individuals who have an injury in the cervical area (C5/C6) that results tetraplegia may be the best candidates for a FES neuroprosthesis for the upper limb. This is because voluntary control of shoulder and elbow flexion movement is preserved in this group. Therefore, they can position their arms appropriately while using the FES device to restore hand grasp functions. Two main objectives in applying FES in these individuals with tetraplegia are either to create a reliable and long-lasting power grasp or to generate a smooth pinch grasp that is used to manipulate small objects. Regardless of the grasp strategy, it is essential that the user can easily command and adjust the grasp and grip strength.

Several well-known grasping neuroprostheses have been designed and used to restore or improve grasping function in individuals with tetraplegia. These include the Freehand System, Handmaster, Bionic Glove, and the systems developed by Rebersk and Vodovink, and Popovie and coworkers. Most of these systems are exclusively used for research purposes; however, the Freehand System is the first neuroprosthesis for grasping approved by the U.S. Food and Drug Administration (FDA) and is commercially available. The Freehand system has eight implanted epimysial stimulation electrodes and an implanted stimulator. The stimulation electrodes are used to generate flexion and extension of the fingers and thumb. The opening and closing of the hand are commanded using an external position sensor that is placed on the shoulder of the patient's opposite arm. The position sensor monitors two axes of shoulder motion: protraction/retraction and elevation/depression. The control strategy can be varied to fit different shoulder motion capabilities of the user. Typically, the protraction/retraction motion of the shoulder is used as a proportional signal for hand opening and closing. The shoulder elevation/depression motion is used to generate logic commands that are used to establish a zero level for the protraction/retraction command and to lock the stimulation levels until the next command is issued. Once an external shoulder sensor translates small shoulder movements into signals that are received by a microprocessor control unit located on the wheelchair, the unit will process the signals into radio waves that travel to an external transmitting coil worn on the skin over the implanted stimulator. The stimulator sends electrical signals to the implanted electrodes, causing the hand muscles to flex and extend. An additional switch is also provided to allow a user to choose between palmer and lateral grasp strategies.

9.6.4 FES for Exercise

A common method of using FES for rehabilitation is cycling. FES cycling can decrease atrophy of skeletal muscle, create muscle tone and provide psychological benefits for the individual's paraplegia and tetraplegia. In addition, this training benefits the cardiovascular system, pulmonary system, immune system, and lower limb circulation. Exercise can reduce edema and increase bone density. Many of the benefits gained through exercise by most unimpaired people can be achieved through FES cycling by individuals with paraplegia and tetraplegia.

Five major components influence the design of FES cycles: pedaling rate, condition of muscles, seating position, electrical stimulation parameters, and electrodes. Pedaling rate is the angular velocity of the crank. This variable greatly affects the required torque as well as the timing of the electrical stimulations. The condition of each patient's muscles is also an important factor and will vary constantly for one person and will vary even more for different people. Because most people who use

FES cycles have disabilities, mechanical design considerations must be made for mounting and dismounting from the bicycle to secure a comfortable and stable setting for users. If the bicycle is to be converted for use by a person with tetraplegia, more constraints will need to be added to stabilize the upper body.

Some serious concerns need to be included during the design process of an FES cycling system. Excessive muscle fatigue or spasms can injure the muscles, thereby inhibiting the user's rehabilitation. In addition, excessive forces can cause joint, tendon, and bone damage. Overall, FES cycling is an optimal method to improve the health and muscle tone of people with SCI, MS, and poststroke symptoms.

9.7 SPINAL CORD STIMULATION

9.7.1 OVERVIEW

A man by the name of Anteros lived in the Roman Era and suffered from an extreme case of gout, suffering from much pain in the joints of his feet and hands. One day, as he was walking along the beach, he inadvertently stepped on an electric eel and was "cured" from the pain his gout was giving him. Scribonius Largus, along with Dioscorides, recommended the application of electric stimulus from fish to relieve headaches. The procedure, known as electrotherapy, continued through the Middle Ages to treat a variety of types of headaches.

The practice of pain relief with electric stimulation has evolved from the use of such sea creatures to the use of subdural implants and hair-thin electrodes. The electrodes, also called channels, are placed in the thoracic or cervical region of the spine and target specific fibers in the spinal cord that are related to the sensation of pain. As these fibers are activated by electric current, the sensation of pain is replaced by a feeling called parasthesia, which is similar to that of the vibration from an electric massager. When electrodes were originally implanted in the human body, the electrodes were placed on the dorsal side of the spinal cord, inspiring the name "dorsal column stimulation," or DCS, for the technique. However, recent success with ventral positioning of the electrodes as well as demonstration that nondorsal fibers are recruited in stimulation has prompted the more generic term "spinal cord stimulation," or SCS, in current practice. Typical stimulation frequencies for pain control are in the range of 33 to 80 Hz, typical pulse widths for stimulations are approximately 210 μsec, and the typical voltage is between 0 and 12 V.

9.7.2 APPLICATIONS

The clinical applications for SCS have also evolved from its original uses. SCS can be used to treat severe angina pectoris; ischemic pain; segmental pain after spinal cord injury, arachnoiditis, failed back surgery syndrome, postamputation pain, or phantom pain, and peripheral vascular disease, such as arteriosclerosis or diabetic vasculopathy. The treatment of severe angina pectoris is relatively new, starting in the mid-1980s, and its exact mechanisms are still being determined. SCS does not appear to be helpful with nociceptive pain, which is the localized sensation of pain, such as arthritis, cancer pain and postoperative pain.

9.7.3 Anatomy and Physiology

The spinal cord is the "information highway" that carries messages between the brain and the rest of the body. It is composed of two different types of neural tissue: gray matter and white matter. Gray matter consists of nerve cell bodies, their unmyelinated processes, and neuroglia. It is located in the center of the spinal column, where its boundary with white matter resembles the shape of a butterfly. White matter, on the other hand, is composed of myelinated and unmyelinated axons that carry messages between different regions of the nervous system. It encompasses the gray matter and is divided into three funiculi, or bands (Figure 9.6), where fibers are grouped according to destination and function. The dorsal funiculus is composed primarily of ascending pathways, which carry afferent (sensory) messages to the brain. The anterior funiculus is composed primarily of descending pathways, which carry efferent (motor) messages away from the brain. And the lateral funiculus contains a mix of ascending and descending pathways. Table 9.1 describes the functions of the spinal cord's major ascending and descending pathways in more detail.

The spinal cord itself is located in the center of the vertebral column, between the foramen magnum of the skull and the first or second lumbar vertebra (Figure 9.7). It is surrounded by four protective layers: the spinal vertebrae, epidural space, dura mater,

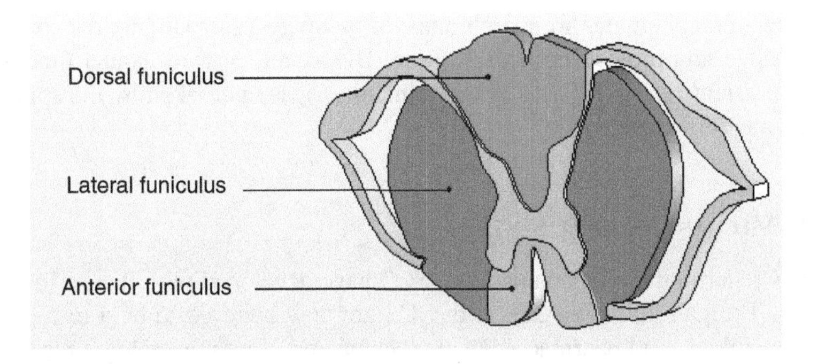

FIGURE 9.6 The three divisions of white matter.

TABLE 9.1
Direction and Function of Major Pathways of the Spinal Cord

Pathway	Direction	Function/destination
Fasciculus gracilis	Ascending	Carries sensory messages to the brain, generally from lower extremities
Fasciculus cuneatus	Ascending	Carries sensory messages to the brain, generally from upper extremities
Anterior spinothalamic tract	Ascending	Carries light touch messages to the brain
Lateral spinothalamic tract	Ascending	Carries pain and temperature messages to the brain
Anterior corticospinal motor pathway	Descending	Carries primarily motor messages from brain, but is related to the substantia gelatinosa

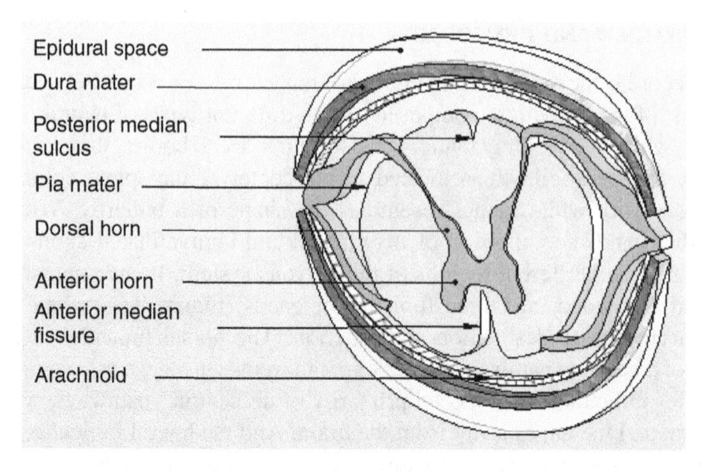

FIGURE 9.7 Protective layers of the spinal cord and major landmarks.

arachnoid membrane, and pia mater. Spinal vertebrae are bony structures that give shape to the spinal column. Between the spinal vertebrae and dura mater is the epidural space, which contains a cushioning layer of fat. The dura mater, arachnoid membrane, and pia mater are membranes, or meninges, surrounding the spinal cord that stabilize and protect the neural tissue. Blood and cerebral spinal fluid (CSF), a clear, nutrient-rich fluid, occupy the remaining space not taken by the spinal cord beneath the dura mater.

9.7.4 Mechanisms of SCS

Theories regarding the mechanisms of SCS have developed with its use in medical practice. The pain relieving effects of SCS are now believed to be a result of both neurophysiologic and pharmacological mechanisms. Neurophysiologic mechanisms refer to the activation of various neural fibers, transmission of messages, and decision-making processes within neurons in response to an electrical stimulation. They are based on the gate control theory of pain, introduced by Melzack and Wall (1965) and Shealy et al. (1967). Pharmacological mechanisms refer to the release of chemicals and response of tissues not in the nervous system.

9.7.4.1 Neurophysiologic Mechanisms

The gate control theory assumes that multiple pathways lead from a noxious stimulus to the brain, where one pathway transmits the pain message and another acts as a "gate" that permits the brain to receive the message. Specifically, small-diameter A-delta and C fibers transmit the pain message to the gate mechanism, located in the substantia gelatinosa. Concurrently, larger-diameter A-alpha and A-beta fibers also transmit messages to the gate mechanism. If more messages are received from the A-delta and C fibers, then the noxious stimulus is transmitted to the brain. If, however,

TABLE 9.2
Conductivities of Spinal Cord Material

Compartment	Conductivity (1/Ω m)
Gray matter	0.23
White matter	0.60 (longitudinal)
	0.083 (transverse)
Cerebrospinal fluid	1.70
Epidural space	0.040
Dura mater	0.020
Vertebral bone	0.040
Surrounding layer	0.0020 (bipolar stimulation)
	0.010 (monopolar stimulation)

Note that electric impulses transmit much easier in the longitudinal direction for white matter than the transverse direction.

more messages are received from the A-alpha and A-beta fibers, then the noxious stimulus is not transmitted to the brain and the pain is not sensed.

While the gate control theory serves as a good baseline of SCS's beneficial effects, the exact mechanisms are still being uncovered with the help of animal studies. Rees, Terenzi, and Roberts (1995) have suggested that, while stimulating the dorsal column of rats, some of the higher centers in the brain (i.e., the anterior pretectal nucleus [APtN] in rats) induce an inhibitory effect on the activity of multireceptive spinal neurons (MSN) deep in the dorsal horn. Chandler et al., however, have suggested with their monkey study that SCS inhibits the activity of the spinothalamic tract by antridromically sending impulses via collateral branches in the spinal cord. Nonetheless, it is believed that targeted electrical stimulation of the A-alpha and A-beta fibers will trigger the gate and cause the noxious sensation to be inhibited. The sense of parasthesia that is felt in its place is caused by the activation of wide dynamic range neurons.

As many of the more specific neurophysiologic mechanisms are still being discovered, the tracts that SCS directly affect are being slowly identified with spinal cord models. Because of the relative conductivities of spinal structures (Table 9.2), approximately 90% of the SCS electrical current flows through the CSF rather than entering the spinal cord itself. The relative conductivities also mandate that the stimulation amplitude must increase significantly to excite a spinal nerve fiber deeper in the spinal cord, while a small increase will stimulate nerve fibers more lateral to the electrode's position. Once a sufficiently high voltage, known as the perception threshold (PT), is reached, the pain is replaced by the feeling of parasthesia. Depths to which the stimulation may penetrate are on the order of 0.20 to 0.25 mm before discomfort is felt, known as the discomfort threshold (DT). The breadth may be as wide as 12 to 13 dermatomes for the T10-11 segment.

Very few A-alpha and A-beta fibers, however, are required for parasthesia to be sensed, as predicted by the computer model at the University of Twente in the Netherlands (see Holsheimer, 1998, for more about their model). The model is used to relate the depth of SCS penetration to the stimulating voltage and to relate the nerve fiber diameter to the stimulation voltage required to trigger it. Assuming an equal distribution of A-alpha and A-beta fibers for the T10-11 segment, only about four or five A-alpha fibers will be stimulated directly under the electrode, whereas just one A-alpha fiber will be stimulated in the most lateral dermatome at 1.4 PT. Hence, it may require only one A-alpha fiber per dermatome to be stimulated for the pain to be replaced by parasthesia.

These results, however, are predicted for just the T10-11 segment in the spinal cord. Results at different levels and mediolateral positions will be different because of the varying thicknesses of the CSF layer and the locations of the dorsal roots. PT is a function of spinal level primarily because the thickness of the CSF changes along the spinal cord: a thicker CSF layer will require a greater PT voltage, which will result in less coverage before DT is reached. PT is a function of mediolateral positioning of the electrodes based on the cervical level. Above the midcervical region, PT is smaller if the electrodes are placed medially. Below the midcervical region, however, PT is higher if the electrodes are placed medially.

9.7.4.2 Pharmacological Mechanisms

Chemical changes in the spinal cord and physiological changes in the rest of the body also contribute to the pain-inhibiting effect of SCS. In the spinal cord, SCS causes the level of gamma-aminobutyric acid (GABA), which inhibits transmission of nociceptive impulses in the spinothalamic tract, to increase. Concomitant with the increase in GABA is the decrease in excitatory amino acids (EAA) glutamate and aspartate, possibly limiting pain perception. Likewise, the release of beta-endorphin is enhanced, which is suggested to play a beneficial role in angina pectoris. Other chemicals, such as adenosine and baclofen, have been shown to enhance the beneficial effects of SCS.

The effects of SCS extend beyond the central nervous system to reduce painful sensations. SCS has been shown to cause vasodilatation and increase capillary density and red blood cell velocity, which improves microcirculatory flow for patients with peripheral vascular disease. Similarly, SCS relieves pain from angina pectoris by improving myocardial blood flow and reducing sympathetic tone. With a reduced sympathetic tone and a reduced perception of pain, cardiac consumption of oxygen decreases and myocardial microcirculatory blood flow improves, further contributing to a diminished sensation of pain. SCS is also believed to suppress neuropathic pain via lowered muscle tension and normalization of posture.

9.7.4.3 SCS Systems

Spinal cord stimulation systems consist of two functional components: the ultralow-power electronics and the leads and electrodes. The ultralow-power electronics are responsible for moderating the stimulation parameters as well as providing a power source for the stimulation. The leads and electrodes deliver the stimulation to the spinal cord.

9.7.5 Ultralow-Power Electronics

There are two approaches to housing the ultralow-power electronics, resulting in two distinct types of SCS systems: implantable pulse generators (IPGs) and radio-frequency-coupled passive implants. All of the electronics of an IPG, including the battery, are implanted in the patient. Although these systems may be more bulky than the radio-frequency-coupled passive implants, they are cosmetically more acceptable because there is no external device to be worn. The user may also shower, bathe or swim while receiving stimulation. The patient uses an external magnet to turn the IPG on or off and to control a few of the SCS system parameters. The most significant disadvantage to an IPG system is that the battery needs to be surgically replaced every 2 to 4 years. As such, the user may be more inclined to keep the system off and the physician may be more likely to adjust system parameters to lengthen the battery life.

Radio-frequency-coupled passive implants were the first SCS systems available and are still in use today. They use an external, battery-powered generator to deliver radio frequency bursts that provide the energy needed for effective stimulation. Because the battery is external, no surgical replacement is required. These systems also give the user significantly more control over stimulation parameters than the IPG. The most significant disadvantage, though, is the inconvenience of wearing an external device.

Four parameters may be adjusted in the SCS system to maximize parasthesia coverage and to lengthen battery life. They are frequency, pulse width, pulse amplitude and mode (Figure 9.8). The amplitude is a measure of the strength of the stimulation. Depending on the user–SCS system's PT and DT, the amplitude can be set anywhere between 0 and 12 V. The pulse width (PW) is a measure of the duration of each electrical pulse, where a larger PW will strengthen the sensation of parasthesia. The frequency is a measure of the number of times per second an impulse is generated. Because frequencies below about 25 Hz induce a painful sensation and high frequencies decrease battery life, typical SCS system frequencies are between 33 and 80 Hz. The mode determines which electrodes in multielectrode systems are activated.

9.7.5.1 Leads and Electrodes

The placement of the leads and electrodes also affects the effectiveness of SCS, and as a result a variety of styles have been designed. As such, a variety of styles have been designed. Leads are typically characterized by the number of electrodes and

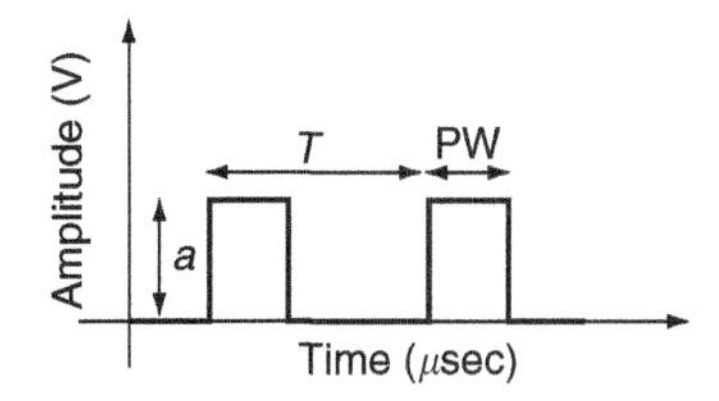

FIGURE 9.8 Typical SCS system waveform: a = amplitude, T = 1/frequency, PW = pulse width.

shape. They may have between one and eight electrodes, where more will increase both the coverage area and DT. The distance between electrodes is usually between 5 and 10 mm. Percutaneous leads are long and cylindrical, and can be surgically implanted with a needle. If four to eight electrodes are exposed and two leads are placed in parallel, the coverage area is increased, reducing the need for surgical repositioning of the electrodes. However, percutaneous leads may stimulate nerves in the ligamentum flavum (LF), resulting in midback pain. Insulated leads resemble a boat paddle, where one side has exposed electrodes. Although they insulate the electrodes from stimulating the LF and are more efficient, their primary disadvantage is that the epidural space must be exposed during surgery.

9.8 DEEP BRAIN STIMULATION

Since the 1960s, researchers have been analyzing the effects of electrically stimulating specific regions of the brain to relieve chronic pain. However, only since the late 1980s and 1990s has the procedure dubbed deep brain stimulation (DBS) become more widespread. DBS has been used to treat obsessive-compulsive disorders; dystonia; dyskinesias; and, more prevalently, essential tremor and Parkinson's disease. The primary benefits for Parkinson's patients include improved unified Parkinson's disease rating scale (UPDRS) scores and reduced usage of medications — medications that, in turn, induce dyskinesias in many patients. In DBS, electrode leads are surgically implanted in the brain, which are attached to insulated extensions that connect to the implantable pulse generator (IPG) located under the collarbone. Stimulation frequencies can range between 2 and 185 Hz, the pulse width may be between 60 μsec and 450 μsec, and the amplitudes can vary between 0 and 10.5 V.

Notwithstanding its benefits, a few risks and problems are associated with DBS. During the surgical implantation, the patient risks intracranial bleeding, infection and loss of function. Side effects from the implantable system can include parasthesia in the head and hands, depression, slight paralysis, dysarthia, loss of balance and impaired muscle tone. DBS also suffers from problems related to diagnosis, battery life and size. Concerning diagnosis, doctors have only a brief interview with patients to assess their level of need; in daily life, the emergence of dyskinesias depends on many factors, such as stress, food, environment, etc., which may not be noticeable during the interview. A finite battery life necessitates surgery every 3 to 5 years to replace the battery. And, although implantable systems are intended to be small (6 to 10 cm^3), the controller may still slide into a fat layer under the skin if it is too big for the patient.

9.8.1 ANATOMY AND PHYSIOLOGY

Deep brain stimulation systems typically target the subthalamic nucleus (STN), internal globus pallidus (GPi) or ventral intermediate nucleus (ViM) of the thalamus. These structures are located in the basal ganglia, deep within the cerebral hemispheres. This area of the brain is primarily concerned with motor movements, from controlling speed and magnitude of movement to sequencing movement for purposeful activity.

9.8.2 Mechanism of DBS

The mechanisms that carry out the beneficial effects of DBS are even more elusive than those of SCS. Nowak and Bullier (1998) and Holsheimer et al. (2000) both propose that the large myelinated axons rather than the cell bodies of the targeted areas are those directly affected by DBS. It is also believed that DBS is able to desynchronize 15–30 Hz firing patterns in neuron pairs in the STN. These 15–30 Hz high-frequency oscillations (HFO) appear common in Parkinson's patients with tremor disorder who are responsive to dopaminergic medications.

9.8.3 DBS Systems

deep brain stimulation systems are very similar to IPG SCS systems with percutaneous leads; their two main functional components are the ultralow-power electronics and the leads and electrodes. As with the IPG SCS systems, the ultralow-power electronics consist of a battery and IPG. Batteries may last between 3 and 5 years depending on use and stimulation parameters. Although DBS systems have a wide range of parameter values as noted earlier, the values during clinical usage are much more constrained. The amplitude is typically less than 3.5 V, the pulse width is between 60 μsec and 90 μsec, and the frequency is between 130 Hz and 185 Hz. Users have the option to turn the system off at night by passing a magnet over the IPG.

The leads and extensions used in DBS systems differ only slightly from those used in SCS systems. The insulated extensions travel subcutaneously to the top of the scalp, where the lead enters the skull through a burr hole. The lead's diameter is in the range of 1.25 mm and 1.27 mm. Four electrodes per lead are typically used, where the end-to-end distance between electrodes is between 7.5 mm and 10.5 mm and the length of each electrode is approximately 1.5 mm. DBS systems may also be either bilateral or unilateral. Bilateral systems will employ two sets of leads to target both sides of the brain. These systems are necessary for most patients diagnosed with Parkinson's and patients with essential tremor with bilateral symptoms. Bilateral stimulation can reduce the dosage of medications for 50% or more patients with Parkinson's. Unilateral systems will target relevant areas on just one side of the brain.

Implanting the electrodes consists of a two-phase process. During the first phase, the physician will use a stereotactic head frame and magnetic resonance imaging (MRI) or computerized topography (CT) to locate the target for the electrodes. The target is dependent on the diagnosis (Table 9.3). The patient is awake in this phase to help the operating team maximize the benefits and minimize the potential side effects. During the second phase, the patient is put under general anesthesia and the extensions and IPG are implanted.

9.8.4 Wearable Technology

Many kinds of wearable sensors, such accelerometers, gyroscopes, goniometers, and magnetic sensors, can be selected to measure dyskinesias. These sensors can be placed on a patient's trunk, wrists, hands, upper arms, and upper legs to measure symptoms

TABLE 9.3
Target Electrode Placement for Various Treatable Conditions

Diagnosis	Lead placement
Chronic pain	Periaqueductal grey of the brain
Dystonia and dyskinesias	GPi
Parkinson's alone	STN or GPi
Parkinson's and Tremor	ViM
Tremor alone	ViM

in daily life. Although distinguishing dyskinesias from voluntary movements such as walking or exercising presents itself as a major difficulty in automatically assessing dyskinesias, using manifold sensors is helpful when solving this problem. For example, when a patient is walking, his voluntary movement can generate an additional component to the amount of morbid movement, affecting the signal observed by an accelerometer. But a gyroscope can measure the angular velocity because it outputs the orientation of the body segment as a function of time. If these two kinds of sensors are used together, more complete and detailed information about the symptom could be captured.

The next step is data analysis. Fourier analysis, wavelet analysis and neural networks are the common methods in this stage. At present, about 95% of the tremor and dyskinesias can be detected accurately in daily life.

Wearable technology has potential in DBS operation: wearable sensor systems can determine the severity of Parkinsonian symptoms, a wearable computer can design an appropriate intervention scheme, and the imbedded controller will practice the scheme. A wearable faradic power supply device will provide energy to the controller. Because the power supply is outside the body, the controller can be made smaller in size.

Parkinson's disease (PD) is a chronic recessive neurological disease. The cause of PD is not clear, and current therapies cannot radically cure this disease. The main pathologic features of PD are the degeneration of neurons in the substantia nigra pars compacta, serious deficiency of dopamine in corpora striata, and the existence of Lewy cytoryctes in the substantia nigra and locus ceruleus. The characteristic symptoms of PD are a rest tremor, rigidity, bradykinesia, and impairment in postural balance. Most patients with PD are competent in their jobs or daily life early in the disease, but they will gradually lose these abilities as they develop motor disorders. During the later years, a patient with PD will have rigidity in his whole body, have difficulty moving, and usually die of pneumonia or other syndromes.

Deep brain stimulation is not the only treatment used for patients with PD, although it is currently emerging as one of the best in clinics. The current treatments for PD involve rehabilitation, medication, physical therapy, cell transplantation, gene therapy, and surgery. Rehabilitation therapy is applied in the earlier period of PD, and its function is to delay the need for medication. Levodopa has been the most successful medication in reducing Parkinsonian symptoms, but serious dependence will appear and patients' sensitivity to the medication will decrease over time. Physical therapy is not curative, but it helps the patient adapt to the illness. Cell transplantation and gene therapy are two ongoing research technologies developed in the 1980s but have not been applied in clinical practice because of problems in technology, efficiency, ethics, and inflammatory reactions.

Surgery has been widely carried out in the field of PD therapy. There are now two kinds of operations: deep brain derogation (DBD) and deep brain stimulation (DBS). The derogation operation has a notable short-lasting curative effect that reduces symptoms; but as a result of regeneration of the central nervous system and the redevelopment of the primary affection, the curative effect will fade and disappear in several years. If the derogation operation is performed in the same patient again, the effect will be worse and many complications may arise. In DBS, the controllers generate electric stimulation signals to enhance the brain's inhibitory functions, therefore eliminating the motor symptoms in patients. DBS also does not suffer from the problem of the CNS regenerating itself as in DBD. The parameters of these signals can be modified according to the severity of the individual's symptoms. Therefore, DBS has been an increasingly used method in the clinical setting when treating PD.

9.9 SUMMARY

In this chapter we discuss historical, clinical, and theoretical aspects of therapeutic electrical stimulation. Specific topics covered include electrodes, functional electrical stimulation (FES), spinal cord stimulation (SCS), and deep brain stimulation (DBS). Electrodes deliver electric energy to the body and can be configured in many ways. FES can restore such functions as walking, gripping with the hand, and bicycle riding. SCS is used to treat pain states including severe angina pectoris, segmental pain after spinal cord injury, postamputation pain, failed back surgery syndrome, and peripheral vascular disease. And, DBS is a new and promising treatment for Parkinson's disease and essential tremor.

9.10 STUDY QUESTIONS

1. Who is considered as the father of electrotherapy?
2. What is the difference between monopolar and the bipolar electrode placement techniques?
3. What is tissue impedance, and why is it important in the application of electrical stimulation?
4. How is FES used to benefit individuals with (a) Dropped foot, (b) Paraplegia, (c) Tetraplegia, (d) Angina pectoris, (e) Peripheral vascular disease, (f) Failed back surgery syndrome, and (g) Parkinson's Disease?
5. What are the main categories of electrodes used in electrical stimulation?
6. Compare and contrast percutaneous leads with insulated leads used in SCS.
7. Why is the depth of penetration of SCS current very small (on the order of 0.25 mm)?

8. What parameters determine the parasthesia coverage of an SCS system? How do they affect the batter life of the system?
9. Name the areas of the brain DBS targets and the cell parts most likely affected.
10. What are the main functional components in a DBS system, and what do they do?

BIBLIOGRAPHY

Agarwal S, Kobetic R, Nandurkar S, Marsolais EB. 2003. Functional electrical stimulation for walking in paraplegia: 17-year follow-up of 2 cases. *Journal of Spinal Cord Medicine* 26:86–91.

Alo KM, Holsheimer J, 2002. New trends in neuromodulation for the management of neuropathic pain. *Neurosurgery* 50:690–703.

Andrews RJ, 2003. Neuroprotection trek — the next generation: neuromodulation I. Techniques — deep brain stimulation, vagus nerve stimulation, and transcranial magnetic stimulation. *Annals of the New York Academy of Sciences* 993:1–13.

Baker L, *Electrical Stimulation to Increase Functional Activity Clinical Electrotherapy*. Eds. Nelson RM, Hayes KW, Currier DP. 3rd ed. Appleton & Lange, Stamford CT. 1999, pp. 355–410.

Baker LL, McNeal DR, Benton LA, Bowman BR, Waters RL. *Neuromuscular Electrical Stimulation: A Practical Guide*. Los Amigos Research & Education Institute, Inc.: Downey, California, 1993.

Basford JR. A historical perspective of the popular use of electric and magnetic therapy. *Archives in Physical Medicine Rehabilitation* 82:1261–1269, 2001.

Bhadra N, Peckham P. Peripheral nerve stimulation for restoration of motor function. *Journal of Clinical Neurophysiology* 14:378–393, 1997.

Bianchi L, Babiloni F, Cincotti F, Arrivas M, Bollero P, Marciani MG, 2003. Developing Wearable Bio-Feedback Systems: A General-Purpose Platform. *IEEE Transactions on Neural Systems and Rehabilitation Engineering* 11:117–119.

Bircan C, Senocak O, Peker O, Kaya A, Tamci SA, Gulbahar S, et al. (2002). Efficacy of two forms of electrical stimulation in increasing quadriceps strength: a randomized controlled trial. *Clinical Rehabilitation* 16:194–199.

Chae J, Bethoux F, Bohine T, Dobos D, Davis T., Friedl A. Neuromuscular stimulation for upper extremity motor and functional recovery in acute hemiplegia. *Stroke* 29:975–979, 1998.

Chandler MJ, Brennan TJ, Garrison DW, Kim KS, Schwartz PJ, Foreman RD, 1993. A mechanism of cardiac pain suppression by spinal cord stimulation: Implications for patients with severe angina pectoris. *European Heart Journal* 14:96–105.

Cui JG, O'Connor T, Ungerstedt U, Linderoth B, Meyerson BA, 1997. Spinal cord stimulation attenuates augmented dorsal horn release of excitatory amino acids in mononeuropathy via GABAergic mechanism. *Pain* 73:87–95.

Daly JJ, Kollar K, Debogorski A, Strasshofer B, Marsolais EB et al. Performance of an intrumuscular electrode during functional neuromuscular stimulation for gait training post stroke. *Journal of Rehabilitation Research and Development* 38:513–26, 2001.

Davis R, Houdayer T, Andrews B, Barriskill A., 2000. *Paraplegia: Implanted Parxis–FES System for Multifunctional Restoration.* Paper presented at the Proceedings of the 5th Annual Conference of the International Functional FES Society, Aalborg, Denmark.

Dimitrijevic MM, Dimirijevic MR. Clinical elements for the neuromuscular stimulation and functional electrical stimulation protocols in the practice of neurorehabilitation. *Artificial Organs* 26:256–259, 2002.

Dubuisson D, 1989. Effect of dorsal column stimulation on gelatinosa and marginal neurons of cat spinal cord. *Journal of Neurosurgery* 70:257–265.

Fisekovic N, Popovic DB. New controller for functional electrical stimulation systems. *Medical Engineering and Physics* 23:391–399, 2001.

Graupe D. An overview of the state of the art of noninvasive FES for independent ambulation by thoracic level paraplegics. *Neurological Research* 24:431–442, 2002.

Graupe D, Davis, R, Kordylewski H, Kohn KH, 1998. Ambulation by traumatic T4–12 paraplegics using functional neuromuscular stimulation. *Critical Reviews in Neurosurgery* 8:221–231.

Graupe D, Kohn KH, 1998. Functional neuromuscular stimulator for short-distance ambulation by certain thoracic-level spinal-cord-injured paraplegics. *Surgical Neurology* 50:202–207.

He J, Barolat G, Holsheimer J, Struijk JJ, 1994. Perception threshold and electrode position for spinal cord stimulation *Pain* 59:55–63.

Helmstadter A. The History of Electrotherapy of Pain. *Pharmazie* 58:151–153. 2003.

Hendelman WJ, 2000. *Atlas of Functional Neuroanatomy*. Boca Raton: CRC Press.

Holsheimer J, 1998. Computer modelling of spinal cord stimulation and its contribution to therapeutic efficacy. *Spinal Cord* 36:531–40.

Holsheimer J, 2002. Which neuronal elements are activated directly by spinal cord stimulation *Neuromodulation* 5:25–31.

Holsheimer J, Barolat G, Struijk JJ, He J, 1995. Significance of the spinal cord position in spinal cord stimulation. *Acta Neurochirurgica* — Supplement 64. 119–124.

Holsheimer J, Demeulemeester H, Nuttin B, deSutter D, 2000. Identification of the target neuronal elements in electrical deep brain stimulation. *European Journal of Neuroscience* 12:4573–4577.

Hossain P, Seetho IW, Browning AC, Amoaku WM, 2005. Artificial means for restoring vision. *BMJ* 330:30–3.

Iacono RP, 2005. Deep Brain Stimulation, http://www.pallidotomy.com/deep_brain_stimulation.html (18 Feb. 2005).

Ingram WR, 1976. *A Review of Anatomical Neurology*. University Park Press: Baltimore.

Jacobs MJHM, Jorning PJ, Beckers RC, Ubbink DT, van Kleef M, Slaaf DW, Reneman RS, 1990. Foot salvage and improvement of microvascular blood flow as a result of epidural spinal cord electrical stimulation. *Journal of Vascular Surgery* 12:354–360.

Jessurun GAJ, DeJongste MJL, Blanksma PK, 1996. Current views on neurostimulation in the treatment of cardiac ischemic syndromes. *Pain* 66:109–116.

Kameyama J, Handa Y, Hoshimiya N, Sakurai M. Restoration of shoulder movement in quadriplegic and hemiplegic patients by functional electrical stimulation using percutaneous multiple electrodes. *Tohuku Journal of Experimental Medicine* 187:329–37, 1999.

Keijsers NLW, Horstink MWIM, and Gielen SCAM, 2003. Online monitoring of dyskinesia in patients with Parkinson's disease. *IEEE Engineering in Medicine and Biology* May/June, 96–103.

Kobetic R, Marsolais, EB, 1994. Synthesis of paraplegic gait with multichannel functional neuromuscular stimulation. *IEEE Trans Rehabil Eng* 2:66–79.

Kobetic R, Triolo RJ, Marsolais EB, 1997. Muscle selection and walking performance of multichannel FES systems for ambulation in paraplegia. *IEEE Transactions on Rehabilitation Engineering* 5:23–29.

Kobetic R, Triolo RJ, Uhlir JP, Bieri C, Wibowo M, Polando G, Marsolais EB, Davis JA Jr., Ferguson KA. Implanted functional electrical stimulation system for mobility in Paraplegia: a follow-up case report. *IEEE Transactions on Rehabilitation Engineering* 7:390–8, 1999.

Kralj A, Bajd T, Turk R, (1988). Enhancement of gait restoration in spinal injured patients by functional electrical stimulation. *Clinical Orthopaedics and Related Research* (233):34–43.

Latif OA, Nedeljkovic SS, Stevenson LS, 2001. Spinal cord stimulation for chronic intractable angina pectoris: A unified theory on its mechanism. *Clinical Cardiology* 24: 533–541.

Levy R, Ashby P, Hutchison WD, Lang AE, Lozano AM, Dostrovsky JO, 2002. Dependence of subthalamic nucleus oscillations on movement and dopamine in Parkinson's disease. *Brain* 125:1196–1209.

Lind G, Meyerson BA, Winter J, Linderoth B, 2004. Intrathecal baclofen as adjuvant therapy to enhance the effect of spinal cord stimulation in neuropathic pain: a pilot study. *European Journal of Pain: Ejp* 8:377–383.

Linderoth B, Foreman RD, 1999. Physiology of spinal cord stimulation: review and update. *Neuromodulation* 2:150–164.

Linderoth B, Stiller C O, Gunasekera L, O'Connor W T, Ungerstedt U, Brodin E. Gamma-aminobutyric acid is released in the dorsal horn by electrical spinal cord stimulation: an in vivo microdialysis study in the rat. *Neurosurgery* 34:484–488.

Marieb EN, 2001. *Human Anatomy and Physiology*, 5th ed. San Francisco: Benjamin Cummings.

Marsolais EB, Kobetic R. Implantation techniques and experience with percutaneous intramuscular electrodes in the lower extremities. *Journal of Rehabilitation Research Development* 23:1–8, 1986.

Melzack RA and Wall PD, 1965. Pain mechanisms: A new theory. *Science* 150:971–979.

Moynahan M, Mullin C, Cohn J, Burns C, Halden E, Triolo R, Betz R. Home use of a functional electrical stimulation system for standing and mobility in adolescents with spinal cord injury. *Archives in Physical Medicine Rehabilitation* 77:1005–1011, 1996.

Murphy DF, Giles KE, 1987. Intractable angina pectoris: management with dorsal column stimulation. *Medical Journal of Australia*. 146:260.

Nandurkar S, Marsolais EB, Kobetic R. Percutaneous implantation of iliopsoas for functional neuromuscular stimulation. *Clinical Orthopaedics and Related Research* (389):210–217, 2001.

North RB, Wetzel TF. 2002. Spinal cord stimulation for chronic pain of spinal origin: a valuable long-term solution. *Spine* 27:2584–2591.

Nowak LG, Bullier J, 1998. Axons, but not cell bodies, are activated by electrical stimulation in cortical gray matter. II. Evidence for selective inactivation of cell bodies and axon initial segments. *Experiment in Brain Research* 118:489–500.

Nuttin B, Cosyns P, Demeulemeester H, Gybels J, Meyerson B, 1999. Electrical stimulation in anterior limbs of internal capsules in patients with obsessive-compulsive disorder. *Lancet* 354:1526.

Popovic D, Tomovic R, Schwirtlich L, 1989. Hybrid assistive system — the motor neuroprosthesis. *IEEE Transactions on Biomedical Engineering* 36:729–737.

Rees H, Terenzi MG, Roberts MH, 1995. Anterior pretectal nucleus facilitation of superficial dorsal horn neurones and modulation of deafferentation pain in the rat. *Journal of Physiology* 489:159–169.

Rintala DH, Loubser PG, Castro J, Hart KA, Fuhrer MJ, 1998. Chronic pain in a community-based sample of men with spinal cord injury: prevalence, severity, and relationship with impairment, disability, handicap, and subjective well-being. *Archives in Physical Medicine Rehabilitation* 79:604–614.

Samuelsson KA, Tropp H, Gerdle B, 2004. Shoulder pain and its consequences in paraplegic spinal cord-injured, wheelchair users. *Spinal Cord* 42:41–46.

Shealy CN, Mortimer JT, Reswick JB, 1967. Electrical inhibition of pain by stimulation of the dorsal columns: Preliminary clinical report. *Anest Analg* 46:489–491.

Shimada Y, Sato K, Abe E, Kagaya H, Ebata K, Oba M, Sato M. Clinical experience of functional electrical stimulation in complete paraplegia. *Spinal Cord* 34:615–619, 1996.

Shimada Y, Sato K, Kagaya H, Konishi N, Miyamoto S, Matsunaga T. Clinical use of percutaneous intramuscular electrodes for functional electrical stimulation. *Archived in Physical Medicine and Rehabilitation*. 77:1014–1018, 1996.

Silverthorn DU, 2001. *Human Physiology: An Integrated Approach*, 2nd ed. New Jersey: Prentice Hall.

Simpson BA. 1997 Spinal cord stimulation *British Journal of Neurosurgery* 11:5–11.

Sinnott KA, Milburn P, McNaughton H, 2000. Factors associated with thoracic spinal cord injury, lesion level and rotator cuff disorders. *Spinal Cord* 38:748–753.

Solomonow M, Aguilar E, Reisin E, Baratta RV, Best R, Coetzee T, D'Ambrosia R, 1997. Reciprocating gait orthosis powered with electrical muscle stimulation (RGO II). Part I: Performance evaluation of 70 paraplegic patients. *Orthopedics* 20:315–324.

Standaert DG, 2003. Wearable technology's applications in Parkinson's disease. *IEEE Engineering in Medicine and Biology* May/June, 25–26.

Stillings D, 1975. A survey of the history of electrical stimulation for pain to 1900. *Medical Instrumentation* 9:255–259.

Struijk JJ, Holsheimer J, Boom HBK, 1993. Excitation of dorsal root fibers in spinal cord stimulation: A theoretical study. *IEEE Transactions on Biomedical Engineering* 40:632–639.

What is deep brain stimulation? http://www.medtronic.com/UK/patients/neuro/brain_stimulation.html (18 Feb. 2005).

10 Wheelchair Transportation Safety

Patricia Karg, Gina Bertocci, Linda van Roosmalen, Douglas Hobson, and Toru Furui

CONTENTS

10.1 LEARNING OBJECTIVES OF THIS CHAPTER

Upon completion of this chapter, the reader will be able to:

Learn the components of the wheelchair transportation safety system
Understand the issues related to safely transporting the wheelchair-seated motor-vehicle occupant
Gain knowledge about regulations and safety standards related to wheelchair transportation

Learn the principles of occupant safety in motor vehicles
Be able to identify the technologies used to transport the wheelchair-seated occupant

10.2 INTRODUCTION

Transportation is a key component for full integration into the community. Accessible public transportation is necessary to provide persons with disabilities the same opportunities as others: access to employment, education, religious worship, and recreation. In the United States, the Individuals with Disabilities Education Act (IDEA) (IDEA-97, 1997) and the Americans with Disabilities Act (ADA) of 1990 (ADA, 1990) have provided people with disabilities the opportunity for accessing schools and public transportation. The ADA transportation requirements mandate accessible fixed-route vehicles, as well as complementary paratransit services for those unable to use the fixed-route service. Other nations also recognize the importance of accessible transportation. For example, big change in the attitude of the Japanese government regarding transportation accessibility was the initiation of the "Law for Promoting Easily Accessible Public Transportation Infrastructure for the Aged and Disabled (2000 law number 68)" in 2000. As accessible vehicles are not as common as in the United States, the goal of this law by year 2010 is to have 20% accessibility in a total of 600,000 public buses, 30% in a total of 51,000 trains, 40% in a total of 420 aircraft, and 50% of a total of 1100 ships. In the United Kingdom, the Disability Discrimination Act of 1995 includes transport provisions that require vehicles to safely accommodate wheelchair users (MDA DB2001 (03), 2001). In addition, advances in technology for adapting vehicles have made personal vehicle transportation available to people with disabilities, either as a passenger or a driver.

A subset of the disabled population is the wheelchair user, whom may or may not be able to transfer to a vehicle seat and might, therefore, remain seated in the wheelchair during transport. A reasonable goal is that the wheelchair user has the same level of safety afforded to passengers sitting in vehicle seats that must meet federal safety standards. The reality is that the wheelchair-seated passenger is at an increased risk of injury in the event of a vehicle collision or even an emergency maneuver. This chapter will explore issues related to the safety of the wheelchair-seated occupant. The chapter presents information from the U.S. perspective; other nations may have different laws, regulations, and device standards. However, the safety principles apply equally to all wheelchair transport situations.

10.3 PRINCIPLES OF OCCUPANT SAFETY IN MOTOR VEHICLES

The primary cause of serious injury to a vehicle occupant is the result of the occupant impacting the interior of the vehicle or structures outside of the vehicle after ejection. Occupant restraint systems are used in motor vehicles to reduce the likelihood of ejection from the vehicle and limit the movement of the occupant to the clear space in the interior of the vehicle. The provision of adequate occupant protection involves the vehicle, vehicle seat, and occupant restraint. Padded or crushable interior components can reduce the effect of impacts with the vehicle interior. The vehicle seat is fixed to the vehicle, assists in dissipating occupant loads, and holds a person in the correct orientation so that the occupant restraints can function properly. The occupant restraint limits movement of the pelvis and torso and applies loads to bony structures, locations that are most likely to be able to handle the high loads without serious or fatal injury. The use of both an upper (shoulder) and a lower torso (pelvic) belt is necessary to prevent "submarining," which occurs when the occupant slides under the pelvic belt, allowing for forces to be applied to the more vulnerable soft tissues of the abdomen. The pelvic belt should be positioned as low as possible on the pelvis and should have an angle between 45° and 75° to the vertical. The shoulder belt should be in contact with the occupant's shoulder and chest, and slack should be removed from the belt in order to be effective.

Regulations and standards focus on protecting vehicle occupants in a frontal impact as they lead to more than half of the serious injuries and fatalities (NHTSA, 2002). Federal Motor Vehicle Safety Standard (FMVSS) 208, Occupant Crash Protection, and FMVSS 222, School Bus Seating Crash Protection, reflect this priority, and FMVSS 222 requires that all school bus seats face forward. Sideways facing is the least safe orientation for a passenger during a frontal crash as it prevents the occupant restraints from being effective, allowing for excessive occupant excursion and potential injury to the torso due to the armrests.

The frequency of distribution of crash severities is a function of vehicle mass. Thus, the distribution of crash severities for larger vehicles, such as large transit vehicles and school buses, will span a significantly lower range of changes in vehicle speed during an impact than smaller passenger vans and cars. This is due to the fact that most vehicle crashes involve impacts between two vehicles. The higher-mass vehicle will have a lower deceleration than the lower-mass vehicle in such an impact. Therefore, it is possible to reduce the risk of injury by riding in a larger vehicle when there is an option (Shaw and Gillispie, 2003).

10.4 COMPONENTS OF THE WHEELCHAIR TRANSPORTATION SYSTEM

To ensure the safety of the wheelchair user during vehicle transportation, the entire wheelchair transportation system must be considered. This system includes three main components: the wheelchair, wheelchair securement, and occupant restraint.

10.4.1 WHEELCHAIR

Historically, the wheelchair industry, largely to reduce liability, chose to place labels on the wheelchairs, stating that they were not suitable for use as a seat in a motor vehicle. IDEA mandates that children with disabilities be transported to their educational settings, and ADA mandates that all individuals with disabilities have access to public transit. Many of these individuals are unable to safely transfer to a vehicle seat and must ride seated in their wheelchair. The ADA and school bus regulations acknowledge this need and require vehicles to provide stations for the wheelchairs that contain wheelchair securement and occupant restraints. Therefore, the wheelchair design must consider the potential function of the wheelchair as a seat in a motor vehicle. This means that the wheelchair must be designed to withstand forces generated in a crash, must provide occupants with effective support during impact loading so that the occupant restraint belts remain properly positioned, consider access and fit of vehicle occupant restraints, and provide attachment points to secure the wheelchair to the vehicle. Prior to the advent of voluntary industry standards for wheelchairs used as seats in motor vehicles, the majority of wheelchairs were not designed to be used as a motor-vehicle seat. Fortunately, this is now slowly changing.

10.4.2 Wheelchair Securement

The wheelchair must not add to the forces on an occupant during an emergency maneuver or a crash. Therefore, the wheelchair must be secured using a means independent from restraining the occupant (e.g., the same strap should not be used to restrain the occupant and secure the wheelchair). The securement system must be designed to attach to the wheelchair frame to prevent excessive movement or tipping during driving events, and withstand crash forces.

10.4.3 Occupant Restraint

A system that prevents occupant ejection from the vehicle and contact with the interior structure of the vehicle is necessary. Both upper and lower torso restraints are needed to limit the excursion of the head, chest, and pelvis. In addition, the restraints must be positioned properly to apply forces to bony structures and not to soft tissue.

10.5 STATUS OF WHEELCHAIR DESIGN

Although legislation established requirements for vehicle accessibility and wheelchair securement locations and equipment in public transportation, it did not ensure compatibility of the equipment with the wheelchair or address the safety of the wheelchair-seated occupant. Transit providers are faced with challenges to safely and effectively transport wheelchairs that do not have securement points, are uncommon shapes and sizes, and are unstable or do not have adequate strength to withstand crash forces (Hardin et al., 2002; NCD, 2005). The ADA defines a "common wheelchair" as a device that does not exceed 76 cm (30 in.) width by 122 cm (48 in.) length and weighs no more than 272 kg (600 lb) when occupied. Even wheelchairs that meet this definition present transport challenges. Many wheelchairs have not been designed or tested for securement and crashworthiness, and may not provide effective seat and back support, which is necessary to ensure proper occupant restraint effectiveness.

Due to the lack of federal safety standards for devices used in the transportation of wheelchair-seated passengers, national and international efforts were made to develop "voluntary" product safety standards for wheelchairs used as seats in a motor vehicle. The standards establish design and performance requirements. Two standards have been published addressing the wheelchair: Section 19 of ANSI/RESNA WC Volume 1

FIGURE 10.1 Securement point on a wheelchair that complies with WC19.

(WC19): "Wheelchairs used as seats in motor vehicles" (ANSI/RESNA, 2000); and ISO 7176/Part 19: "Wheeled mobility devices for use in motor vehicles" (ISO, 2001).

Wheelchairs that comply with these standards provide increased occupant protection in motor vehicle crashes and improved usability. The wheelchair is crash tested at a speed of 48 kph (30 mph) and acceleration of 20g in a simulated frontal crash. The wheelchair contains four crash-tested securement points and crash-tested anchor points on the frame to which a crash-tested pelvic belt can be attached. The securement points are clearly marked and easily accessible with one hand (Figure 10.1), and the wheelchairs have increased compatibility with vehicle-anchored occupant restraints.

The standards evaluate the complete wheelchair, frame/base, and seating system. However, seating systems are often added after-market. Wheelchairs utilizing after-market seating systems are likely not to be tested for evaluation of their ability to withstand crash forces or will invalidate previous Part 19 testing. The integrity of the seat and back of the wheelchair is important to the safety of the seated individual in a motor vehicle. Voluntary standards that address the need to independently evaluate wheelchair-seating-system crashworthiness are under development at the national and international levels. The standards included under development are Section 20 of ANSI/RESNA WC Volume 4 and ISO 16840-4.

10.6 SECUREMENT SYSTEMS

10.6.1 Four-Point Tiedown System

A motor vehicle seat must be effectively anchored to the vehicle floor to prevent excessive movement and ensure that its mass does not add to the restraint loads on the occupant. When an occupant remains seated in the wheelchair, the wheelchair becomes the vehicle seat. After-market securement systems must be utilized to anchor the wheelchair to the vehicle floor. Again, due to the lack of federal safety standards

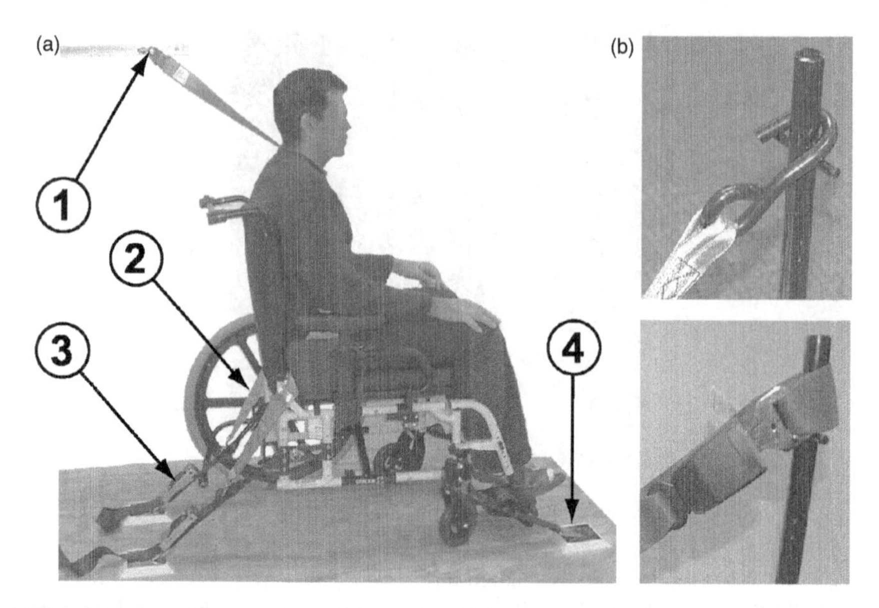

FIGURE 10.2 (a) Example of a four-point tiedown system: (1) vehicle-anchored shoulder belt, (2) lap belt anchored to the tiedown assembly, (3) over-center cam buckle tensioning mechanism on rear tiedown strap, (4) vehicle floor anchor for front tiedown straps; (b) hook-type and loop-type end fittings.

for devices used in the transportation of wheelchair-seated passengers, national and international efforts were made to develop voluntary product safety standards for wheelchair tiedown and occupant restraint systems (WTORS). In the US, the Society of Automotive Engineers' (SAE) Adaptive Devices Subcommittee developed SAE J2249: "Wheelchair tiedown and occupant restraint systems for use in motor vehicles (SAE, 1999)." Shortly after, an international standard was published — ISO 10542 Parts 1 and 2 (ISO, 2001). Further details are provided in Section 10.8.

The most common securement system found in public transportation is the four-point tiedown system (Figure 10.2). This system consists of four tiedown straps that anchor to the vehicle floor and four points on the wheelchair (two front and two rear). The end fittings that attach to the wheelchair can be hooks or loops that wrap around the wheelchair frame or designated securement attachment point. The main advantage of this type of system is that it is one of the most universal systems available on the market and can be used with a wide variety of wheelchairs, without the need for special hardware to be added to the wheelchair. In addition, there are a number of systems on the market that comply with WTORS safety standards. There are also disadvantages for this type of system; several have been documented (Hardin et al., 2002). First, someone other than the wheelchair user must attach the straps to the wheelchair. On a public transit vehicle, the vehicle operator is usually that person. The attachment of the four straps is also time consuming and can be intrusive to the wheelchair user. Another disadvantage is that it is often difficult to identify appropriate locations to attach the straps to the wheelchair, especially when the frames are encased in a plastic housing. These disadvantages often result in disuse or misuse of the systems, such as utilizing less than four tiedowns, which can compromise safety during emergency maneuvers or a crash. Use of wheelchairs that comply with WC19 standards helps to eliminate this problem by having four clearly marked, easily accessible securement points.

Transit providers seek ways to limit the demand for paratransit services and, at the same time, attract more riders and avoid costs of new rail systems by developing bus rapid transit (BRT) services. Both transit driver unions and transit management would welcome wheelchair securement technologies that did not require drivers to perform time-consuming, physically awkward, and sometimes confusing securement procedures.

If we could restart the clock today and begin with a clean state, would we choose a four-point strap tiedown to be the *de facto* industry standard? Probably not. If one were to design a wheelchair with the transport needs of wheelchair users in mind, would one completely cover the structural frame in cosmetic, plastic cowlings? Probably not. If the wheelchair user desires to independently secure their wheelchair, without the time-consuming intervention of a transit driver or attendant, would they opt for strap-type tiedowns? Probably not. If you wanted to increase the independence of wheelchair users and reduce their dependency on transit drivers/attendants, would you use a docking-type technology? Perhaps. If a safe, cost-effective, and user-friendly passive securement system could be developed for use in fixed-route transit vehicles, would this make good sense? Perhaps. Therefore, it is important to work toward alternative approaches for securement that are more compatible with the desires of wheelchair users and the operational needs of transit systems while still offering wheelchair users a reasonable level of transportation safety.

10.6.2 Docking Systems

A second method of securement is an automatic docking system. This type of securement utilizes a wheelchair adaptor, special hardware mounted on the wheelchair, which engages with a docking station or receptacle mounted to the vehicle floor or side wall. The advantages of wheelchair-docking technology have been demonstrated by proprietary systems that have been used for independent securement of wheelchairs in private vans for many years. The EZ Lock is one such device (EZ LOCK Inc., Model 6290) (Figure 10.3). The EZ Lock device uses a wheelchair-model-specific docking adaptor that is bolted to the bottom of the wheelchair frame. The docking adaptors are designed to mate with EZ Lock's proprietary vehicle-mounted docking device. The two devices (adaptor and receptacle) are adjusted so that the two components will align and autoengage when the wheelchair user drives into the wheelchair station of their van, either as a driver or a passenger. A switch control, located for activation by the wheelchair user, disengages the docking device when exiting the vehicle. EZ Lock has demonstrated crashworthiness at a 30-mph/20-g crash severity level. Although it has been used successfully in private vehicle applications, it has not evolved as a viable solution for use in public transit applications, largely because a specially designed adaptor would be required for each wheelchair size and model that would work only with EZ Lock's proprietary docking device. Also, it requires fittings on the vehicle floor that would be problematic in public transit vehicles.

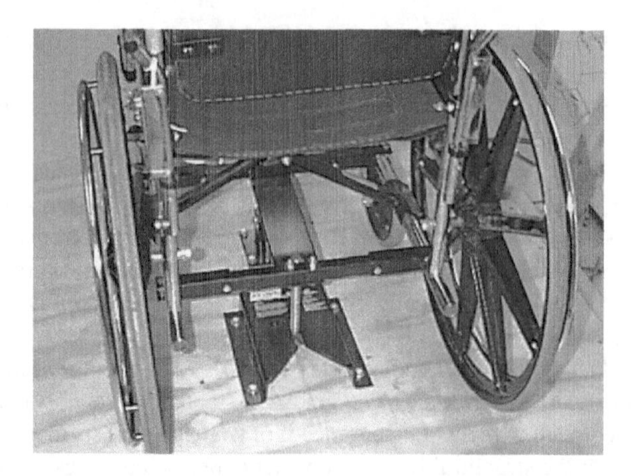

FIGURE 10.3 Rear view of EZ Lock docking device showing docking adapter (black square tubing) bolted to lower wheelchair frame and the vehicle floor mount receptacle (V shape).

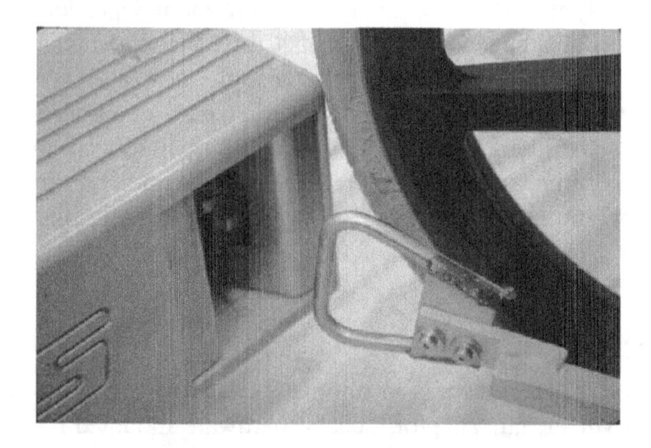

FIGURE 10.4 Oregon State University–Mobil–Tech/ILS docking device showing a vehicle-mounted docking device and mating proprietary wheelchair adaptors.

Another autodocking device, developed at Oregon State University in the early 1990s, demonstrated potential for the use of autodocking securement technology in public transit vehicles (Hunter-Zaworski, 1992). This device also requires that model-specific wheelchair adaptors be added, in this case, to the lower rear of the wheelchair frame. This adaptor interfaces or docks with the docking receptacle permanently mounted on the floor of the wheelchair station of a public transit vehicle (Figure 10.4). Autodocking occurs when the wheelchair user backs into the receptacle, and disengagement occurs when the user or transit driver activates an electric switch control similar to EZ Lock. Efforts to commercialize this technology for use in public transportation over the past ten or more years have been only partially successful as only one transit agency has the device in routine operation today.

Both docking devices have demonstrated the key advantages of the autodocking wheelchair securement approach: increased independence by the wheelchair user and reduced driver/attendant intervention. These systems also remove the need for human judgment, resulting in more consistent and proper use of the systems. However, both these systems have inherent limitations from both the usage and safety perspectives. For example, they require that a unique proprietary adaptor be designed and retrofitted to each and every model of wheelchair, of which there are hundreds of different makes and models. Because wheelchair manufacturers have not marketed these adaptors as accessories to their products, due to liability concerns, it means that the adapters must be sold and applied as after-market add-on accessories. As it is impractical to crash test each and every after-market installation, it is very difficult to ensure that an adaptor installation will meet industry safety standards. Because proprietary adaptors retrofitted to wheelchairs would only work with one type of proprietary docking receptacle, this limits the technology that could be developed for widespread use in public transit. Therefore, if a competitor developed a docking system, complete with wheelchair-mounted adaptors, wheelchair users could secure their wheelchair in the transit system in one city but would find incompatibility when using the transit system in another. In addition, the adaptors may add unwanted weight to the wheelchair, impact ground clearance if mounted underneath, or add to overall length if mounted in the rear. Docking systems will also tend to cost more to purchase and maintain than the traditional four-point tiedown system.

Despite the limitations, it appears that docking technology, especially for use in high-g vehicle environments, holds promise as a future solution. Examples in which agreement has been reached as to how two or more technologies will effectively mate, interconnect, interface, or dock with each other exist in other fields. Obvious examples are international space exploration, electronic connectors, and 18-wheel tractor/trailer units. Similarly, if a nonproprietary specification or standard for universal docking interface geometry could be agreed upon and implemented by the involved industries, the door would remain open for the development of competing docking technologies using this interface geometry. This would ensure the potential for universal docking wheelchair securement. The concept is demonstrated in Figure 10.5.

The industry standard ISO10542-3 specifies how docking technologies interface (ISO, 2005). This standard permits development of technologies to automate the wheelchair securement process. One requirement of ISO10542-3 is the specification for the hardware interface. The interface specification, termed *universal docking interface geometry* (UDIG), specifies three key design requirements for the hardware or UDIG adapter that is to be located on the rear of a wheelchair intended for universal docking. These are: (1) the geometry or physical shape and dimensions of the adapter, (2) its location in space relative to a known structure on the rear of the wheelchair, and (3) the dimensions of the clear spaces or zones around the adapter geometry needed to allow unimpeded access by the docking device or used for attachment of the adapter to the wheelchair. Figure 10.6 shows an experimental UDIG adapter mounted on the rear of a power wheelchair. Supporting research studies have demonstrated the feasibility of constructing and successfully crash testing UDIG adapters mounted to both manual and powered wheelchairs (RERC on Wheelchair Transportation Safety, 2004).

A limitation of the universal docking approach is that wheelchair manufacturers must be willing to incorporate UDIG adaptors into new designs, and existing

FIGURE 10.5 Conceptual graphic of universal docking concept based on a common nonproprietary hardware interface between transit-ready wheelchairs and docking devices on transport vehicles.

FIGURE 10.6 UDIG adapter (arrow) on power wheelchair.

wheelchairs will need add-on adaptors. Therefore, deployment for general use will take time, and four-point tiedowns will still be needed in public transit vehicles during the phase-in period. However, broad implementation of the UDIG standard will stimulate the development of competing docking devices that have the potential to resolve many of the shortcomings of today's securement technology.

10.6.3 Rear-Facing Wheelchair Passenger Space

Another wheelchair securement approach, widely demonstrated in Canada and Europe, and having significant promise for problem solving on large, accessible transit vehicles (LATVs), is rear-facing passive containment or compartmentalization. Rear-facing passive wheelchair containment is the most promising near-term solution for use in LATVs in which the securement forces very rarely exceed 1 g. Current four-point tiedown devices are all nominally tested to the 48-kph/20-g crash pulse. This is necessary when manufacturers are uncertain as to the size of the vehicle into which their products are being installed. Therefore, the testing replicates the worst-case scenario — small transport vehicles. However, these devices are more robust than required in the LATV environment, which makes other securement options possible, such as the passive, rear-facing containment.

Accident data analysis of LATVs used on fixed urban routes clearly indicates that the frequency of fatalities per million passenger miles due to frontal crashes is very small and that most fatalities or injuries occur during collision-avoidance driving maneuvers (braking/swerving) or during nonmoving events, such as, entering or exiting the vehicle (Shaw and Gillispie, 2003). These findings support a wheelchair securement approach that is concerned primarily with the forces associated with collision-avoidance driving and, thereby, accept the very small risk that a frontal collision will ever occur. Other research has shown that the acceleration forces associated with avoidance driving maneuvers (maximum braking or rapid swerving) all have values lower than 1g (Adams et al., 1995; Mercer and Billing, 1990). Based on these and other findings (Blennemann, 1995; Rutenberg U and Association, 2000; TCRP-50, 2003), many European countries, Canada, and Australia have been introducing rear-facing wheelchair passenger spaces in LATVs since the early 1990s. The advantage claimed is that of allowing wheelchair users to enter and leave without the need of active involvement by an attendant or driver for wheelchair securement. In the securement station, during forward travel, the rear-facing passenger is protected in response to vehicle braking or frontal impact by a vehicle-mounted, padded restraint that fits in close proximity to the passenger's head and back. As seen in Figure 10.7, lateral movement or rotation of the wheelchair is limited on one side by the wall of the LATV and on the aisle side by a vertical or horizontal barrier. The brakes of the wheelchair, and/or the user's ability to grasp handholds located within the passenger space, limit movement toward the rear of the vehicle.

In practice, most wheelchairs users and transit providers favor the passive rear-facing approach because it is more user friendly than the four-point strap tiedowns (TCRP-50, Chapter 3, 2003). From the perspective of injury risk, the passive system approach needs refinement in several important areas. Many wheelchair designs,

FIGURE 10.7 Rear-facing wheelchair passenger space installed in a Belgian LATV.

combined with varying sizes of personal bags hung on the rear of the backrest, prevent the head and back restraint from coming in close contact with the occupant's head and back — a requirement for a safe passive system (TCRP-50, 2003). Additionally, manual wheelchairs can have unreliable brakes, occupants vary in their ability to use handrails, and lateral barriers on the aisle side do not work effectively with some scooters. Fixed aisle-side stanchions and other barriers do not allow for wide variation of wheelchair shapes and sizes, and create impediments to ease of movement within LATVs and driving wheelchairs into the space. These deficiencies have allowed, for example, tipping or swerving into the aisle during vehicle cornering. Attempts to resolve these problems have resulted in the addition of various types of secondary securement straps. For example, a 2002 survey of six Canadian transit systems reported that one-half of the systems use secondary straps and two-thirds provide wheelchair positioning and/or securement assistance (TCRP-50, 2003). Although these approaches are an improvement over four-point tiedowns, this continued need for driver intervention nullifies a main advantage of passive containment — independence for the wheelchair user.

To date, there are no international or U.S. standards for the design, testing, and installation of a rear-facing wheelchair passenger space in a LATV. Therefore, misapplication of the concept can easily occur, and confusing and conflicting variations within a country are likely. A review of existing regulations and industry standards across implementing countries indicates significant variations in both strength and dimensional requirements of rear-facing stations and approaches to aisle-side anti-tipping protection (TCRP-50, 2003). Therefore, the development of harmonized industry standards for a rear-facing wheelchair passenger space (RFWPS), at both the international and national levels, is the logical next step.

As indicated earlier, introduction of the RFWPS on LATVs offers many attractive benefits to both wheelchairs users and transit providers. As in the past, the key to having RFWPS installations meet a nominal level of safety and design requirements is to seek agreement among stakeholders within the standards development forum. Over the past decade, most countries that have experimented with RFWPSs have established national regulations or voluntary industry standards. Countries such as England, France, and Germany have regulations that are now being integrated into a European Community (EC) Directive (TCRP-50, 2003). In Canada, Canadian Standards Association (CSA) has published a national standard (CSA-435) with plans for review and possible revisions based on continuing testing (CSA, 2002). The US and ISO, both lag behind on this initiative. However, proposals for new work items (NWI), complete with initial working drafts for a standard on rear-facing containment in LATVs, have been accepted for implementation by ISO and ANSI/RESNA (US).

10.6.4 Other Securement Systems

Prior to legislation and the development of product safety standards, other types of securement systems were more commonly used, and continue to be used. These systems include single and double wheel clamps, T-bars, and frame clamps. The systems have not been demonstrated to be safe under crash conditions and were often used to secure wheelchairs in a side-facing position, the least safe orientation in a frontal crash. Continued use of these devices is discouraged, as much safer products exist on the market today.

10.7 OCCUPANT RESTRAINTS

The increased risk of injury to the wheelchair-seated passenger is largely due to the fact that these individuals are not able to benefit from the manufacturer-installed, federally regulated occupant restraints. The wheelchair user must use after-market restraints that are not required to comply with federal safety standards. Wheelchair occupant restraint systems (WORS) are safety belt systems designed to provide upper torso and pelvic restraint to the occupant. During normal driving as well as emergency maneuvers, a WORS prevents the occupant from being ejected or sliding out of the wheelchair and impacting the motor vehicle interior.

Figure 10.8 shows three belt-restraint configurations in use or proposed for WORS: (a) pelvic and torso belts with all three anchorages on the vehicle or on the tiedown system, (b) a wheelchair-integrated pelvic belt with a vehicle-mounted torso belt that connects to the latch plate of the lap belt, and (c) a fully integrated system with both lap and shoulder anchorages on the wheelchair.

Configuration (a) is the most common system in use today in public and private vehicles. Configuration (b) is made possible by the integrated pelvic belts required by the voluntary ANSI/RESNA WC19 standard. Compliance with WC19 requires wheelchair manufacturers to provide crash tested pelvic belt anchor points as part of

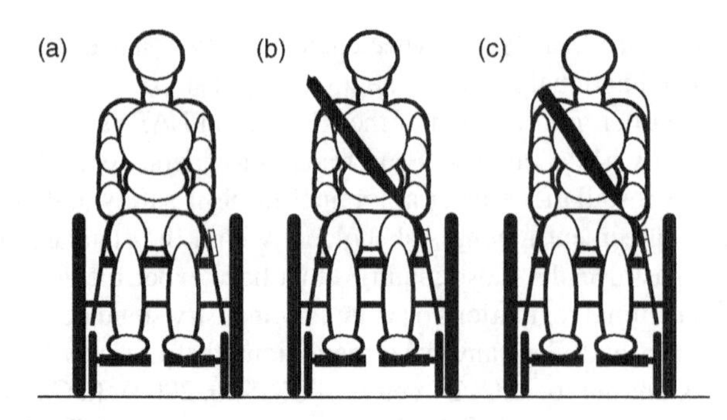

FIGURE 10.8 Wheelchair occupant restraint system configurations: (a) two-part vehicle-mounted system, (b) wheelchair-integrated pelvic belt and vehicle-mounted torso belt, and (c) wheelchair-integrated pelvic and torso belts.

the wheelchair frame (ANSI/RESNA, 2001). Configuration (c) has been shown to be feasible for use on some wheelchairs (van Roosmalen et al., 2002a), but is not in common use for adult individuals.

Occupant restraints for the wheelchair-seated occupant can be further categorized by where they anchor. An integrated restraint is anchored to the rear tiedowns of the securement system, whereas an independent restraint is anchored to the vehicle floor. An integrated restraint will increase the force on the rear tiedowns, which must handle both the load of the wheelchair and the occupant. An independent restraint has the potential to allow the load of the wheelchair to be borne by the occupant in a frontal crash or emergency stop if the wheelchair is able to travel forward further than the occupant, thus increasing the risk for injury.

Configuration (a) typically does not allow for independent use and often provides poor belt fit. Due to variations in wheelchair design and height, and the fixed location of the vehicle anchor points, the pelvic and torso belts may not be optimally positioned with respect to the occupant (van Roosmalen et al., 1998). Another common problem is that the armrests, postural supports, or controller on the wheelchair prevent the pelvic belt from being properly positioned low on the pelvis. Poor belt fit, discomfort, and nonuse of the currently available safety belts result in decreased occupant protection during motor vehicle incidents or normal driving situations (Aibe et al., 1982; Bertocci et al., 1996, 1997; Bertocci and Evans, 2000; Bunai et al., 2001; van Roosmalen et al., 1998, 2002b; Linden et al., 1996; Sprigle et al., 1994).

One solution to poor belt-fit and problems with independent use is to partially or completely integrate the occupant restraint system with the wheelchair. Wheelchairs that comply with the voluntary ANSI/RESNA WC19 standard include anchorages for a pelvic belt that is integrated with the wheelchair. Torso restraint is obtained by attaching a vehicle-mounted belt to the latch plate of the pelvic belt (configuration (b) in Figure 10.8). In addition to improved restraint performance, the integrated pelvic belt could increase the number of wheelchair users who can don safety belts independently. Preliminary data from surveys of wheelchair users show that safety belts that are completely wheelchair integrated, restraining both torso and pelvis (configuration (c) in Figure 10.8), may benefit pediatric wheelchair users as well as adult power wheelchair users (Armstrong et al., 2003; van Roosmalen et al., 2003).

A study by van Roosmalen et al. (2005) showed evidence of user problems with common WORS designs for use by wheelchair-seated individuals with significant functional limitations. Results from this pilot study suggested redesign of latch plates and buckles used in WORS to enable wheelchair-seated individuals with functional limitations to don and buckle belt restraints more easily and independently.

Postural support devices used by wheelchair users are often confused for or used as safety restraints in a motor vehicle. Postural support devices should not be used as a substitute for a safety restraint. These devices have not been designed to provide protection in a vehicle crash and may be placed over more vulnerable soft tissue. The SAE J2249 and WC19 standards address this by requiring labeling on the occupant restraint belts, indicating they are suitable for use in a motor vehicle.

10.8 REGULATIONS AND STANDARDS

Vehicle regulations resulting from the American with Disabilities Act (ADA 1990) and Federal Motor Vehicle Safety Standard 222 (FMVSS 222), School Bus Passenger Seating and Crash Protection, include requirements for wheelchair tiedowns and occupant restraint systems (WTORS) and, thereby, include provisions for the safety of wheelchair-seated travelers in public and school transportation. However, most of the progress over the past two decades toward improving occupant protection and transportation safety for wheelchair users who do not transfer to the vehicle seat has been made through national and international efforts to development voluntary recommended practices and standards. The following sections summarize the key points of published regulations and standards with respect to safety of the wheelchair-seated passenger.

10.8.1 REGULATIONS

The Americans with Disabilities Act (ADA, 1990) states that at least two securement stations shall be provided on vehicles in excess of 6.7 m (22 ft.) in length and at least one securement station shall be provided on vehicles 6.7 m (22 ft.) in length or less. In vehicles longer than 6.7 m (22 ft.), at least one securement station shall secure the wheelchair in a forward facing orientation, others can be either forward or rearward facing. In vehicles 6.7 m (22 ft.) long or less, the securement station may secure the wheelchair either forward or rearward facing. In those cases where wheelchairs will be oriented towards the rear of the vehicle, a padded barrier shall be provided at the rear of the wheelchair.

Securement systems complying with ADA must be capable of securing a "common wheelchair" and mobility aids. A "common wheelchair" is defined as a three- or four-wheeled device being no larger than 76 cm (30 in.) wide by 122 cm (48 in.) long, and weighing no more than 272 kg (600 lb) when occupied. Securement system performance is dependent upon gross vehicle weight. In those vehicles weighing 13,608 kg (30,000 lb) or more, securement systems must be capable of withstanding a forward longitudinal static load of 907 kg (2000 lb) per leg or a minimum or 1814 kg (4000 lb) per wheelchair. On vehicles weighing up to 13,608 kg (30,000 lb), each leg of the securement device must withstand a 1134 kg (2500 lb) forward static longitudinal load and a minimum of 2268 kg (5000 lb) per wheelchair. Additional securement system performance criteria include limiting the movement of the occupied wheelchair to no more than 5 cm (2 in.) in any direction under normal vehicle operating conditions. No test method is provided to evaluate whether the wheelchair meets this performance criteria. ADA also requires that a lap and shoulder belt be provided at each securement station. Occupant restraints are required to meet the applicable provisions of 49 CFR 571 (Federal Motor Vehicle Safety Standards) as a part of ADA.

FMVSS 222, *School bus passenger seating and crash protection*, establishes occupant protection requirements for wheelchair users traveling on school buses (FMVSS 222). This regulation focuses primarily on vehicle requirement and requires that each wheelchair station on a school bus have not less than four wheelchair securement anchorages, arranged such that the wheelchair can be positioned in a forward-facing orientation. The wheelchair must be secured by a device that attaches to the wheelchair at two locations in the front and two in the rear. The securement device must be able to limit wheelchair movement, although limitations are not specified. Each of the vehicle anchorages must be able to withstand a statically applied load of 1376 kg (3033 lb). Wheelchair stations must also provide a lap and shoulder belt along with an anchorage for the shoulder belt and not less than two floor anchorages for the lap and shoulder belt combination. Each lap belt anchorage point must also be capable of withstanding a 1376 kg (3033 lb) static load, and the shoulder belt vehicle anchorage point must withstand a 688 kg (1516 lb) static load.

The requirements for wheelchair securement under ADA apply only to public transit vehicles, and the requirements of FMVSS 222 apply only to WTORS installed by the original equipment manufacturer (OEM) of the vehicle. Additionally, these federal standards require only static strength testing, which do not account for strength under dynamic loading that occurs in motor-vehicle crash environments. In most cases, WTORS are after-market products that are purchased and installed after the vehicle has been purchased. Furthermore, neither ADA nor FMVSS 222 address wheelchair design and performance when used as a vehicle seat, thereby leaving a significant gap in the occupant-protection "system" for people who remain seated in their wheelchairs when traveling in motor vehicles. Finally, there is no federal standard that applies to private vehicles, such as modified vans and minivans, used by wheelchair-seated drivers and passengers.

10.8.2 Voluntary Standards

The lack of federal regulations for after-market wheelchair securement and occupant restraint systems and for the wheelchair used as a motor vehicle seat has resulted in the development of voluntary industry standards. The voluntary nature of these standards means that their implementation by manufacturers is market driven; if there is a demand for products that meet the standards, these products will be designed and marketed to meet the demand. The standards follow the basic principles of occupant protection, strive to provide equipment that is comparable to manufacturer-installed federally regulated devices, and allows for use in all types of vehicles.

In the United States, efforts toward developing voluntary standards for WTORS began in the mid-1980s with the establishment of the Restraint Systems Task Force of the Society of Automotive Engineers (SAE) Adaptive Devices Subcommittee (ADSC). This Subcommittee and its various Task Forces were, in fact, the direct result of efforts by NIDRR to set up a mechanism for establishing requirements for the after-market equipment installed in vans that are modified for use by wheelchair-seated drivers and passengers. After more than ten years of effort, much of which involved working toward the development of similar WTORS standards within the International Standards Organization (ISO 10542 series) as well as in Canada (CSA Z605), SAE J2249 Wheelchair Tiedowns and Occupant Restraints for Use in Motor Vehicles was approved and published as an SAE recommended practice in 1999. Comparable WTORS standards have now been completed in ISO and Canada, and today there are many WTORS on the market that comply with the requirements of these standards.

Common to all of the WTORS standards is the requirement to dynamically test wheelchair tiedowns and associated occupant belt restraints using an 85 kg (187 lb) rigid surrogate wheelchair and a 76 kg (170 lb) midsize adult male ATD (crash dummy) to load the tiedowns and restraints in a 48 kph (30 mph), 20-g sled impact test. Figure 10.9 shows an SAE J2249 sled test utilizing the surrogate wheelchair and 50th percentile male ATD. Compliance with the standard requires that wheelchair and occupant excursions are within established limits and that there is no sign of WTORS failure. This level of testing is patterned after tests required by federal safety standards for passenger vehicles and is a "worst-case" frontal-impact crash pulse. The use of this worst-case crash pulse in these initial WTORS standards has been

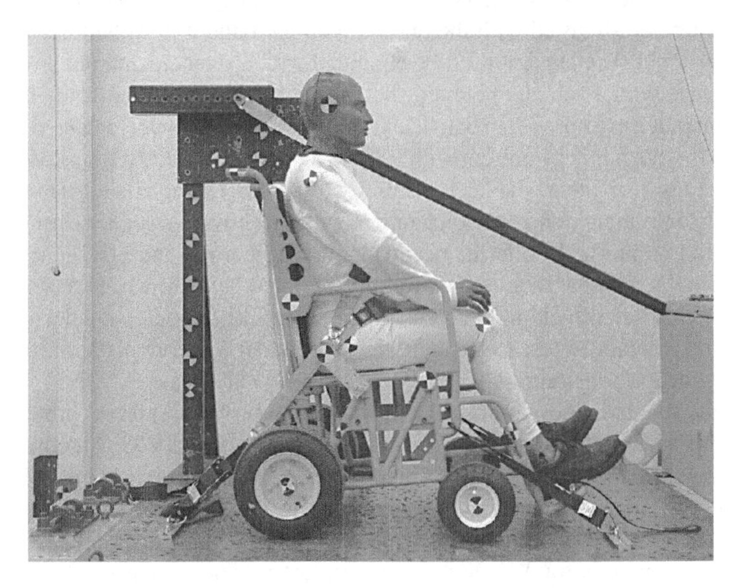

FIGURE 10.9 SAE J2249 frontal impact crash test.

based on the assumption that the WTORS manufacturer does not, and cannot, control the types and sizes of vehicles in which the WTORS equipment will be installed. It is important to understand, however, that SAE J2249 applies to WTORS that use all types of wheelchair securement systems, and does not specifically encourage or endorse any method of wheelchair securement. It does, however, require that a belt-type restraint system be included or specified by the WTORS manufacturer, and that occupant restraint must include both upper and lower torso restraints.

As SAE J2249 was nearing completion, it became clear to the members of the SAE Restraints Systems Task Force that the remaining missing link in the "system" to providing safe transportation to wheelchair-seated occupants of motor vehicles was the wide variety of wheelchairs that function as vehicle seats, but that have not been designed or tested for that application. As a result, a new standards-development effort was initiated within the wheelchair standards committee of ANSI/RESNA. Four years later, the group developed the first ANSI/RESNA wheelchair standard for wheelchairs used as a seat in a motor vehicle, Section 19 ANSI/RESNA WC/Vol 1 (ANSI/RESNA 2000), otherwise known as WC19. Harmonized wheelchair standards were later developed in ISO (7176/19) and Canada (CSA Z604).

The wheelchair standards require that a wheelchair provide four easily accessible securement points on the wheelchair frame, and that the wheelchair be dynamically tested in a simulated 48 kph (30-mph), 20-g frontal impact test while secured by a surrogate four-point strap-type wheelchair tiedown system. To achieve a successful dynamic test, ATD and wheelchair excursion must be within specified limits and the wheelchair must not show any visible evidence of failure. ATD excursion measures include forward and rearward head excursion, as well forward knee excursion. Seat integrity is assessed through a comparison of the ATD's pre-test to post-test H-point (hip) vertical displacement, with a limit of no more than a 20% difference. WC19 also requires that the wheelchairs be equipped with integrated lap belts. The wheelchair must be crash tested with the on-board lap belt and it must be made available to the consumer. The lap belt must be equipped with a standardized pin bushing that can interface with a vehicle-mounted shoulder belt. The integrated lap belt requirement is intended to improve lap belt fit and to avoid poor belt fit and reduce wheelchair interference that might be present with vehicle mounted restraint systems. WC19 also includes a test to evaluate the lateral stability of a transit wheelchair. The purpose of the test is to promote transit wheelchair designs that provide a stable seating support surface and comfortable ride for occupants during normal and emergency driving maneuvers. Despite the requirement for an integrated lap belt, wheelchairs may still require the use of vehicle-mounted occupant restraints in some vehicles, therefore WC19 also contains a test to evaluate the wheelchair's accommodation of vehicle mounted occupant restraints.

Future standards development efforts are underway to improve protection for wheelchair users in side and rear impacts. In addition, since WC19 applies to only complete wheelchairs with a base frame and seating system, efforts are underway to develop a standard to evaluate wheelchair seating system crashworthiness independent of the wheelchair frame using a surrogate wheelchair base. Work is also being done to develop standards for WTORS for use in large vehicles that see low-g impact conditions.

Throughout these standard development efforts, and subsequent to the completion of these initial WTORS and wheelchair standards, one thing had become very clear. In the absence of a serious effort on the part of federal governments to establish safety regulations for wheelchair-seated travelers, the process of developing voluntary industry standards is one of the most important and effective ways to achieve real-world improvements in safety and crash protection for wheelchair users. Furthermore, the development of these standards on an international level has been a key factor toward achieving worldwide harmonization of requirements for WTORS and wheelchairs, and has greatly benefited manufacturers and consumers on a global basis.

It has also become evident that developing a voluntary product standard is only the first step in getting products that conform to the requirements of the standard into the market place, and into the real world where they will benefit the wheelchair user and transportation providers. Once a voluntary standard has been completed and approved, a significant education and promotion effort is required to ensure effective interpretation and implementation of the requirements of the standards by manufacturers, primary users, clinicians, and others. In particular, because the impact of a voluntary standard is largely consumer driven, it is critical that wheelchair users, clinicians, and transit providers clearly understand the standards and the benefits offered by products that conform to them, in terms of reduced injury risk and increased operational efficiency. Only then, will they request, and even insist on purchasing, products that comply with these standards, and only then will the manufacturers be willing to design and test products to the requirements of the standards.

10.9 SUMMARY

The goal for wheelchair transportation is that all wheelchair users be able to attain a level of safety equivalent to that afforded to passengers and drivers seated in federally regulated, original equipment manufacturer's (OEM) vehicle seats. To achieve this goal requires that wheelchairs be capable of serving as a vehicle seat, that the wheelchair be secured to the vehicle and that the occupant be restrained with a crashworthy occupant restraint. This systems approach dictates that after-market adaptive devices are needed to provide safe wheelchair transport. Voluntary standards have been developed by ANSI/RESNA, SAE, and ISO to guide manufacturers in the design, development and testing of these devices. Clinical rehabilitation engineers should become familiar with these standards so that they are able to prescribe compliant products that for their clients. Utilizing products that comply with applicable product safety standards will provide consumers with the highest level of safety possible given current technology. In addition, rehabilitation engineers will then be able to educate consumers so that they are able to assure that proper wheelchair securement and occupant restraint techniques are employed. Despite the availability of products that comply with safety standards, challenges remain in achieving independent, quick and effective wheelchair securement and occupant restraint. Rehabilitation engineers can play a vital role in defining the technical problems that wheelchair users experience in transport and to developing new technologies to further improve current wheelchair transportation practices. The application of rehabilitation engineering principles to wheelchair transportation issues is imperative to assuring continued improvements in the safe and efficient transport for all wheelchairs users.

10.10 STUDY QUESTIONS

1. Postural support devices used by wheelchair users are often confused for or used as safety restraints in a motor vehicle. Provide two reasons why postural support devices may not be suitable as safety restraints in motor vehicles.
2. What is the primary cause of serious injury to a vehicle occupant?
3. The use of both an upper (shoulder) and a lower torso (pelvic) belt is necessary to prevent "submarining," which occurs when the occupant slides under the lap belt, allowing for forces to be applied to the more vulnerable soft tissues of the abdomen. What is the optimal position of the pelvic belt and shoulder belt in order to be effective?
4. What speed and rate of acceleration do wheelchair standards call for in a simulated frontal crash test?
5. What is the most common securement system found in public transportation and its disadvantages?
6. What are some of the barriers in the implementation of a universal docking securement system for wheelchairs used as seats in motor vehicles?
7. Describe some of the problems associated with using standard automotive occupant restraint systems that are installed in motor vehicles for wheelchair users.
8. Occupant restraints for the wheelchair-seated occupant can be categorized by where they anchor. Describe three common configurations.
9. Rear-facing passive wheelchair containment is a promising near-term solution for use in large, accessible transit vehicles (LATVs). What are some of the study findings that support this statement?
10. List at least four key provisions of ANSI/RESNA WC19.

BIBLIOGRAPHY

Americans with Disabilities Act of 1990 (PL101-336), 49CFR37: Transportation Services for Individuals with Disabilities (Act). Washington, D.C.: Federal Register.

Adams, T. C., Reger, S. I., Linden, M. A., & Kamper, D. G. 1995. Wheelchair user stability during simulated driving maneuvers. Paper presented at the RESNA Annual Conference, Vancouver, Canada.

Aibe, T., Watanabe, K., Okamoto, T., & Nakamori, T. 1982, Influence of occupant seating posture and size on head and chest injuries in frontal collision. SAE.

ANSI/RESNA 2000, Wheelchairs/Volume 1: Requirements and Test Methods for Wheelchairs (including Scooters). Section 19: Wheelchairs Used as Seats in Motor Vehicles, Arlington: American National Standards Institute (ANSI)/Rehabilitation Engineering Society of North America (RESNA).

Armstrong, C., Van Roosmalen, L., Manary, M., & Bertocci, G. E. 2003, User preferences on wheelchair integrated restraint design. Paper presented at the submitted to: Annual RESNA Conference, Atlanta, GE.

Bergman, E. & Johnson, E. 1995, Towards accessible human–computer interaction. In J. Nielsen (Ed.), *Advances in Human–Computer Interaction* (5th ed.): Ablex Publishing Co.

Bertocci, G. E., Digges, K., & Hobson, D. A. 1997, The effects of wheelchair securement point location on occupant injury risk. Paper presented at the Annual RESNA Conference.

Bertocci, G. E., Digges, K., & Hobson, D. A. 1996, Shoulder belt anchor location influences on wheelchair occupant crash protection. *Journal of Rehabilitation Research and Development*, 33, 279–289.

Bertocci, G. E., & Evans, J. 2000, Injury risk assessment of wheelchair occupant restraint systems in a frontal crash: a case for integrated restraints. *Journal of Rehabilitation Research and Development*, 37, 573–589.

Blennemann, F. 1995, German Experiences of Carrying Wheelchairs in Low Floor Buses. Paper presented at the 7th TRANSED International Conference on Mobility and Transport for Elderly and Disabled People, Reading, UK.

Bunai, Y., Nagai, A., Nakamura, I., & Ohya, I. 2001, Blunt pancreatic trauma by a wheelchair user restraint system during a traffic accident. *Journal of Forensic Science*, 46, 965–968.

CSA 2002, D435-02 Accessible transit buses (Standard No. D435-02). Mississauga, Ontario, L4W 5N6. Canada: Canadian Standards Association.

CSA Z605, Mobility Aid Securement and Occupant Restraint Systems for Motor Vehicles — Draft, CSA International Toronto, Canada, January, 2001.

CSA Z604, Transportable Mobility Aids — Draft, CSA International Toronto, Canada, January, 2001.

Dalrymple, G. D., Hsia, H., Ragland, C. L., & Dickman, F. B. 1990, Wheelchair and Occupant Restraint on School Buses, U.S. DOT-SC-NHTSA-90-1, Final Report NHTSA.

EZ LOCK, Inc. Model 6290. Baton Rouge, LA.

FMVSS 201, National Highway Traffic Safety Administration, Department of Transportation. Code of Federal Regulations, 49 CFR 571.201, Standard No. 201. U.S. Government Printing Office, Washington, D.C.

FMVSS 208, National Highway Traffic Safety Administration, Department of Transportation. Code of Federal Regulations, 49 CFR 571.208, Standard No. 208: Occupant Crash Protection (10-1-00 Edition), pp. 480–554. U.S. Government Printing Office, Washington, D.C.

FMVSS 222, National Highway Traffic Safety Administration, Department of Transportation. Code of Federal Regulations, 49 CFR 571.222, Standard No. 222: School Bus Passenger Seating and Crash Protection (10-1-00 Edition), pp. 670–681. U.S. Government Printing Office, Washington, D.C.

Hardin, J. A., Foreman, C., & Callejas, L. 2002, Synthesis of securement device options and strategies (Technical report No. 416-07). Tampa: National Center for Transit Research (NCTR).

Hunter-Zaworski, K. 1992, The development of an independent locking securement system for mobility aids on public transportation vehicles: Volume 2 (No. FTA-OR-11-0006-92-2): NTIS.

IDEA-97 Individuals with Disabilities Education Act of 1997, (PL105-17) 34CFR.300 Federal Register March 12, Volume 64, Number 48.

ISO 2001, 10542-1, Technical systems and aids for disabled or handicapped persons — Wheelchair tiedown and occupant-restraint systems — Part 1: Requirements and test methods for all systems. Geneva, Switzerland: International Organization for Standardization.

ISO 2001, 10542-2, Technical systems and aids for disabled or handicapped persons — Wheelchair tiedown and occupant-restraint systems — Part 2: Four-point strap-type tiedown systems. Geneva, Switzerland: International Organization for Standardization.

ISO 2001, 7176-19, Wheelchairs: Wheeled mobility devices for use in motor vehicles, Geneva, Switzerland: International Organization for Standardization.

ISO 2005. ISO 10542-3: Technical systems and aids for disabled or handicapped persons — Wheelchair tiedown and occupant-restraint systems — Part 3: Docking type tiedown systems. Geneva, Switzerland: International Organization for Standardization.

Law for Promoting Easily Accessible Public Transportation Infrastructure for the Aged and Disabled (a 2000 law number 68). (http://www.mlit.go.jo/barrierfree/transport-bf/transport-bf.html) (http://www.mlit,go,jp/kisha/kisha04/01/011026/ 02.pdf).

Linden, M. A., Kamper, D. G., Reger, S. I., & Adams, T. C. 1996, Transportation needs: Survey of individuals with disabilities. In: Proceedings of the RESNA 1996 Annual Conference; June 7–12; Salt Lake city, Utah. Arlington, VA: RESNA Press.

MDA DB2001(03): *Device Bulletin: Guidance on the Safe Transportation of Wheelchairs*, 2001, Medical Devices Agency, Department of Health.

Mercer, P. W., & Billing, J. R. 1990, Assessment of a transportable mobility aid in severe driving conditions: An exploratory test (No. CV-90-03). Ontario: Vehicle Technology Office, Transportation Technology and Energy Branch.

NCD 2005, The current state of transportation for people with disabilities in the United States, National Council on Disability, Washington, D.C.

NHTSA 2002, Traffic safety facts 2002: A compilation of motor vehicle crash data from the fatality analysis reporting system and the general estimates system, Washington, D.C., 2004, U.S. Department of Transportation.

RERC on Wheelchair Transportation Safety 2004. Tasks SP3a&b: Universal Interface and Docking Technologies (Progress Report). Pittsburgh, PA: University of Pittsburgh (RST), University of Michigan (UMTRI). www.rercwts.pitt.edu.

Rutenberg, U., & Association C. U. T. 2000, Accommodating mobility aids on Canadian low-floor buses using the rear-facing position design: experience, issues and requirements. (STRP Report 13, Nov. 2000). STRP Report 13.

SAE 1999, Recommended Practice J2249, Wheelchair Tiedown and Occupant Restraint Systems for Use in Motor Vehicles, Society of Automotive Engineers, Warrendale, PA.

Shaw, G., & Gillispie, T. 2003, Appropriate protection for wheelchair riders on public transit buses. *Journal of Rehabilitation Research and Development*, 40, 309–320.

Sprigle, S., Morris, B., Nowacek, G., & Karg, P. 1994, Assessment of adaptive transportation technology: a survey of users and equipment vendors. *Assistive Technology*, 6, 111–119.

TCRP-50 2003, Transit Cooperative Research Program (TCRP): Use of Rear-Facing Position for Common Wheelchairs on Transit Buses. Washington, D.C.: Transportation Research Board.

van Roosmalen, L., Bertocci, G. E., Ha, D., & Karg, P. 2002a, Wheelchair integrated occupant restraints: feasibility in frontal impact. *Medical Engineering and Physics*, 23, 685–696.

van Roosmalen, L., Bertocci, G. E., Hobson, D. A., & Karg, P. 2002b, Preliminary evaluation of wheelchair occupant restraint system usage in motor vehicles. *Journal of Rehabilitation Research and Development*, 39.

van Roosmalen, L., Bertocci, G. E., Karg, P., & Young, T. 1998, Belt fit evaluation of fixed vehicle mounted shoulder restraint anchors across mixed occupant populations. In: Proceedings of the RESNA 1998 Annual Conference; June 26–30; Minneapolis, MN. Arlington, VA: RESNA Press.

van Roosmalen, L., Manary, M., & Armstrong, C. 2003, Wheelchair occupant restraint issues for adult and pediatric wheelchair users. Paper presented at the 19th International Seating Symposium, Hyatt Orlando, Orlando, FL.

van Roosmalen, L., Reed, M, Bertocci, G. E. 2005, Wheelchair occupant restraint usability issues, *Assistive Technology*, 17, 23–36.

[illegible]

11 Rehabilitation Robotics

Dan Ding, Richard Simpson, Yoky Matsuoka, and Edmund LoPresti

CONTENTS

11.1 LEARNING OBJECTIVES OF THIS CHAPTER

Upon completion of this chapter, the reader will be able to:

- Be able to differentiate between an industrial robot and a rehabilitation robot
- Be able to discuss how the sensors (e.g., laser rangefinder, infrared, and sonar) function in obstacle avoidance and navigation of smart wheelchairs
- Be able to discuss the importance of shared control strategy vs. autonomous control in smart mobility aids
- Be able to list the advantages of robot-assisted therapy over the conventional therapy
- Understand the position control strategy of assistive robot manipulators
- Be able to discuss various robotic manipulation aids
- Be able to discuss the challenges that face implementation of rehabilitation robots
- Be able to discuss various therapeutic robot systems
- Be able to discuss the results of clinical studies on selected therapeutic robotics such as the Lokomat and their implications.

11.2 INTRODUCTION

Rehabilitation robotics is a combination of industrial robotics and medical rehabilitation, which encompasses many areas including mechanical, electrical, and biomedical engineering, prosthetics, autonomous mobility, artificial intelligence, and sensor technology (Prior and Warner, 1990). The success of robotics in the industrial arena opens up opportunities to significantly improve the quality of life of people with disabilities, including integration into employment, therapy augmentation, and so on. These opportunities could be realized to the full, provided rehabilitation robotic systems can be developed to meet the needs of people with disabilities in their daily living activities. Rehabilitation robotics can be categorized mainly as three kinds: mobility aids, manipulation aids, and therapeutic robots.

The use of robots in rehabilitation is quite different from industrial applications where robots normally operate in a structured environment with predefined tasks and are usually separated from human operators. Many tasks in rehabilitation robots cannot be preprogrammed, for example, pick up a newspaper or open a door. Furthermore, industrial robots are operated with specially trained personnel with a certain interest in the technology, whereas rehabilitation robots are usually used by people who have physical limitations. Rehabilitation robots have more in common with service robots which integrate humans and robots in the same task, requiring certain safety aspects and special attention to man–machine interfaces. Therefore, more attention must be paid to the user's requirements, as the user is a part of the process in the execution of various tasks (Bolmsjo et al., 1995).

The use of robots in rehabilitation occurred in 1960s, when a powered orthosis with four degrees of freedom (DOF) was used to move the user's paralyzed arm (Kim and Cook, 1985). There have been more efforts on investigating specific areas of rehabilitation robotics in recent years, and the field is technically well advanced.

However, rehabilitation robotics has been recognized overall as a technology playground for universities and academia, and it is penetrating the market very slowly and is still considered to be a "future technology." The major problem is the lack of evidence for usability and benefits of rehabilitation robots (Biihler, 1997). Although some evaluations and studies have been undertaken, the real benefits and disadvantages of systems in service need to be analyzed to better understand who the users are and what they actually need. In this chapter, we will focus on three kinds of rehabilitation robots, i.e., intelligent mobility aids, robotic manipulation aids, and therapeutic robots. We will discuss the latest technology in the field and analyze the benefits and disadvantages of each system.

11.3 INTELLIGENT MOBILITY AIDS

Although the needs of many individuals with disabilities can be satisfied with traditional mobility aids (e.g., canes, walkers, manual wheelchairs, power wheelchairs, scooters, etc.), there exists a segment of the disabled community who find it difficult or impossible to use traditional mobility aids independently. This population includes, but is not limited to, individuals with low vision, visual field neglect, spasticity, tremors, or cognitive deficits. Individuals in this population often lack independent mobility and are dependent on a caregiver to push them in a manual wheelchair.

To accommodate this population, researchers have used technologies originally developed for mobile robots to create *intelligent mobility aids* (IMAs). These devices typically consist of either a traditional mobility aid to which a computer and a collection of sensors have been added or a mobile robot base to which a seat and/or handlebars have been attached. IMAs have been designed based on a variety of traditional mobility aids and provide navigation assistance to the user in a number of different ways, e.g., assuring collision-free travel, aiding the performance of specific tasks, and autonomously transporting the user between locations.

11.3.1 SMART POWER WHEELCHAIRS

A useful way of classifying smart wheelchairs is based on the degree to which the components of the smart wheelchair are integrated with the underlying mobility device. The majority of smart wheelchairs that have been developed to date have been tightly integrated with the underlying power wheelchair, requiring significant modifications to function properly (Bourhis et al., 1998; Katevas et al., 1997; Levine et al., 1999; Simpson et al., 2002). A smaller number of smart wheelchairs have been designed as "add-on" units that can be attached and removed from the underlying power wheelchair (Fehr et al., 2000; Gomi and Ide, 1996).

Despite a long history of research in this area, there are very few smart wheelchairs currently on the market. Two North American companies, Applied AI and ActivMedia, sell smart wheelchair prototypes for use by researchers, but neither system is intended for use outside of a research lab. The CALL Centre smart wheelchair is sold in the UK and Europe by Smile Rehab, Ltd. (Berkshire, UK), as the "Smart Wheelchair." The "Smart Box," which is also sold by Smile Rehab in the UK and Europe, is compatible with wheelchairs using either Penny + Giles or Dynamics control electronics and includes bump sensors (but not sonar sensors) and the ability to follow tape tracks on the floor.

The CALL Centre has, by far, the most clinical experience in using smart wheelchairs (Nisbet et al., 1996). The CALL Centre uses a standard power wheelchair equipped with bump sensors and line-tracking sensors as an instructional tool for children who learn to operate a power wheelchair. Clients use the smart wheelchair to increase their navigation skills, until they reach the limit of their control potential (at which point they continue to use the smart wheelchair as a mobility aid) or reach the point where they are fully independent.

11.3.2 Smart Wheeled Walkers

There are several IMAs based on wheeled walkers (i.e., rollators) that are currently being developed. The goal of these devices is to provide the basic support of a traditional rollator coupled with the obstacle-avoiding capability of a mobile robot. Ideally, these devices function as a normal rollator most of the time, but provide navigational and avoidance assistance whenever necessary. An IMA that is currently making the transition from research project to commercial product is the PAM-AID (Lacey and MacNamara, 2000). The PAM-AID, which is marketed as the Guido (Figure 11.1), consists of a mobile robot base to which sonar sensors, a laser rangefinder, and a pair of handles (oriented like bicycle handles) have been added. The PAM-AID was developed to assist elderly individuals who have both mobility and visual impairments, and has two different control modes. In the *manual mode*, the user has complete control over the walker. Voice messages describing landmarks and obstacles are given to the user. In the *automatic mode*, the device uses the sensor information along with the user's input to negotiate a safe path around obstacles. The central processing unit controls motors that can direct the front wheels of the walker away from obstacles.

11.3.3 Form Factors

One obvious way to classify IMAs is form factor. Early smart wheelchairs (Bourhis et al., 1993; Connell and Viola, 1990) were actually mobile robots to which seats were added. The majority of smart wheelchairs that have been developed to date have been based on (heavily modified) commercially available power wheelchairs (Levine et al., 1999; Borgolte et al., 1998; Prassler et al., 2001; Katevas et al., 1997), whereas a smaller number of smart wheelchairs (Simpson et al., 2002, 2003, 2004; Miller and Slack, 1995; Mazo, 2001) have been designed as "add-on" units that can be attached and removed from the underlying power wheelchair. Similarly, smart walkers are still at the point in their evolution where several (Ulrich et al., 2001) are robots to which handles have been added, but the majority are based on commercially-available rollators (Wasson et al., 2003) or custom-built form factors (Lacey and MacNamara, 2000).

There are currently only two smart wheelchairs that are based on manual wheelchairs. The Collaborative Wheelchair Assistant controls the direction of a manual

FIGURE 11.1 The robotic walker Guido.

wheelchair with small motorized wheels that are placed in contact with the wheelchair's rear tires to transfer torque (Boy et al., 2002). The smart power assist module (SPAM) uses pushrim-activated power-assist wheelchair hubs in place of traditional rear wheels and equips with an obstacle avoidance function (Figure 11.2) (Cooper et al., 2001; Simpson et al., 2003).

11.3.4 Input Methods

Smart wheelchairs have been used to explore a variety of alternatives to the more "traditional" input methods associated with power wheelchairs (for example, joysticks and pneumatic switches). Voice recognition has often been used for smart wheelchairs (Levine et al., 1999; Katevas et al., 1997; Balcells et al., 1998) because of their low cost and the wide availability of commercial voice recognition hardware and software. More exotic input methods that have been implemented include detecting where the wheelchair user is looking through electro-oculogram (EOG) activity

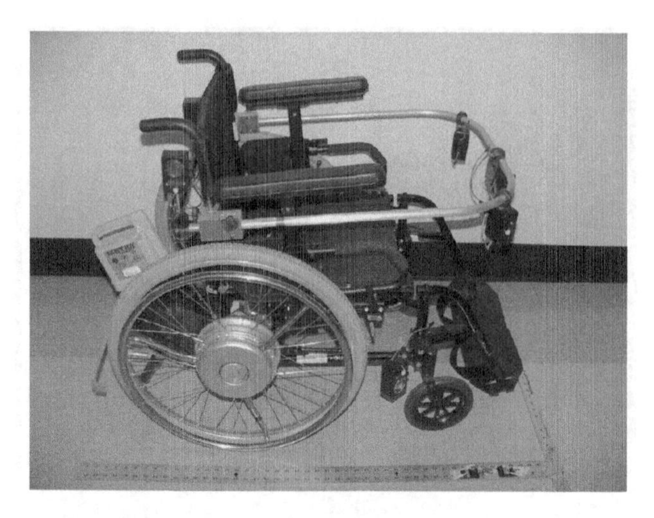

FIGURE 11.2 The smart power assist module (SPAM).

(Yanco, 2000; Mazo, 2001) or by using machine vision to calculate the position and orientation of the head (Kuno et al., 2003; Matsumoto et al., 2001).

Smart wheelchairs are excellent test beds for novel input methods because, unlike standard wheelchairs, smart wheelchairs have an onboard computer on which the interface can be implemented. More importantly, the obstacle avoidance provided by smart wheelchairs provides a safety net for input methods that are inaccurate or have limited bandwidth. Voice control, for example, has proven very difficult to implement successfully on standard wheelchairs (Miller et al., 1985; Clark and Roemer, 1977). On the NavChair, however, the obstacle avoidance capabilities built into the control software protect the user from the consequences of unrecognized (or misrecognized) voice commands (Simpson and Levine, 2002).

11.3.5 Sensors

To avoid obstacles, IMAs need sensors to perceive their surroundings. By far, the most frequently used sensor is the ultrasonic acoustic rangefinder (i.e., sonar). Sonar sensors are very accurate when the sound wave emitted by the sensor strikes an object head on. As the angle of incidence increases, however, the likelihood that the sound wave will not reflect back towards the sensor increases. This effect is more pronounced if the object is smooth or sound absorbent. Sonar sensors are also susceptible to "cross talk," in which the signal generated by one sensor produces an echo that is received by a different sensor.

Another frequently used sensing modality is the infrared rangefinder (i.e., IR). IR sensors emit light, rather than sound, and can therefore be fooled by dark (light absorbent) material rather than sound absorbent material. IR sensors also have difficulty with transparent or refractive surfaces. Despite their limitations, however, sonar and IR sensors are often used because they are small, inexpensive, and well understood.

Neither sonar nor IR sensors are particularly well suited to identifying drop-offs, such as stairs, curbs, or potholes. It is not uncommon for floors to be dark and smooth, meaning that both a sonar and an IR sensor would need to be facing almost straight down towards the ground in order to receive an echo. In this case, the IMA would not have enough warning time to stop.

More accurate obstacle and drop-off detection is possible with laser rangefinders, which provide a 180° scan within a two-dimensional plane of the distance to obstacles in the environment. Examples of IMAs that use a laser rangefinder include PAM-AID (Lacey and MacNamara, 2000), Rolland (Roefer and Lankenau, 2000), Maid (Prassler et al., 2001), and SENARIO (Katevas et al., 1997). Unfortunately, laser rangefinders are expensive, large, and consume lots of power, which make it difficult to mount enough rangefinders on an IMA to provide complete coverage.

Another option is a "laser striper," which consists of a laser emitter coupled with a CCD camera. The image of the laser stripe returned by the camera can be used to calculate distance to obstacles and drop-offs based on discontinuities in the stripe. The laser striper hardware is less expensive than a laser rangefinder, but can return false readings when the stripe falls on glass or a dark surface. To date, this system has not been used within an IMA.

A significant obstacle to bringing IMAs to the market is the need for sensors that are accurate, inexpensive, small, lightweight, impervious to environmental conditions (e.g., lighting, precipitation, and temperature), and require little power. Because no single sensor exists that meets these needs, many IMAs (e.g., VAHM, TAO, OMNI, and Rolland) combine information from multiple sensors to locate obstacles. In this way, the limitations of one sensor can be compensated by other sensors. Sonar and IR sensors are frequently used in combination for precisely this reason. When other sensors fail, the last line of defense is often the bump sensor, which is triggered when an IMA comes in contact with an obstacle.

Perhaps the most promising sensing modality is machine vision. Cameras are much smaller than laser rangefinders and, thus, much easier to mount in multiple locations on an IMA. Cameras can also provide much greater sensor coverage. The cost of machine vision hardware has fallen significantly — what used to require special cameras and digitizing boards (frame grabbers) can now be accomplished with a $20 USB Web cam — and machine vision software continues to improve, making successful implementation of an IMA based on computer vision increasingly likely. Some IMAs already use computer vision for landmark detection (e.g., Rolland, MAid, and CPWNS) and as a means of head- and eye-tracking for wheelchair control (Matsumoto et al., 2001; Moon et al., 2003; Mazo, 2001).

11.3.6 Control Software

Investigators have taken a variety of approaches to implementing control software for IMAs based on the functions supported by the IMA and the sensors it uses. The University of Plymouth (Bugmann et al., 1998) and the Chinese University of Hong Kong (Chow et al., 2002), for example, both developed smart wheelchairs that use neural networks to reproduce pretaught routes. The NavChair (Levine et al., 1999), on the other hand, uses an obstacle density histogram to combine information from its sonar sensors with joystick input from the user, and the SWCS (Simpson et al., 2004) and SPAM (Simpson et al., 2003) use rule-based approaches.

Several IMAs use subsumption control architectures (Brooks, 1986), in which primitive "behaviors" are coupled to produce more sophisticated emergent behavior (Gomi and Griffith, 1998; Connell and Viola, 1990; Li et al., 2000). Reactive control methods, such as subsumption, are occasionally used as the lowest layer of a multilayered control architecture. The reactive control layer provides sense–react behaviors that interact directly with the underlying hardware, whereas the upper layers of the architecture provide deliberative reasoning and control. Examples of IMAs that use a multilayered control architecture include VAHM, which uses a subsumption control approach at its lowest level (Bourhis et al., 2001), OMNI (Borgolte et al., 1998), and Rolland (Roefer and Lankenau, 2000).

11.3.7 Operating Modes

Some IMAs (Balcells et al., 1998) operate in a manner very similar to autonomous robots — the user gives the system a final destination and supervises as the smart wheelchair plans and executes a path to the target location. To reach their destination, these systems typically require either a complete map of the area through which they navigate or some modifications to their environment (e.g., tape tracks placed on the floor or markers placed on the walls), and they are usually unable to compensate for unplanned obstacles or to travel in unknown areas. IMAs in this category are most appropriate for users who (1) lack the ability to plan and/or execute a path to a destination and (2) spend the majority of their time within the same controlled environment.

Other IMAs confine their assistance to collision avoidance, and leave the majority of planning and navigation duties to the user (Levine et al., 1999; Miller and Slack, 1995). These systems do not normally require prior knowledge of an area or any specific alterations to the environment. They do, however, require more planning and continuous effort on the part of the user and are only appropriate for users who can effectively plan and execute a path to a destination. Finally, a third group of IMAs offers both autonomous and semiautonomous navigation (Bourhis et al., 2001; Katevas et al., 1997; Parikh et al., 2003).

Within the group of IMAs that provide semiautonomous navigation assistance, a subset offers multiple behaviors, each designed for a specific set of tasks and input methods. For example, the NavChair offers three distinct operating modes for (1) traversing a room while avoiding obstacles, (2) passing through doorways, and (3) following a wall down a hallway (Levine et al., 1999). Other IMAs that offer task-specific behaviors include Wheelesely (Yanco, 2000), Mister Ed (Connell and Viola, 1990), OMNI (Borgolte et al., 1998), and Rolland (Roefer and Lankenau, 2000). IMAs in this category are able to accommodate a wider range of needs and abilities but present the added requirement of selecting the most appropriate configuration for a given task.

The responsibility for selecting the most appropriate operating mode can be performed by the user (manual adaptation to changing task requirements) or the IMA (automatic adaptation). The TinMan smart wheelchair provides an example of manual adaptation (Miller and Slack, 1995). Users can change the setting of a dial to specify the amount of obstacle avoidance assistance provided by the chair. The NavChair (Levine et al., 1999) and the TAO systems (Gomi and Griffith, 1998), on the other hand, use automatic adaptation. The NavChair uses probabilistic reasoning techniques to combine information from the sonar sensors and a topological map to make adaptation decisions, whereas the TAO system uses a subsumptive reasoning system to allow the most appropriate behavior to emerge from a collection of potential behaviors.

11.3.8 Internal Mapping and Landmarks

IMAs that navigate autonomously to a destination often do so based on an internal map. The map can encode distance (in which case it is referred to as a *metric map*) or can be limited to specifying the connections between locations without any distance information (a *topological map*). There are, of course, other approaches to autonomous navigation that do not require an internal map, such as following tracks laid on the floor (Wakaumi and Nakamura, 1992).

A significant problem with using an internal map is unambiguously determining where the IMA is located within the map. A small number of IMAs (Gomi and Griffith, 1998; Li et al., 2000) use machine vision to identify naturally occurring landmarks in the environment, but the majority modify the environment to create "artificial" landmarks that can be easily identified and linked with a unique location. Most IMAs use machine vision to locate artificial landmarks, but other IMAs use radio beacons (Prassler et al., 2001; Balcells et al., 1998).

Several IMAs also make use of a "local" map that moves with the device (Levine et al., 1999; Roefer and Lankenau, 2000; Katevas et al., 1997). This map is often referred to as an "occupancy grid" or "certainty grid" and stores the location of obstacles in the IMA's immediate vicinity. Occupancy grids are used as the basis for many obstacle avoidance methods.

11.3.9 Future Mobility Aid Research

IMAs will remain a fertile ground for technological research for many years to come. They are excellent test beds for research in sensors, particularly machine vision, and also provide an opportunity to study human–robot interaction, adaptive and/or shared control, and novel input methods, such as voice, electroencephalogram (EEG), and eye gaze. Furthermore, IMAs will continue to serve as test beds for robot control architectures.

Although there has been a significant amount of effort devoted to the development of IMAs, there has been scant attention paid to evaluating their performance. Very few IMAs have involved people with disabilities in their evaluation activities. Furthermore, no IMA has been subjected to a rigorous, controlled evaluation that involves extended use in real-world settings. Conducting user trials is difficult for several reasons. Some users do not show any immediate improvement in navigation skills (measured in terms of average velocity and number of collisions) when using an IMA on a closed course in a laboratory setting. This could happen because the

IMA does not work very well, or because the user was already good enough that little improvement was possible. Users who have the potential to show large performance gains, on the other hand, often have little or no experience with independent mobility and may need a significant amount of training before they are ready to participate in valid user trials.

The primary obstacle to conducting long-term studies is the prohibitive hardware cost associated with constructing enough devices for such a study. Long-term studies are necessary, however, because the actual effects of using an IMA for an extended period of time are unknown. Some smart wheelchairs (Nisbet et al., 1996) are intended to be used as a means of developing the skills necessary to use standard wheelchairs safely and independently. Most investigators, however, have intended their IMA to be a person's permanent mobility solution or not addressed the issue at all. It is possible that using an IMA could actually *diminish* an individual's ability to use a standard wheelchair or wheeled walker, as that individual comes to rely on the navigation assistance provided by the IMA. Ultimately, it is likely that, for some users (particularly children), IMAs will be effective "training wheels" that can be used to teach the most basic mobility skills (e.g., cause and effect, starting and stopping on command), and for other users, IMAs will be permanent solutions.

The distinction between using an IMA as a *mobility aid*, a *training tool*, and an *evaluation instrument* is also worthy of study. Each of these functions is unique, and requires very different behaviors. As a mobility aid, the IMA's goal is to help the user reach a destination as quickly and comfortably as possible. Feedback to the user is kept to a minimum to avoid distractions, and collisions are to be avoided. As a training tool, on the other hand, the goal is to develop specific skills. In this case, feedback is likely to be significantly increased, and the extent to which the IMA complies with the user's input will be a function of the actual training activity. Finally, as an evaluation instrument, the IMA's goal is to record activity without intervention. In this case, there would possibly be no feedback or active navigation assistance to the user.

11.3.10 Summary of Robotic Mobility Aids

There are several barriers that must be overcome before IMAs can become widely used. A significant technical issue is the cost–accuracy trade-off that must be made with existing sensors. Until there is an inexpensive sensor that can detect obstacles and drop-offs over a wide range of operating conditions and surface materials, liability concerns will limit IMAs to indoor environments. To date, only a few IMAs have made efforts at operating outdoors, and most IMAs focus their efforts entirely on indoor environments. Only the Wheelesley (Yanco, 2000) has separate operating modes for indoor and outdoor navigation.

Another technical issue is the lack of a standard communication protocol for wheelchair input devices (e.g., joysticks and pneumatic switches) and wheelchair motor controllers. There have been several efforts to develop a standard protocol, for example, multiple masters multiple slave (M3S) (Linnman, 1996), but none has been adopted by the industry. A standard protocol would greatly simplify the task of interfacing smart wheelchair technology with the underlying wheelchair.

Even if these technical barriers are overcome, there still remain issues of clinical acceptance and reimbursement. Third-party payers are unlikely to reimburse clients for the expense of IMAs until they have been proven to be efficacious, if not cost-effective. Unfortunately, the evidence needed to prove efficacy will not exist until sufficient numbers of IMAs have been prescribed, and this will not be possible without adequate numbers of clinicians with training and expertise in the use of this technology. IMAs are expensive and complicated, and so the familiarization and training effort will require the extensive resources and infrastructure that only the durable medical equipment manufacturers possess.

This is not to imply, however, that IMAs cannot be commercialized. IMAs are ready, today, for use in indoor environments that have been modified to prevent access to drop-offs. These modifications can take the form of "baby gates," doors in front of stairwells, or ramps placed over single steps. The first IMA that may be commercially successful in North America is likely to be marketed as a device that can be operated independently indoors, but must be controlled by an attendant outdoors or in unmodified indoor environments. However, as sensor technology improves, the environments in which IMAs can safely operate will continue to expand.

11.4 ROBOTIC MANIPULATION AIDS

Robotic manipulation aids provide people with disabilities the tools to perform activities of daily living (ADL) and vocational support tasks that would otherwise require assistance from others. Such a manipulation aid is usually under control of its operator by a joystick, keypad, or voice or other input devices. Table 11.1 lists a number of robotic manipulation aids that were actively being developed at the time this chapter was written. Robotic manipulation aids can be classified into three groups:

Task-specific devices: Simple electromechanical devices used to perform simple tasks, e.g., powered feeders and page turners.

Workstation-based manipulation aids: The robotic manipulator could be built into a workstation and is suitable for use in a vocational environment.

Wheelchair mounted manipulation aids: With the robotic manipulator being mounted on an electric power wheelchair, such system can augment both mobility and manipulation, allowing the user to accomplish ADL tasks and vocational activities.

11.4.1 CONTROL INTERFACE

There are two aspects of control in a robot manipulation aid that affect its functionality: the interface between the user and the robot, and the interface between the robot and the objects being manipulated (Honzik, 2003).

The control interface between the user and the manipulation aid is a critical factor in the acceptance of such a system. Robot manipulation aids are usually controlled by the user through typical adaptive interfaces, e.g., joystick, keypad, or sip-and-puff

TABLE 11.1
Robotic Manipulation Aids

System	Distributor	Description
PROVAR (Van Der Loos et al., 1999)	VA Palo Alto	Desktop vocational assistive robot for high-level tetraplegia
Raptor	Applied Resources Corp., NJ	Wheelchair-mounted simple arm, joint-by-joint control from keypad
Assistive Robotic Manipulator (ARM) (it was called MANUS)	Exact Dynamics, The Netherlands	Seven-DOF arm mounted on a rotating and telescoping base, could be wheelchair mounted, has been extensively evaluated (Kwee et al., 2002), and is now being manufactured and marketed by Exact Dynamics
Winsford Feeder	Winsford Products Inc., NJ	Feed at a selected rate, controlled by either the chin switch or the plug-in rocker switch
My Spoon	SECOM Co., Ltd., Japan	Five-DOF manipulator arm and 1-DOF end-effector (spoon and fork) to assist eating. It can be operated in manual, semi-automatic mode via a joystick
RAID (Jones, 1999)	European TIDE (technology for the socio-economic integration of disabled and elderly people) program	Workstation-based arm, assists in office environments, manual mode, and autonomous grasping
Heidelberg Manipulator	The University of Heidelberg, Germany	Workstation-based arm, general-purpose pneumatic end effector and a separately controlled vacuum finger
HANDY 1	Rehab Robotics, Ltd., UK	Desktop robot arm to assist eating and other ADL tasks such as drinking, shaving, and teeth cleaning

device. However, some users may be limited in the control he/she is able to exert due to dexterity of the body parts that operate the device or the complexity of the operation. For example, a conventional 2D joystick is insufficient to adequately control the end-effector on a manipulator in a 3D space. Nonetheless, advanced technologies such as voice control, gesture recognition, brain computer interface, and teleoperation may minimize the user's burden and increase the quality of the user interface significantly. The robotic system FRIEND comprised of an electric wheelchair and the ARM (Exact Dynamics, the Netherland) can be commanded through a speed interface. The recognized words are shown on the screen attached to the wheelchair, which allows the user to monitor the system's behavior and react quickly if an error occurs (Martens et al., 2001). The user interface in PROVAR is a 3D graphical representation of the workspace including the robotic arm on a computer screen where the user can plan and examine tasks in the virtual world before the tasks are actually executed (Van Der Loos, 1999). Users could also simply use gestures (e.g., pointing with a laser pointer) to indicate a location or a desired object and uses speech to activate the system (Chen et al., 1996). Though these advanced human-machine interface technologies have demonstrated their advantages, they are usually subject to problems such as reliability and complexity (e.g., calibration and setup). It remains a challenge to design intuitive and effective interfaces between the user and the robot manipulation aid.

The interface between the robot manipulator and the objects being manipulated is usually a simple pincerlike gripper. Thus, the types of manipulations that can be performed could be limited. However, a large portion of activities we perform at school, work, and daily living involve pick-and-place tasks, which may be carried out with a simple end-effector. The end-effector could be equipped with sensors such as force-sensitive resistors and optical emitter/detector pairs to help detect objects and allow more robust manipulation (Van Der Loos, 1999).

11.4.2 Control Modes

The robotic manipulation aid can usually be controlled in two modes: joint and Cartesian modes. For the joint control mode, each joint of the arm will be controlled individually to achieve the destination. We can take the RAPTOR arm as an example (see Figure 11.3). It has four DOFs: flexion/extension of the shoulder, and rotation, flexion/extension of the elbow, and rotation of the wrist. The arm kinematical model should be developed so that the RAPTOR can reach objects on the floor, on a table, turn on switches or elevator buttons, etc. The arm kinematics includes forward kinematics and inverse kinematics. Forward kinematics deals with the problem of determining the end-effector position given all joint angles. To derive forward kinematics, Denavit–Hartenberg (D–H) parameters including link length, twist angle, offset, and joint angle are to be identified and transformation matrix describing the relationships between the manipulator adjacent links are to be formulated. Through forward kinematics, the joint control mode control can be realized. Users can specify the values of one or more joints and watch how the end of the arm moves. Currently, each joint of the RAPTOR is controlled independently in the joint space, which is easy to implement, but the user has to constantly adjust one or more joints to position the end-effector exactly where they want to have it. The joint control mode allows the

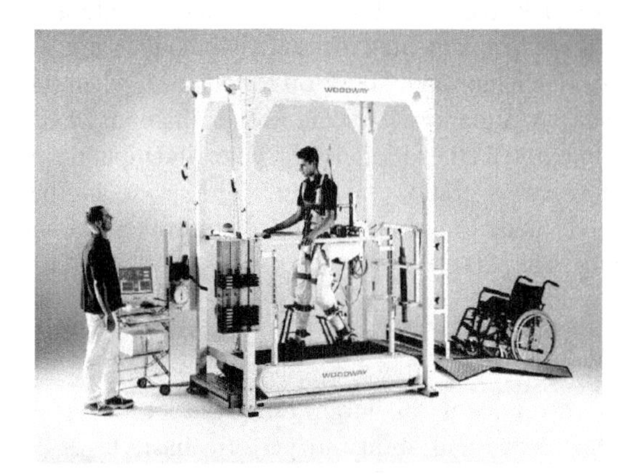

FIGURE 11.3 Components of the RAPTOR arm.

RAPTOR to adopt a simple open-loop control scheme without requiring encoders to measure joint positions.

For the Cartesian control, the position of the end-effector is controlled using a Cartesian coordinate system, and the orientation of the end-effector is controlled by rotations about the vertical, lateral and longitudinal axes of the end-effector. An inverse kinematical model is necessary for the continuous control of joints in the task space, where the required joint angles leading to a desired end-effector position and orientation are calculated. Users can simply set the end effector's position and orientation and let the arm follow to the destination point automatically. The ARM from the Netherland allows both task space control where the user can directly control the motion of the end-effector in Cartesian coordinates, and joint space control where joints are controlled independently. It also has a drinking mode where a drinking movement can be mimicked, and a folding mode where the arm can be unfolded before use and folded after use. In addition, several advanced modes have been implemented including a pilot mode, a cylindrical mode, a position mode, and a collaborative mode. In the collaborative control mode, a small camera is mounted on the end-effector of the arm and the image coming from the camera is used to control the arm (Tijsma et al., 2005). This approach could dramatically reduce the time to approach the object, but the required time to select the object in the camera view will be increased (Driessen et al., 2005).

11.4.3 Integrated Control

An integrated control system is a means whereby a single control interface (e.g., joystick, head switches, voice recognition system, keypad) is used to operate two or more assistive devices. Connecting one input device to two or more end-effectors (e.g., a wheelchair and a robotic manipulation aid mounted on a wheelchair) to form an integral aid may offer people with disabilities convenience to function as independently as possible. In an evaluation study of an early version MANUS, it was reported that users had difficulty in using two joysticks to control the wheelchair and the robot arm individually (Eftring and Boschian, 1999). The current MANUS system allows the same joystick to control both the wheelchair and the arm. An integrated control scheme was also implemented in a study where the same joystick was used to control an electric wheelchair and a desktop robotic arm when the wheelchair stays close to the workstation (Stefanov, 1999). However, there may exist an all-or-nothing problem with the integrated controller, that is, when the input device is broken, users may lose access to every operation mode. In addition, it could be cognitively more demanding to switch mode to use an integrated controller than simply use separate input devices (Ding et al., 2003).

11.4.4 Clinical Studies

Researchers at the University of Pittsburgh carried out some experiments to determine if functional independence and ability to perform basic tasks could be improved with the use of the RAPTOR. Level of independence with and without the device was recorded on 16 basic activities in 11 participants with tetraplegia. Significant improvement in task independence as well as in the time it took to complete tasks were seen in seven activities including picking up an item from the floor, reaching for items, and object manipulation (e.g., drinking, pouring water). Subjects became more proficient in using the RAPTOR after 13 h of training and practice for over half of the ADL activities (Chaves et al., 2003).

The ARM from the Netherland has been extensively evaluated in a number of studies. In one study, the researchers adapted the manipulator for seven children and young adults, and the control modes and procedures have been elaborated individually. They demonstrated the interactive development approach where the concept development, implementation, training and evaluation have been realized with the participation of the user. The result was satisfactory and two subjects have applied for such a manipulator for personal use (Kwee et al., 2002).

11.4.5 Safety

Since a robotic manipulation aid operate in close proximity to the user, the issue of safety and system reliability becomes very important. Existing manipulation aids are usually designed to have limit workspace volume and operate at a low velocity with low force, which inevitably results in limitations with these manipulation aids. For example, a recent survey among MANUS users showed that most of them find that the robot moves too slowly, is unable to handle heavy objects, and is not physically long enough to accomplish certain movement tasks (Gelderblom, 1999). However, designing much faster and more powerful manipulation aid is not a trivial problem that can be solved by merely implementing more powerful motors. Sensors such as machine vision and proximity sensors have been used to distinguish human and objects and automatically stop the robot when it moves too closely toward the user (Kawamura et al., 2000). A fail-safe force sensor was used to realize a safe force/torque feedback and enable the end-effector to touch the user with adequate force when necessary (Tejima et al., 2005). Zinn et al. (2004) have developed a prototype manipulator that reduces the impact loads associated with uncontrolled collision without compromising performance. The major source of actuation effort was relocated from the joint to the base to reduce inertia, and small, low inertia servomotors collocated at the joints helped maintain the torque capability.

11.4.6 Summary of Robotic Manipulation Aids

Similar to intelligent mobility aids, there are several barriers that must be overcome before the robotic manipulation aids can become widely used. Ease of use, economic considerations, safety, and effectiveness are critical elements for the acceptance. Robotic manipulation aids are still rather expensive. Though hardware such as computers, sensors, and vision systems are becoming affordable, robotic manipulation aids usually need to be customized to individual users, which could significantly increase the cost. There also exists considerable difficulty to have these robotic aids accepted by health care providers and insurance companies for reimbursement. Existing manipulation aids could help people with disabilities in various important daily living tasks, but they are still far from the stage of assisting users in the same range of tasks with the same speed and efficiency as humans can. Performance is usually compromised in favor of safety in a robotic manipulation aid, and users may feel that the robotic aid is not as effective as expected. Another issue is the lack of intuitive and effective user interface with learning capability and a better understanding of users' intentions. Nevertheless, the field of robotic manipulation aids has good technical and commercial potential. Much research and development on mechanism design, use of materials, large-scale clinical trials and cost-effectiveness analysis, and intelligent interface design is needed.

11.5 THERAPEUTIC ROBOTS

Traditional human-assisted therapy for mobility impairments has been augmented with therapeutic robots designed to administer different modes of therapy to the impaired limb. These robots are haptic interfaces, meaning that they allow a patient to interact with a computer by displacing or exerting force on the robot and by feeling forces exerted on them by the robot (force feedback). Patients also usually receive visual information about their performance from a computer monitor or another display. Robotic therapy for the upper limb can focus on movements of the shoulder, elbow, wrist, and/or hand, and robotic therapy for the low limb concentrates on restoration of normal gaits.

11.5.1 Therapeutic Robots for Upper-Limb Movements

Many of the robotic systems that have been used for rehabilitation of the upper limb have been designed to help patients relearn the skills needed to make reaching movements with the arm and the hand. These skills can be lost when someone experiences a stroke or other motor impairments. We will discuss nine different robotic systems in this section: MIT-MANUS, MIME, ARM Guide, GENTLE/s, HapticMASTER, WAM, BAM, PHANTOM™, and Rutgers Master II.

As of 2005, the MIT-MANUS (see Figure 11.4) is the most widely tested robot under therapeutic conditions. It is a 2-DOF robotic arm that consists of a five-bar-linkage SCARA (selective compliance assembly robot arm) (Krebs et al., 1998). A stroke patient moves a cursor on a computer screen to a target by moving the end point of the robotic arm in a horizontal plane. The robot is designed so that the patient experiences minimum impedance when moving the end point of the arm. The force provided by two brushless motors can be used to help the patient to complete the target reach if necessary (Krebs et al., 2000). This mode of operation is known as an active assisted mode. An extension to the MIT-MANUS allows about 19 in. of vertical movement in addition to movement in the horizontal plane (Krebs et al., 2004). During trials of MIT-MANUS, 56 stroke patients received about 4 hours per week of reaching practice with the robot in addition to a traditional human-assisted rehabilitation program (Volpe et al., 2001). After the four weeks, robot-treated patients showed greater gains in muscle control and muscle strength in the shoulder and elbow as compared to control patients who received standard rehabilitation with 1 to 2 h per week of exposure to the robot (robot motors were never turned on) (Volpe et al., 2001).

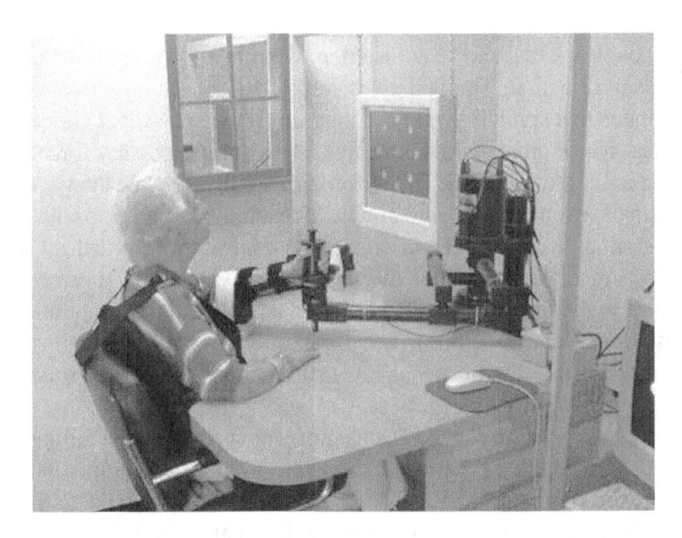

FIGURE 11.4 The MIT-MANUS (© 2004 Krebs et al; licensee BioMed Central Ltd.).

The "Mirror-Image Motion Enabler" (MIME) is another haptic interface designed for rehabilitation of the elbow and shoulder. The MIME system uses a 6-DOF industrial robot, the PUMA-560 robot arm (Lum et al., 2002). The forearm of the impaired limb rests in a splint that prevents movement of the wrist and hand; the splint is coupled to the robot arm via a six-axis force sensor. Using the MIME system, a patient can make reaching movements to targets in one of four modes (Burgar et al., 2000). In the passive mode, the patient relaxes while the robot moves the arm, and the active assisted mode is similar to that described earlier for the MIT-MANUS. In the active constrained mode, the patient moves in the target direction against a viscous (velocity-dependent) resistive force. The system uses springlike forces normal to the target direction to encourage the patient to move only in the desired direction. The fourth mode, the bilateral mode, is designed for hemiparetic patients or stroke patients who have one impaired limb and one normal limb. In this mode, a position sensor is attached to the unimpaired limb, and the system measures the movements that the patient makes with the unimpaired limb. The robot is then used to guide the movements of the impaired arm so that its movements mirror those of the normal arm. Burgar and coworkers hypothesized that these mirror image movements will stimulate activity in neural pathways important to stroke recovery (Burgar et al., 2000). As a test of the MIME system, 13 experimental subjects received 24 1-h reaching sessions with the robot, while 14 control subjects devoted an equal amount of time to practice of daily tasks with the impaired arm (Lum et al., 2002). The robot-treated group showed significantly greater gains in shoulder and elbow strength and reach.

The Assisted Rehabilitation and Measurement Guide (ARM Guide) is a simpler, relatively inexpensive system designed to provide therapy to improve reaching performance after stroke. This system has also been used to study various factors that impair reaching in stroke patients (Reinkensmeyer et al., 2000). The ARM Guide features a linear track that can be adjusted in two dimensions to allow reaching practice in a chosen direction. The patient makes 1-DOF reaching movements with the forearm strapped to a splint that moves along this track. The ARM Guide can operate in either a passive or an active assisted mode, with the necessary force provided by a chain drive and motor attached to the forearm splint (Reinkensmeyer et al., 2000). When a group of seven patients who practiced reaching movements with ARM Guide for 24 1-h sessions were compared with seven controls who practiced reaching for the same amount of time without the robot, the two groups showed comparable improvements in the time to complete functional tasks and the straightness of reaching movements (Kahn et al., 2001). The robot group also showed an improvement in trajectory smoothness, possibly because the robot imposed a smooth trajectory during practice.

The last robotic system for providing shoulder and elbow rehabilitation that will be discussed is the GENTLE/s system. In this system, a commercially available 3-DOF robot, the HapticMASTER from FCS Robotics, is attached to the wrist via a gimbal with three passive DOFs (Amirabdollahian et al., 2003). This system suspends the upper limb from a sling to eliminate gravity. The system can provide therapy in three possible modes: passive, active assisted, and an active mode similar to the active constrained mode of the MIME system (Amirabdollahian et al., 2002). When compared with task practice with the arm suspended in the sling, the GENTLE/s system seems to yield a greater rate of improvement for some physical variables such as shoulder range of motion and the ability to perform certain arm movements.

Just as the HapticMASTER from FCS Robotics is used for the GENTLE/s system, there are increasing numbers of commercially available haptic interfaces that are used for upper extremity rehabilitation. For example, WAM Arm from Barrett Technology, Inc. (http://www.barretttechnology.com), is a highly dexterous backdrivable 4- or 7-DOF manipulator. It is known for its dexterity, fast dynamics, zero backlash, and near-zero friction. Because of these superior capabilities that exceed any of the robotic devices used for rehabilitation, WAM Arm will be used frequently for rehabilitation in the future.

As the robotic devices become more accepted in the rehabilitation, especially domestic uses, safety will become a great concern. The robots listed here have active actuators (meaning the robot can move faster and apply forces larger than humanly possibly) that make them potentially unsafe in the rehabilitation environment. To overcome this safety issue, a rehabilitation robotic device called Brake Actuated Machine (BAM) has been developed (Matsuoka et al., 1999, 2001). BAM is a 6-DOF haptic robot with a workspace of approximately 2 m^3, large enough for full extension of the arm while taking a few steps. It utilizes dissipative brake actuators (it cannot store energy and thus cannot exceed the strength of the user) for motion control in order to safeguard against user injury.

Moving away from systems for the rehabilitation of the shoulder and elbow, Hesse and coworkers have constructed a robot designed to train pronation/supination of the forearm (twisting of the forearm about its long axis) and flexion/extension of the wrist (back and forth movement of the hand) (Hesse et al., 2003). Both of the patient's arms interact with the device. This haptic interface can move both limbs passively, or the unimpaired limb can be used to move the impaired limb in a mirrorlike fashion, similar to the bilateral mode of the MIME system. This device also has a bilateral mode in which the impaired arm must overcome an initial resistance before it follows the path of the unimpaired arm. The robot has been tested with 12 people with hemiparesis. The therapy reduced excessive tone (tightness of the muscles) in the impaired wrist and fingers, but these effects were not sustained at a 3-month follow-up. This may be because patients spent less time interacting with the system than in comparable studies.

Robotic systems for hand rehabilitation have not yet been widely investigated. Boian and coworkers have constructed a system that provides force feedback to the hand via a Rutgers Master II haptic glove (Boian et al., 2002). The Rutgers Master II uses four pneumatic actuators to apply force to the thumb and the index, middle, and ring fingers (Jack et al., 2001). This device also contains infrared and Hall effect sensors that can be used to estimate the posture of the hand. The system has been tested with four stroke patients, and improvements in range of motion and independence of movement were seen. These preliminary results also indicate that these improvements seem to cause improvements in patients' performance in a reaching and grasping task.

Moving forward, there are three areas that are important to address for therapeutic robotics:

1. Create a rehabilitation environment that is able to target as many patients as possible (including those with perceptual or cognitive deficits)
2. Attempt to take advantage of robots beyond their ability to repeat precise movements repetitively without fatigue
3. Provide an immersive task-oriented environment

Little has been done toward these goals thus far. Matsuoka and coworkers are creating a therapeutic environment called "rehabilitation by distortion" that satisfies all these three points (Brewer et al., 2005). In this environment, a patient with a paralyzed limb could immerse himself/herself in a virtual reality environment, while a comfortable and safe robot coupled to his/her paralyzed limb would exercise and monitor his/her progress. This environment takes advantage of the fact that virtual reality can provide visual feedback that is slightly different from reality. Due to the fact that people, in general, believe what they see over what they feel on the skin or the muscles, if the patient sees her virtual limb on the visual display moving more slowly than she had intended, she exerts more effort to move her actual limb faster. As a result, the patient moves further and more forcefully than she might have otherwise. When the virtual environment is removed, the patient's perception returns to normal but her muscles stay stronger and more coordinated.

This group uses the PHANTOM from Sensable Technologies, Inc. with six DOFs (three active and three passive). A preliminary therapy was conducted with an 8-year postinjury traumatic brain-injured patient (who has not improved with traditional therapy for several years) in a 6-week therapeutic session. This environment extended her finger range of motion from 40° to 68° and her maximum strength from 3 to 5 N. Hypertone was decreased from 2 to 1 on the Ashworth scale, and the results in standard therapeutic tests (such as AMAT and ARAT) were clinically noticeable.

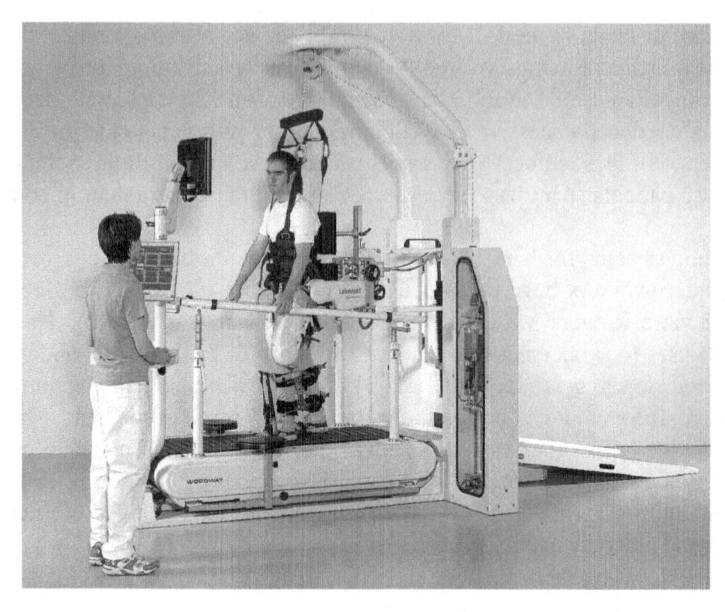

FIGURE 11.5 The Lokomat Robotic Gait Orthosis (Lokomat®, Hocoma AG: www.hocoma.com).

11.5.2 Therapeutic Robots for Low-Limb Movements

The Lokomat system (Hocoma AG, Volketswil, Switzerland), is the most developed commercially available robot-assisted therapy system for lower limbs (Figure 11.5). It is a bilateral robotic orthosis that incorporates computers, a treadmill, and a body-weight support system to control the leg movements in the sagittal plane. The Lokomat's hip and knee joints are actuated by linear back-drivable actuators integrated into an exoskeletal structure. Passive foot lifter assists ankle dorsiflexion during the swing phase. The legs of the patient, which are fixed to the exoskeleton by straps, are moved according to a position control strategy with predefined hip and knee joint trajectories. The built-in position and force sensors can help assess physiological and neurological status of patients, providing objective assessment of the patients in a repeatable manner and with minimal efforts from therapists. One of the limitations of the Lokomat is that it only focuses on the leg movements and restricts pelvic rotation, pelvic obliquity, and horizontal translation of the pelvis, which plays an important role in normal gait.

Mirbagheri and coworkers studied the effects of Lokomat training on the neuromuscular mechanical properties and voluntary movement of the spastic ankle in persons with incomplete Spinal Cord Injury. They found that reflex stiffness was significantly reduced following Lokomat training, and active range of motion, peak velocity, and peak acceleration of voluntary movement increased (Mirbagheri et al., 2005). Dietz and Muller conducted a study where three uninjured individuals and 16 with complete paraplegia or quadriplegia received gait training on a Lokomat for 2–3 times 1-h session per week for 3 months. It was found that people without injury and people with an injury less than 1 year have more muscle activity after the training.

Gait Trainer developed in Germany is based on a double crank and rocker gear system. A subject stands on the two footplates while being suspended on a weight-relief support system. The center of mass of the subject is controlled along fixed vertical and horizontal trajectories (Hesse and Uhlenbrock, 2000). Gait Trainer imposes gait-like, repetitive practice of lower limbs through exercising fixed trajectories of feet by means of both footplates. A manual assistance from therapist is needed to assist in proper movement of the knee. So the Gait Trainer might not be suitable for non-ambulatory people with weak muscles but only for those that have some degree of control of the knee joints.

11.5.3 Comparison between Conventional and Robotic Therapy

Robotic therapy may improve the quality of upper limb rehabilitation both by addressing current problems in therapy, such as the limited amount of time patients can participate in rehabilitation, and by making new types of therapy possible. For stroke patients, previous research has shown that intensive therapy for an impaired limb can improve the amount and quality of use of the impaired limb in daily life, and patients can make improvements even years after the stroke has occurred (Taub et al., 1999). Augmenting the limited number of human therapists with robotic devices can allow more patients to receive sufficient therapy to reach optimal outcomes. The use of robots in therapy could reduce the potential for therapist injury that exists when therapists manually assist patients in weight-bearing activities such as walking. Robotic devices could also enable rehabilitation to occur in the home with remote supervision by a human therapist.

In addition, robots can facilitate new types of therapy not feasible for a human therapist to provide. Robots offer a unique opportunity to minutely measure a patient's movement, allowing therapists to better evaluate an impairment and track changes in performance during recovery. Robots can also provide precise force perturbation profiles to help patients relearn specific target movements (Patton et al., 2004). Researchers are investigating whether the feedback given to patients during robotic therapy can be manipulated to improve the outcome of rehabilitation (Wei et al., 2005; Brewer et al., 2005). For example, the rehabilitation by distortion method described earlier manipulates the visual feedback to the patients. This is something that cannot be done in conventional therapy; and something that takes advantage of the robot's capabilities beyond the ability to repeat precise movements. In the future, rather than having robots act as automated human therapists, they should become invaluable assistants to the therapists in the way that another human therapist could not be.

11.5.4 Challenges ahead for Therapeutic Robots

Although robotic systems have been clinically shown to improve such parameters as patient strength and range of motion, many challenges remain before the use of robots in therapy will become widespread. Because some robots have the capability to exert large amounts of force, their ability to interact safely with impaired patients must be proved before anything else. In addition, to gain large-scale acceptance in therapy, robotic therapy must also be shown to be effective for large numbers of patients. The MIT-MANUS, the most widely tested robot for upper limb rehabilitation, has been evaluated in several trials with a total of over 250 stroke subjects over a period of 10 years and has shown gains in muscle strength and control with robotic therapy (Krebs et al., 2004; Volpe et al., 2001), but more evaluation is necessary before the robot could be approved for use in standard therapy clinics. Other robots have been tested with fewer subjects and face a longer period of time before widespread clinical use.

To compound these difficulties, most robots have been evaluated based on changes in low-level motor abilities, such as strength, rather than the patient's ability to function in everyday life. Such a holistic, functional goal is ultimately more important from a rehabilitation perspective. Robotic systems have not yet been shown to significantly impact a patient's functional ability. Evaluation of robotic therapy is also complicated by the wide variety of techniques used in conventional therapy; this variety makes it hard to compare robotic therapy to "standard" rehabilitation. Finally, most testing to date has been done with stroke subjects, and the effects of robotic therapy on other impairments remain unclear.

Even after researchers show the safety and efficacy of robotic therapy in general, they must persuade therapists to adopt the new technologies. Many therapists may not consistently follow current developments in the field, and the safety and benefits of robotic therapy will have to be convincingly argued at the level of individual therapists and clinics. Economic arguments will have to be made to insurance companies justifying the expense of robotic systems. After these hurdles are overcome, robotic therapy has the potential to improve the outcome of rehabilitation for many impaired patients.

Robotic technology is playing an important role in rehabilitative medicine. Rehabilitative locomotor training for patients with spinal cord injuries or stroke-related paralysis can be very labor-intensive, and may require three or four physical therapists for one patient intervention. Robot-assisted therapy might offer additional benefits compared to conventional physical therapy or locomotor training using physical therapists. The real perceived advantage of the robotic systems is that they can alleviate the labor-intensive aspect of this rehabilitation and ultimately cut treatment costs and avoid potential staffing shortages. Different robotic systems are now in various stages of development and testing to determine whether use of robots offers an additional clinical benefit beyond traditional physical therapy.

11.6 STUDY QUESTIONS

1. Which populations are most likely to benefit from IMAs?
2. Discuss the advantages, disadvantages, and applications of potential sensing technologies appropriate for obstacle avoidance in an IMA.
3. Why are valid trials of IMAs so difficult to run?
4. Compare and contrast the hardware of two therapeutic robots for the upper extremity.

5. What kind of clinical results support the use of robotics in rehabilitation?
6. What are the advantages of therapeutic robots in comparison with conventional therapy?
7. Discuss the safety measures that are used or can be used in a robotic manipulation aid.
8. Discuss the control mode of MANUS (or ARM) from the Netherland.

BIBLIOGRAPHY

Amirabdollahian F, Gradwell E, Loureiro R, Collin C, Harwin W. Effects of the GENTLE/s robot mediated therapy on the outcome of upper limb rehabilitation post-stroke: analysis of the Battle Hospital data. *The 8th International Conference on Rehabilitation Robotics*, 2003; 55–58.

Amirabdollahian F, Loureiro R, Harwin W. Minimum jerk trajectory control for rehabilitation and haptic applications. *IEEE International Conference on Robotics and Automation*, 2002; 3380–3385.

Balcells AC, Gonzalez J. TetraNauta: a wheelchair controller for users with very severe mobility restrictions. *Proceedings of the 3rd Annual TIDE Congress*, 1998.

Biihler C. Robotics for rehabilitation: factors for success from a European perspective. *Rehabilitation Robotics Newsletter of the Applied Science & Engineering Laboratories at the University of Delaware*, 1997. Accessed on October 12, 2005 at http://www.asel.udel.edu/robotics/newsletter/.

Boian R, Sharma A, Han C, Merians A, Burdea G, Adamovich S, Recce M, Tremaine M, Poizner H. Virtual reality-based post-stroke hand rehabilitation. *Studies in Health Technology Information*, 2002;85:64–70.

Bolmsjo G, Neveryd H, Eftring H. Robotics in rehabilitation. *IEEE Transactions on Rehabilitation Engineering*, 1995;3:77–83.

Borgolte U, Hoyer H, Buehler C, Heck H, Hoelper R. Architectural concepts of a semi-autonomous wheelchair. *Journal of Intelligent and Robotic Systems: Theory and Applications*, 1998;22:233–253.

Bourhis G, Agostini Y. The VAHM robotized wheelchair: System architecture and human-machine interaction. *Journal of Intelligent and Robotic Systems*, 1998;22: 39–50.

Bourhis G, Horn O, Habert O, Pruski A. An autonomous vehicle for people with motor disabilities. *IEEE Robotics & Automation Magazine*, 2001;8:20–28.

Bourhis G, Moumen K, Pino P, Rohmer S, Pruski A. Assisted navigation for a powered wheelchair. *IEEE International Conference on Systems, Man and Cybernetics*, 1993; 553–558.

Boy ES, Teo CL, Burdet E. Collaborative wheelchair assistant. *IEEE/RSJ International Conference on Intelligent Robots and Systems*, 2002; 1511–1516.

Brewer B, Klatzky R, Matsuoka Y. Perceptual limits for a robotic rehabilitation environment using visual feedback distortion. *IEEE Transactions on Neural Systems and Rehabilitation Engineering*, 2005;13:1–11.

Brooks R. A robust layered control system for a mobile robot. *IEEE Journal of Robotics and Automation*, 1986;2(1):14–23.

Bugmann G, Koay KL, Barlow N, Phillips M, Rodney D. Stable encoding of robot trajectories using normalised radial basis functions: application to an autonomous wheelchair. *International Symposium on Robotics Automation and Robotics*, 1998; 232–235.

Burgar CG, Lum PS, Shor PC, Van der Loos MHF. Development of robots for rehabilitation therapy: the Palo Alto VA/Stanford experience. *Journal of Rehabilitation Research and Development*, 2000;37:663–673.

Chaves E, Koontz AM, Garber S, Cooper RA, Williams AL. Clinical evaluation of a wheelchair mounted robotic arm. *RESNA Proceedings*, 2003.

Chen S, Kazi Z, Beitler M, Salganicoff M, Chester D, Foulds R. Gesture-speech based HMI for a rehabilitation robot. *Proceedings of IEEE Southeastcon*, 1996; 29–36.

Chow HN, Xu Y, Tso SK. Learning human navigational skill for smart wheelchair. *IEEE/RSJ International Conference on Intelligent Robots and Systems*, 2002; 996–1001.

Civit-Balcells A, del Rio FD, Jimenez G, Sevillano JL, Amaya C, Vicente S. SIRIUS: improving the maneuverability of powered wheelchairs. *Proceedings of the 2002 IEEE International Conference on Control Applications*, 2002; 790–795.

Clark JA, Roemer RB. Voice controlled wheelchair. *Archives of Physical Medicine and Rehabilitation*, 1977;58:169–175.

Connell J, Viola P. Cooperative control of a semi-autonomous mobile robot. *IEEE International Conference on Robotics and Automation*, 1990; 1118–1121.

Cooper R, Corfman T, Fitzgerald S, Boninger M, Spaeth D, Ammer W, Arva J. Performance assessment of a pushrim activated power assisted wheelchair. *IEEE Transactions of Control Systems Technology*, 2001;10(1):121–126.

Ding D, Cooper RA, Kamniski BA, Kanaly JR, Allegretti A, Chaves E, Hubbard S. Integrated control of assistive devices and related technology. *Assistive Technology*, 2003;15: 89–97.

Driessen B, Liefhebber F, Kate TT, Woerden KV. Collaborative control of the MANUS manipulator. *IEEE International Conference on Rehabilitation Robotics*, 2005; 247–251.

Dubowsky S, Genot F, Godding S, Kozono H, Skwersky A. Yu H, Yu LS. PAMM — a robotic aid to the elderly for mobility assistance and monitoring: a ‘helping hand’ for the elderly. *IEEE International Conference on Robotics and Automation*, 2000; 570–576.

Eftring H, Boschian K. Technical results from MANUS user trials. *IEEE International Conference on Rehabilitation Robotics*, 1999; 136–141.

Fehr L, Langbein WE, Skaar SB. Adequacy of power wheelchair control interfaces for persons with severe disabilities: a clinical survey. *Journal of Rehabilitation Research and Development*, 2000;37:353–360.

Gelderblom GJ. Manus manipulator use profile. *Proceedings on AAATE SIG1 Workshop of Rehabilitation Robotics*, 1999; pp. 8–14.

Gomi T, Griffith A. Developing intelligent wheelchairs for the handicapped. *Assistive Technology and Artificial Intelligence*, 1998.

Gomi T, Ide K. The development of an intelligent wheelchair. *Proceedings of IEEE Symposium on Intelligent Vehicles*, 1996; 70–75.

Gribble WS, Browning RL, Hewett M, Remolina E, Kuipers BJ. Integrating vision and spatial reasoning for assistive navigation. *Assistive Technology and Artificial Intelligence*. New York, NY: Springer; 1998; pp. 179–193.

Hans M, Graf B, Schraft RD. Robotic home assistant Care-O-bot: Past–present–future. *IEEE International Workshop on Robot and Human Communication*, 2002; 380–385.

Hesse S, Schulte-Tigges G, Konrad M, Bardeleben A, Werner C. Robot-assisted arm trainer for the passive and active practice of bilateral forearm and wrist movements in hemiparetic subjects. *Archives in Physical Medicine Rehabilitation*, 2003;84:915–920.

Hesse S, Uhlenbrock D. A mechanized gait trainer for restoration of gait. *Journal of Rehabilitation Research and Development*, 2000;37:701–708.

Honzik B. Inverse kinematics and control of the assistive robot for disabled. Accessed on August 21, 2006 at http://www.uamt.feec.vutbr.cz/robotics.

Jack D, Boian R, Merians AS, Tremaine M, Burdea GC, Adamovich SV, Recce M, Poizner H. Virtual reality-enhanced stroke rehabilitation. *IEEE Transactions on Neural Systems and Rehabilitation Engineering*, 2001;9:308–318.

Jones T. RAID — Toward greater independence in the office & home environment. *ICORR*, 1999; 201–206.

Kahn LE, Zygman ML, Rymer WZ, Reinkensmeyer DJ. Effect of robot-assisted reaching and unassisted exercise on functional reaching in chronic hemiparesis. *Annual International Conference of the IEEE Engineering in Medicine and Biology Society*, 2001; 1344–1347.

Katevas NL, Sgouros NM, Tzafestas SG, Papakonstantinou G, Beattie P, Bishop JM, Tsanakas P, Koutsouris D. The autonomous mobile robot SENARIO: A sensor-aided intelligent navigation system for powered wheelchairs. *IEEE Robotics and Automation Magazine*, 1997;4:60–70.

Kawamura K, Peters II RA, Wilkes MW, Alford WA, Rogers TE. ISAC: foundations in human-humanoid interaction. *IEEE Intelligent Systems and their Applications*, 2000; 38–45.

Kim Y, Cook AM. Manipulation and mobility aids, *Electronic Devices for Rehabilitation* (J.G. Webster et al., Eds). London, Chapman Hall, 1985.

Krebs HI, Ferraro M, Buerger SP, Newbery MJ, Makiyama A, Sandmann M, Lynch D, Volpe BT, Hogan N. Rehabilitation robotics: pilot trial of a spatial extension for MIT-Manus. *Journal of Neuroengineering Rehabilitation*, 2004;1:5.

Krebs HI, Hogan N, Aisen ML, Volpe BT. Robot-aided neurorehabilitation. *IEEE Transactions on Rehabilitation Engineering*, 1998;6:75–87.

Krebs HI, Volpe BT, Aisen ML, Hogan N. Increasing productivity and quality of care: robot-aided neuro-rehabilitation. *Journal of Rehabilitation Research and Development*, 2000;37:639–652.

Kuno Y, Shimada N, Shirai Y. Look where you're going. *IEEE Robotics and Automation Magazine*, 2003: 26–34.

Kwee H, Quaedackers J, van de Bool E, Theeuwen L, Speth L. Adapting the control of the MANUS manipulator for persons with cerebral palsy: an exploratory study. *Technology and Disability*, 2002;14:31–42.

Lacey G, MacNamara S. Context-aware shared control of a robot mobility aid for the elderly blind. *International Journal of Robotics Research*, 2000;19:1054–1065.

Levine SP, Bell DA, Jaros LA, Simpson RC, Koren Y, Borenstein J. The NavChair Assistive Wheelchair Navigation System. *IEEE Transactions on Rehabilitation Engineering*, 1999;7:443–451.

Li X, Zhao X, Tan T. A behavior-based architecture for the control of an intelligent powered wheelchair. *IEEE International Workshop on Robot and Human Communication*, 2000; 80–83.

Linnman S. M3S: The local network for electric wheelchairs and rehabilitation equipment. *IEEE Transactions on Neural Systems and Rehabilitation Engineering*, 1996;4(3): 188–192.

Lum PS, Burgar CG, Shor PC, Majmundar M, Van der Loos M. Robot-assisted movement training compared with conventional therapy techniques for the rehabilitation of upper-limb motor function after stroke. *Archives in Physical Medicine Rehabilitation*, 2002;83:952–959.

Martens C, Ruchel N, Lang O, Ivlev O, Graser A. A FRIEND for assisting handicapped people. *IEEE Robotics and Automation Magazine*, 2001; 57–65.

Matsumoto Y, Ino T, Ogasawara T. Development of intelligent wheelchair system with face and gaze based interface. *IEEE International Workshop on Robot and Human Communication*, 2001; 262–267.

Matsuoka Y, Miller LC. Domestic rehabilitation and learning of task-specific movements. *Sixth International Conference on Rehabilitation Robotics*, 1999; 177–182.

Matsuoka Y, Townsend WT. Design of life-size haptic environments. *Experimental Robotics VII*, 2001; 461–470.

Mazo M. An integral system for assisted mobility. *IEEE Robotics & Automation Magazine*, 2001;8:46–56.

Miller GE, Brown TE, Randolph WR. Voice controller for wheelchairs. *Medical and Biological Engineering and Computing*, 1985;23:597–600.

Miller DP, Slack MG. Design and testing of a low-cost robotic wheelchair prototype. *Autonomous Robots*, 1995;2:77–88.

Min CT, Luo RC. Multilevel multi-agent based team decision fusion for mobile robot behavior control. *The 3rd World Congress on Intelligent Control and Automation*, 2000; 489–494.

Mirbagheri MM, Tsao C, Pelosin E, Rymer WZ. Therapeutic effects of robotic-assisted locomotor training on neuromuscular properties. *IEEE International Conference on Rehabilitation Robotics*, 2005; 561–564.

Moon I, Lee M, Ryu J, Mun M. Intelligent robotic wheelchair with EMG-, gesture-, and voice-based interfaces. *IEEE/RSJ International Conference on Intelligent Robots and Systems*, 2003; 3453–3458.

Mori H, Kotani S. A robotic travel aid for the blind. *Intelligent Autonomous Systems*, 1998.

Murakami Y, Kuno Y, Shimada N, Shirai Y. Collision avoidance by observing pedestrians' faces for intelligent wheelchairs. *IEEE/RSJ International Conference on Intelligent Robots and Systems*, 2001; 2018–2023.

Nisbet PD, Craig J, Odor JP, Aitken S. "Smart" wheelchairs for mobility training. *Technology and Disability*, 1996;5:49–62.

Parikh SP, Rao RS, Jung S-H, Kumar V, Ostrowski JP, Taylor CJ. Human–robot interaction and usability studies for a smart wheelchair. *IEEE/RSJ International Conference on Intelligent Robots and Systems*, 2003; 3206–3211.

Patton JL, Mussa-Ivaldi FA. Robot-assisted adaptive training: custom force fields for teaching movement patterns. *IEEE Transactions on Biomedical Engineering*, 2004;51: 636–646.

Pires G, Nunes U. A wheelchair steered through voice commands and assisted by a reactive fuzzy-logic controller. *Journal of Intelligent and Robotic Systems: Theory and Applications*, 2002;34:301–314.

Prassler E, Scholz J, Fiorini P. A robotic wheelchair for crowded public environments. *IEEE Robotics & Automation Magazine*, 2001;8:38–45.

Prior SD, Warner PR. A review of world rehabilitation robotics research. *IEE Colloquium on High-Tech Help for the Handicapped*, 1990; 1–3.

Reinkensmeyer DJ, Kahn LE, Averbuch M, McKenna-Cole A, Schmit BD, Rymer WZ. Understanding and treating arm movement impairment after chronic brain injury: progress with the ARM guide. *Journal of Rehabilitation Research and Development*, 2000;37:653–662.

Roefer T, Lankenau A. Architecture and applications of the Bremen autonomous wheelchair. *Information Sciences*, 2000;126:1–20.

Schilling K, Roth H, Lieb R, Stutzle H. Sensors to improve the safety for wheelchair users. *Proceedings of the 3rd Annual TIDE Congress*, 1998.

Seki H, Kobayashi S, Kamiya Y, Hikizu M, Nomura H. Autonomous/semi-autonomous navigation system of a wheelchair by active ultrasonic beacons. *International Conference on Robotics and Automation*, 2000; 1366–1371.

Simpson RC, Levine SP. Voice control of a powered wheelchair. *IEEE Transactions on Neural Systems and Rehabilitation Engineering*, 2002;10:122–125.

Simpson RC, LoPresti EF, Hayashi S, Guo S, Ding D, Cooper RA. Smart power assistance module for manual wheelchairs. *Proceedings of Rehabilitation Engineering and Assistive Technology Society of North America*, 2003.

Simpson RC, LoPresti EF, Hayashi S, Nourbakhsh IR, Miller DP. The Smart Wheelchair Component System. *Journal of Rehabilitation Research and Development*, 2004;41:429–442.

Simpson RC, Poirot D, Baxter MF. The Hephaestus Smart Wheelchair system. *IEEE Transactions on Neural Systems and Rehabilitation Engineering*, 2002;10:118–122.

Stefanov D. Integrated control of desktop mounted manipulator and a wheelchair. *IEEE International Conference on Rehabilitation Robotics*, 1999; 207–215.

Taub E, Uswatte G, Pidikiti R. Constraint-induced movement therapy: a new family of techniques with broad application to physical rehabilitation — a clinical review. *Journal of Rehabilitation Research and Development*, 1999;36:237–251.

Tejima N, Stefanov D. Fail-safe components for rehabilitation robots — a reflex mechanism and fail-safe force sensor. *IEEE International Conference on Rehabilitation Robotics*, 2005; 456–460.

Tijsma HA, Liefhebber F, Herder JL. A framework of interface improvements for designing new user interfaces for the MANUS robot arm. *Proceedings of the IEEE International Conference on Rehabilitation Robotics*, 2005; 235–240.

Ulrich I, Borenstein J. The GuideCane — Applying mobile robot technologies to assist the visually impaired. *IEEE Transactions on Systems Man and Cybernetics*, 2001;31:131–136.

Van Der Loos HFM, Wagner JJ, Smaby N, Chang K, Madrigal O, Leifer LJ, Khatib O. ProVAR assistive robot system architecture. *Proceedings of the IEEE International Conference on Robotics and Automation*, 1999; 741–746.

Volpe BT, Krebs HI, Hogan N. Is robot-aided sensorimotor training in stroke rehabilitation a realistic option? *Current Opinion in Neurology*, 2001;14:745–752.

Wakaumi H, Nakamura K. Development of an automated wheelchair guided by a magnetic ferrite marker lane. *Journal of Rehabilitation Research and Development*, 1992;29(1):27–34.

Wasson G, Sheth P, Alwan M, Granata K, Ledoux A, Huang C. User intent in a shared control framework for pedestrian mobility aids. *IEEE/RSJ International Conference on Robots and Systems*, 2003; 2962–2967.

Wei Y, Bajaj P, Scheidt R, Patton JL. A real-time haptic/graphic demonstration of how error augmentation can enhance learning. *IEEE-International Conference on Robotics and Automation*, 2005.

Yanco HA. Shared user-computer control of a robotic wheelchair system. PhD Dissertation, Cambridge, MA: Massachusetts Institute of Technology, 2000.

Yoder J-D, Baumgartner ET, Skaar SB. Initial results in the development of a guidance system for a powered wheelchair. *IEEE Transactions on Rehabilitation Engineering*, 1996;4:143–151.

Zinn M, Khatib O, Roth B, Salisbury JK. Towards a human-centered intrinsically-safe robotic manipulator. *IEEE Robotics and Automation Magazine*, 2004;11(2):12–21.

12 Major Limb Prosthetic Devices

Diane M. Collins, Brad Dicianno, Amol Karmarkar, Paul F. Pasquina, Rick Relich, and Annmarie Kelleher

CONTENTS

12.1 LEARNING OBJECTIVES OF THIS CHAPTER

Upon completion of this chapter, the reader will be able to:

- Classify and name amputations based on anatomical location
- Describe the modern prosthetic fabrication processes
- Characterize types of prostheses used for various levels of amputation
- Understand fundamental biomechanical principles used in the field
- Determine how prosthetic needs of those with acquired and congenital limb loss differ
- Understand the fundamentals of advanced prosthetic control features such as myoelectric control for the upper extremity and microprocessor control and magnetorheologic (MR) fluid actuators for transfemoral prostheses.

12.2 INTRODUCTION

In the developing world, trauma is the leading cause of amputation in 80% of cases, often due to inadequately treated fractures (Esquenazi, 2004). In the US, 57,000 new amputations are seen per year (Edmond and James, 1990; Esquenazi, 2004). Of these, 68% are acquired from disease, mainly lower limb vascular complications of diabetes, whereas 30% are acquired from trauma. Table 12.1 lists the incidence of amputations by anatomical level, and Table 12.2 provides the Classification of different amputations and the prostheses typical for each level (Dillingham et al., 2002).

Up to 3% of limb losses are congenital, or present at birth, and often have unclear etiology (Edmond and James, 1990). However, several genetic syndromes and congenital exposures are known to include skeletal deficiencies. As of 1990, the incidence of congenital skeletal deficiencies involving the upper limb was 1.58 per 10,000 births, almost double that of congenital lower limb deficiencies (Edmond and James, 1990).

TABLE 12.1
U.S. Population Incidence Rates per 100,000 (1996)

Lower Extremity	Foot/Toe	21.2
	Ankle	0.07
	Transtibial	12.9
	Knee Disarticulations	0.32
	Transfemoral	12.6
	Hip Disarticulations	0.21
Upper Extremity	Hand/Finger	3.79
	Wrist Disarticulations	0.04
	Transradial	0.22
	Elbow Disarticulations	0.02
	Transhumeral	0.22
	Shoulder Disarticulations	0.02

TABLE 12.2
Classification of Amputation Types and Prescribed Prostheses

Amputation	Anatomical structures involved	Commonly prescribed prosthesis
	Upper Limb	
Fore quarter	Entire upper limb including scapula and clavicle removed	Shoulder disarticulation prosthesis
Shoulder disarticulation	Shoulder joint removed (scapula and clavicle are preserved)	
Transhumeral	Through humerus	Transhumeral prosthesis
Elbow disarticulation	Elbow joint removed	Transhumeral prosthesis
Transradial	Through radius	Transradial prosthesis
Wrist disarticulation	Through wrist joint	Transradial prosthesis
Partial hand	Through metacarpals	Cosmetic hand
	Lower Limb	
Hemipelvectomy or transpelvic	Through pelvic bone	Hip disarticulation prosthesis
Hip disarticulation	Hip joint removed	
Transfemoral	Through femur	Transfemoral prosthesis
Knee disarticulation	Knee joint removed	Transfemoral prosthesis
Transtibial	Through tibia	Transtibial prosthesis
Ankle disarticulation	Ankle joint removed	Transtibial prosthesis
Syme's amputation	Removal of both medial and lateral malleoli	Cosmetic foot/ankle foot prosthesis
Partial foot or chopart	Through metatarsal	

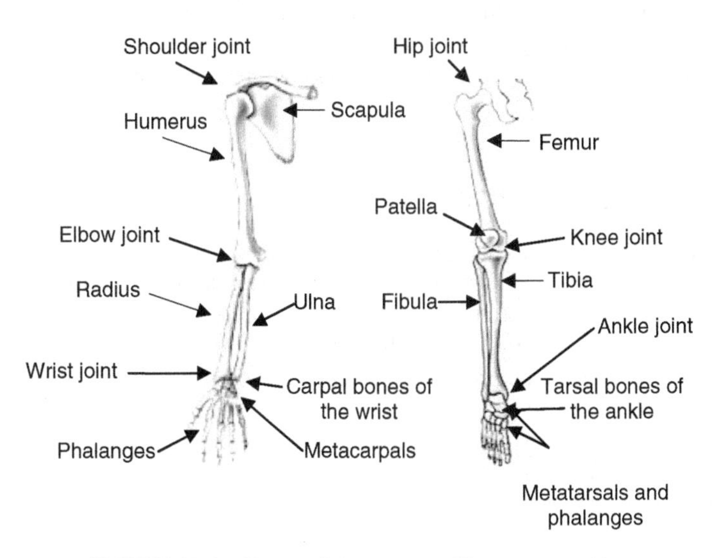

FIGURE 12.1 Bones of the upper and lower extremities.

When upper limb amputations are acquired, it is usually due to trauma on the dominant arm of young males, and less often to cancer, tumors, or diabetic complications.

According to the National Limb Loss Information Center, as of 1996, nearly 1.3 million American live with major limb loss (Amputee Coalition of America [ACA], 2006). Most lower limb amputations occur in individuals 65 years of age or older, who typically have dysvascular disease (ACA, 2006). Trauma is the second most common cause of lower limb amputations and most often occurs in young males.

12.3 ANATOMY AND CLASSIFICATION

The classification of amputations is based on the level at which the amputation occurs. When it occurs across a bone shaft, an amputation is considered transverse, as in a "trans"radial amputation of the forearm or a "trans"tibial amputation occurring through the tibia. When an amputation occurs across a joint, it is considered a disarticulation, as in an elbow or hip disarticulation. See Table 12.3 for anatomical landmarks.

No universal system exists for naming and classifying congenital limb loss. However, the system advocated by the International Society of Prosthetics and Orthotics (ISPO), called the International Terminology for the Classification of Congenital Limb Deficiencies (Kay et al., 1975), is the most widely used. Extremities that develop normally to a level, past which no further bony elements exist are called *terminal* limbs. For example, in a terminal transradial deficiency, the radius and ulna are shortened and no hand or fingers are present. If bony elements exist distal to the affected portion of the limb, these are called *intercalary* limbs. For example, the ulna may be shortened with a normal radius, hand, and fingers present. Congenital deficiencies are often surgically revised to optimize functional use of the limb and prosthesis. In these cases, the deficiency is usually named as if the amputation were acquired. The most common type of congenital limb loss is left transradial amputation (Dillingham et al., 2002).

12.4 CLINICAL MANAGEMENT

Level of amputation is the one of the most important characteristics in determining postamputation function. In upper limb amputations, preservation of the thumb to allow opposition with the remaining fingers will preserve some fine motor skills. A residual limb that is too long or too short may hinder the ideal fitting and prosthesis use. Longer residual limbs provide an extended lever arm to power prosthesis and allow for more area of contact to secure the prosthesis. In addition, longer limbs have more *proprioception*, or the ability to sense where joints are in space. Pediatric amputations are often performed such that joints are disarticulated. This preserves the *epiphyses*, or growth plates of the bones, so extremities can grow to their normal lengths.

Surgeons must keep biomechanical principles in mind when choosing surgical techniques. *Myoplastic closure*, the technique of drawing the muscle and fascia over

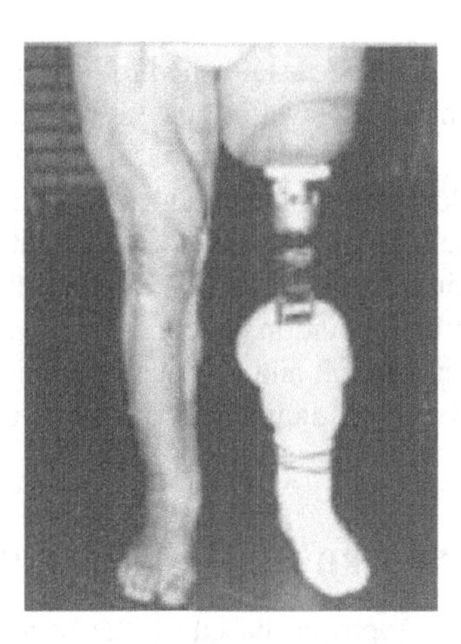

FIGURE 12.2 Body-powered transradial prothesis. Photo courtesy of R. Myers, Ph.D.

the end of the severed bone, creates a cylindrical shape desirable for prosthetic fitting, anchors the lever arm of the limb, and provides adequate padding for the distal portion of the residual limb (Edmonds and James, 1990). Other surgical techniques are osseous integration, in which the prosthetic device is anchored directly into bone (Edmonds and James, 1990) (Figure 12.2), bone-lengthening procedures such as the Ilizarov procedure to extend the lever arm of residual limbs, and the Ertl procedure, where a bone bridge is formed between the tibia and fibula in a lower limb amputation (Ertl, 1949).

12.4.1 Postsurgical Management

Rehabilitation after surgery is often necessary to prepare the patient for wearing a prosthesis. This phase helps shape the residual limb so that it is amenable to fitting into a prosthesis. Muscle atrophy contributes to changes in limb shape. Volume of the residual limb also fluctuates due to fluid accumulation and reabsorption. A postsurgical cast, removable rigid dressing (RRDs) (Edmonds and James, 1990), ace bandages, or a shrinker, which is a gradient pressure sock garment, may be used to help prevent fluid accumulation. Terminal devices such as hands or pylons with prosthetic feet can be added directly to RRDs for immediate postsurgical prosthetic training. Prefabricated fittings and adjustable, postoperative preprosthetic systems (APOPS) are available for immediate postsurgical fittings and volume control. The APOPS offers the rigidity of a postoperative cast but can be opened to visualize the residual limb to evaluate volume and wound healing and to perform dressing changes. Studies have shown that up to 60% of individuals with upper limb amputations and 85% of individuals with lower limb amputations may experience phantom pain or sensation after surgery (Rehabilitation Medicine, 1998). Desensitization techniques can be used for overly sensitive limbs. In addition, symptoms of overuse injury may begin to occur in the unaffected limb (Datta et al., 2004).

Multiple rehabilitation goals are established following surgery, with emphasis on pain management and controlling swelling of the residual limb to promote wound healing and skin integrity. Patient education is vital at this point and may include proper positioning and volume-control techniques in addition to maintaining strength and range of motion, developing independence in activities of daily living (ADLs) and mobility, and promoting psychological well-being. The most successful rehabilitation of individuals with amputations is completed with a team approach which includes a prosthetist, occupational and physical therapists, doctors specializing in rehabilitation, such as physical medicine and rehabilitation doctors, and biomedical engineers.

12.5 PROSTHETIC PRESCRIPTION AND FABRICATION

This phase involves a team approach based on the needs and the surgical level of the individual. A proper prescription includes all components of the prosthesis, the suspension, and the type of control to be used. Prostheses can be fabricated once the prescription is completed and the wound will tolerate weight bearing. In traditional fabrication (Rehabilitation Medicine, 1998), the limb is wrapped with plaster to create a *negative impression* of the residual limb. Pressure-sensitive and pressure-tolerant areas are delineated. This negative impression is then filled with plaster to create a *positive mold.* Plaster is added to the positive mold over the pressure-sensitive areas, such as the anterior tibia or fibular head. Major pressure-tolerant areas, or areas that participate in weight bearing, such as the patellar tendon or medial tibial flare, are places where plaster is removed. This modified mold is called a *positive model.* Then, a clear, diagnostic socket is fabricated out of a thermoplastic material, which is heated up to 300 to 400°F and vacuum-formed over the positive model. The clear, diagnostic socket is then usually evaluated for proper fit under weight-bearing conditions. Using a clear material allows visualization of the fit of the socket over the residual limb.

Technology now allows much of this process to be carried out using computers (Zheng et al., 2001; Houston et al., 1995). Computer-aided design/computer-assisted manufacturing (CAD/CAM) technology can also be used to secure a residual limb measurement for socket fabrication. The residual limb is scanned to input its dimensions, and a prosthetist then uses software to create a digital positive model of the residual limb. He can then alter the model by adding or subtracting volume. A computer milling machine automatically generates a positive mold from which a diagnostic socket can be fabricated. Prostheses made by CAD/CAM may be of the same quality as those made by traditional methods (Oberg et al., 1993).

The final socket is usually made with acrylic epoxy or polyester resins that impregnate various fabrics such as graphite, Kevlar, nylon, or fiberglass through the use of negative or vacuum pressure (Rehabilitation Medicine, 1998). The final socket can be *exoskeletal*, or composed of a strong, outer plastic lamination with a hollow wood or foam center, or *endoskeletal*, in which the strong, inner components are covered by a cosmetic cover. Endoskeleton designs are generally lighter and easier to adjust because of their modular componentry.

In choosing prostheses, the social situation, employment, and hobbies of the person with limb loss must be considered. For example, the prescription for an elderly person with an amputation who cares for her grandchildren would be very different from the prescription for a steelworker who operates heavy machinery. Moreover, many with amputations benefit from using more than one type of prosthesis, depending on their functional needs (Millstein et al., 1986).

In both adults and children, the first prosthesis may require multiple revisions before fabrication of a definitive prosthesis. After initial fitting, rehabilitation begins and may last for weeks. Close follow-up is needed after this initial fitting, usually within four to six weeks after amputation, to ensure functional goals are being met and medical complications such as skin breakdown are avoided. Follow-up appointments are usually conducted with the prescribing physician in the prosthetic clinic, and immediate problems are referred to the prosthetists. Follow-up visits are needed every 3 months until a final prosthesis is prescribed. Afterwards, yearly visits are usually adequate unless new problems arise. As children grow, they require multiple revisions of their prostheses, not only to accommodate their growing size but also to accommodate the new activities they will perform to meet their developmental milestones and personal interests. Generally, a growing child needs a new prosthesis every year.

12.6 COMPONENTS OF THE UPPER LIMB PROSTHESIS

12.6.1 Sockets and Liners

Socket design will depend on residual limb length, muscle strength, and joint stability. The socket is the firm shell that encases the residual limb and contains the interface between the prosthesis and the residual limb. Hard and soft socket designs are commonly used (Millstein et al., 1986). Hard and soft sockets may have a single- or double-wall design. In the double-wall design, the first, i.e., inner, lamination creates the socket's interface, and the second lamination provides an attachment for the rest of the components of the prosthesis. Single-wall design incorporates the interface and the attachment area for components in one lamination. Sockets may also contain a flexible plastic inner liner. Sockets can be split such that two separate shells are used, one above the elbow joint and the other below. They can also partially encase the joint, as in Muenster or Northwestern University sockets.

Sometimes, the socket or inner flexible liner is a direct interface with the limb. However, other liners can provide another interface and a suspension mechanism for the prosthesis. Cushion liners are made from gel materials with varying thickness and are mainly designed to increase comfort (Figure 12.3). Locking liners use lanyards or shuttle pin mechanisms along with a soft silicon liner not only to cushion but also to suspend the prosthesis. Occasionally, layers of socks are used between a locking or cushion liner and the socket, especially if volume has been lost in the residual limb.

12.6.2 Suspension

The suspension system secures the prosthesis to the residual limb despite the weight of the prosthesis and the forces associated with its use. Three types of suspension are used, with harnesses being the most commonly used system (Datta et al., 2004). A

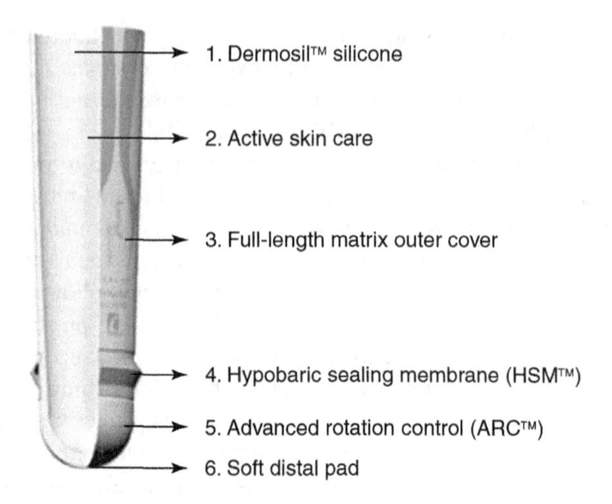

FIGURE 12.3 Iceross Seal-in Liner with a hypobaric sealing membrane in conjunction with an air-expulsion valve (from Ossur, www.ossur.com/template110.asp?pageID=15230).

FIGURE 12.4 Figure-8 arm prosthesis attachment system.

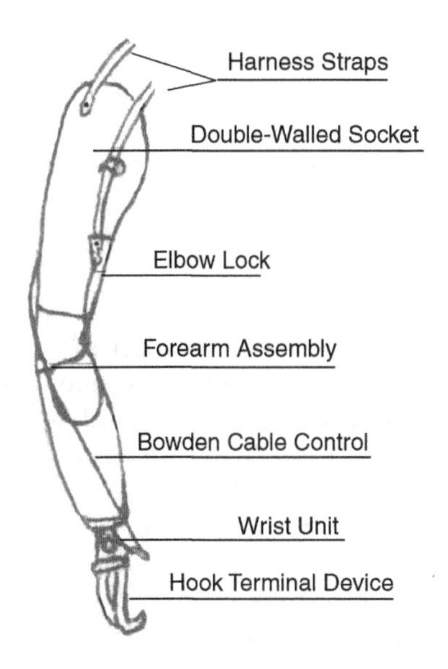

FIGURE 12.5 Body-powered transradial prosthesis.

figure-8 is the most commonly used harness. One loop of the figure-8 harness goes underneath the axilla on the unaffected side. The strap then crosses over the upper back and loops over the shoulder. The anterior portion connects to either the socket in transhumeral limb loss or to a Y-strap and triceps pad in a transradial amputation. These straps provide most of the suspension force. The posterior portion of the strap attaches to a control cable, which is affixed to the elbow in a transhumeral amputation or to an axilla loop in a transradial amputation, as well as to the terminal device (Figure 12.4).

Another type of harness is a shoulder saddle with chest strap (Figure 12.5). This is used when the figure-8 strap causes discomfort or when heavy lifting is expected. The chest strap loops around the chest wall on the unaffected side and connects to a leather or plastic saddle that sits on the opposite shoulder. The saddle is then connected to the prosthesis using a posterior and anterior strap that anchors in a fashion similar to the figure-8 strap.

Self-suspension is used when the socket design is sufficient to secure the prosthesis to the residual limb. This type of suspension is typically used in wrist disarticulations or short transradial amputations. Muenster and Northwestern University sockets are self-suspending.

Suction suspension can be used in two ways. First, a total-contact socket with a one-way air valve can be donned by using a lubricant or a pull-sock. The valve creates a negative pressure inside the socket to secure the prosthesis. Second, the prosthesis can be suspended by connecting to a silicon sleeve that rolls onto the residual limb and locks via a pin or lanyard mechanism. Skin friction is also an important consideration in this suspension mechanism.

A less commonly recognized form of suspension is the aforementioned osseous integration procedure (Figure 12.2). A titanium bone implant directly attaches to the prosthesis, eliminating the need for a socket. Being a relatively new attachment system, the ramifications of long-term use are unknown.

12.6.3 Control Systems

In upper limb prostheses, two types of control systems currently exist: body-powered control and myoelectric control. Body-powered control relies on the effectiveness of the harness to capture body movements to control the terminal device or elbow joint (Rehabilitation Medicine, 1998). To open a transradial terminal device such as a simple hook device, 2.5 in. of body movement is required. Specific movements are used to control upper limb prostheses. Scapular abduction and shoulder flexion typically operate the terminal device or flex the elbow when it is unlocked, while shoulder depression with shoulder extension and abduction lock and unlock the elbow (see Figure 12.4 and 12.5) (Esquenazi and Meier, 1996). When movements activate a single component of the prosthesis, typically the terminal device, the cable is called a *single-control cable*, or a *Bowden cable*. When movements activate two components, typically elbow flexion and the terminal device, the cable is called a *dual-control cable*.

Myoelectric control uses the electrical activity of muscles to control the actions of the prosthesis. Muscle electricity, which is recorded by electromyography (EMG), is produced by the activation of particular muscle groups. In this control system, the EMG signals are recorded by electrodes in the socket that come into contact with the residual limb of the user and get amplified and transmitted for activation and operation of the prosthetic system. The use of this control system has been effectively implemented as the control mechanism of transradial prostheses. Myoelectric controls may provide stronger grasp forces and more fine movements, with less energy expenditure than the traditional cable-controlled prosthesis system (Datta et al., 1989). This can be used to operate the elbow, wrist, and terminal device.

12.6.4 Shoulder Components

Three types of shoulder joint units are commonly used for high-level upper limb loss. *Nonarticulated shoulder joints* do not permit any motion at the shoulder joint. The major advantages of using this type of unit are that it reduces weight and complexity of the prosthesis significantly and, therefore, is often preferred by individuals with unilateral amputation. The *passive friction-loaded shoulder joint* allows the user to preposition his or her prosthesis for basic activities such as dressing. This unit is usually used by unilateral amputees because assistance from the other upper limb is required for positioning the prosthesis. This type of unit is classified by the degrees of freedom of motion permitted. A single axis allows positioning of the prosthesis only in shoulder abduction, whereas a double axis allows both shoulder flexion and abduction. The *locking shoulder joint* permits active locking of the shoulder

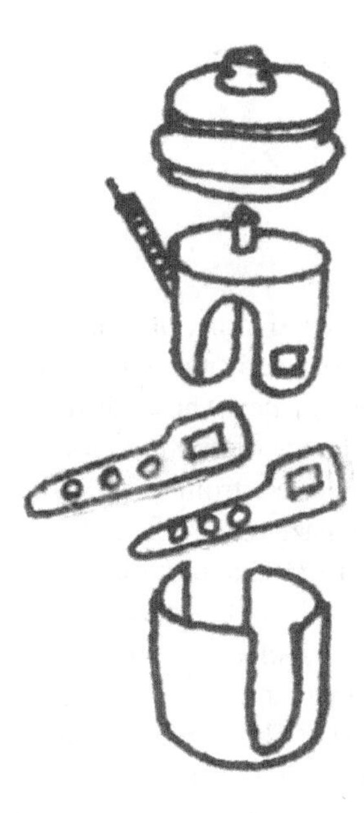

FIGURE 12.6 A popular mechanical elbow unit.

joint in different positions, allowing bilateral activities. This lock can be operated through chin control, scapular abduction, or scapular elevation (Esquenazi and Meier, 1996).

12.6.5 Elbow Components

Prosthetic elbow joints are needed when an amputation occurs at or proximal to the elbow (Figure 12.6). They allow internal and external rotation of the upper arm via a turntable located just superior to the mechanical joint. In order to make the prosthetic limb the same length as the other limb, the joint sometimes has to be external to the prosthesis, as in an elbow disarticulation. However, if the amputation is shorter, the elbow mechanism can be internal. A spring-loaded device provides counterbalance of the forearm, and a flexion-assist mechanism is available for internal elbow joints. Manual control is an additional way to lock elbow joints, using the contralateral hand or chin. This is often termed *nudge control* (Esquenazi, 2004; Esquenazi and Meier, 1996).

12.6.6 Wrist Components

Mechanical wrists connect the prosthesis to the terminal device (Rehabilitation Medicine, 1998). They allow flexion and rotation, which are controlled using the unaffected hand or by applying force to the device through a surface. The wrist is held in place by friction, or is locked in place. The latter is preferable when heavy loads are expected.

Often, a quick-release feature is preferable when multiple terminal devices are to be interchanged.

12.6.7 Terminal Devices

The devices used to replace hand function can be either active active or passive (cosmetic). When the user activates the device to open it, the device is termed a *voluntary opening device*. These devices (Figure 12.7) close by the action of springs or rubber bands (Rehabilitation Medicine, 1998). About 1 to 1.5 lb of pinch force is generated from rubber bands, and 5 to 10 lb is produced by each spring (Esquenazi and Meier, 1996), with a total force of 5 to 10 lb generated in this closure (Rehabilitation Medicine, 1998). For most activities of daily living, 3 lb of pinch force is adequate (Carandall and Tomhave, 2002). Conversely, a voluntary closing device provides indirect sensory feedback through the control cable and can provide 20 to 25 lb of closing force (Crandall and Tomhave, 2002). The continued grip in this design requires constant pull on the cable, and can be very difficult. Hooks provide a lateral-pinch grip, and myoelectric hands can provide three-point chuck prehension patterns. Customized adaptation in the terminal device can enable users to perform certain tasks. Finally, terminal devices able to hold such objects as basketballs, hockey sticks, and bows in archery can be offered for sports and recreation (Figure 12.8).

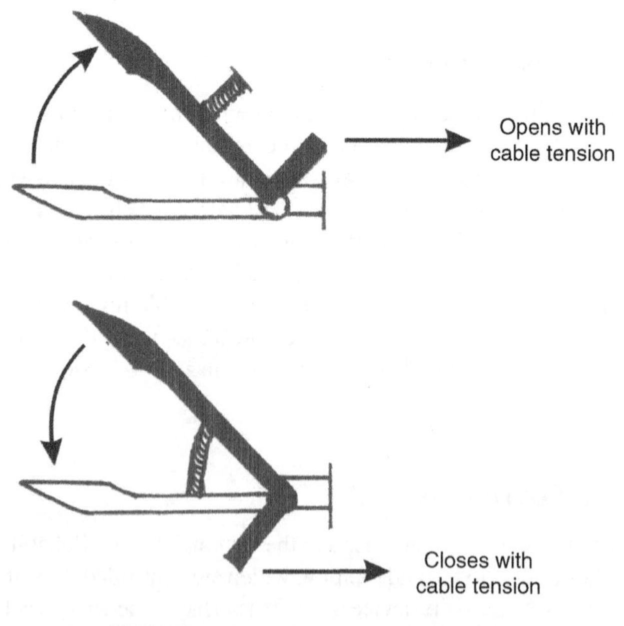

FIGURE 12.7 Voluntary opening devices.

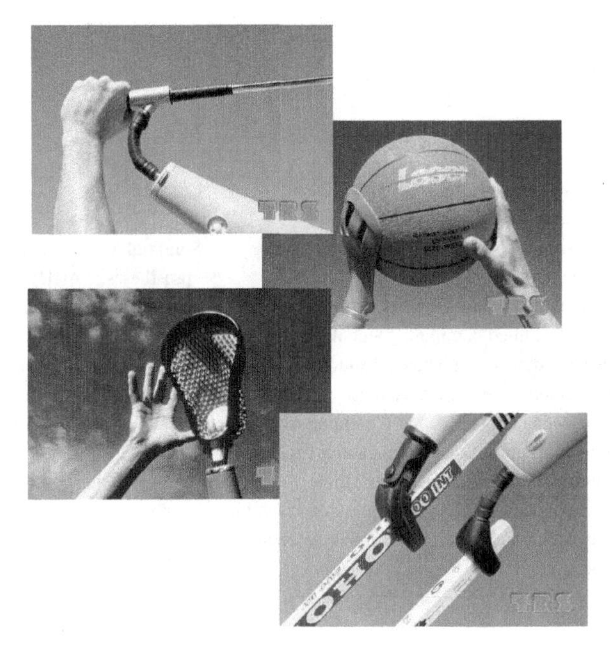

FIGURE 12.8 Upper extremity prosthetic attachment for sports activities. (*Source*: TRS Inc., http://www.oandp.com/products/trs/sports-recreation/.)

12.7 COSMETIC PROSTHESES

Some prostheses are made to enhance the appearance of consumers concerned with cosmesis, such as prostheses that give the appearance of an intact shoulder and upper limb. They are custom-molded and shaped to look similar to the contralateral side with matching intricate details of finger size and shape, nail structure, vascularity, and even skin tone. Growing evidence suggests people use these types of prostheses as frequently as other types of prostheses intended for function, and often, this passive prosthesis is the prosthesis of choice if users are given multiple options (Crandell and Tomhave, 2002; Frasen, 1998). This supports the argument that individuals with amputations benefit from having multiple prosthetic choices. Bimanual tasks such as holding objects or use of the cosmetic prosthesis to secure an object such as a sheet of paper while writing are examples of ways in which individuals with amputations use these prostheses.

12.8 COMPONENTS OF THE LOWER LIMB PROSTHESIS

Individuals who use lower limb prosthetic devices are assessed for body weight, activity levels in vocational and avocational activities, and type of activity for which they

TABLE 12.3

	Functional level	Device
K0	No ability to ambulate or transfer safely; prosthesis does not enhance mobility	Cosmesis
K1	Able to transfer and ambulate on level surfaces; household use	Solid-ankle cushioned-heel prostheses (SACH)
K2	Able to negotiate over low-level environmental barriers; limited community ambulator	Low-level energy storage feet
K3	Prosthetic demands are beyond simple ambulation; able to traverse most environmental barriers and is community ambulatory; used for vocation and avocation	Multiaxial/energy storage prosthesis
K4	Able to perform prosthetic ambulation exceeding basic skills (i.e., high impact); child, active adult and athlete; used for vocation and avocation	Multiaxial/energy storage prosthesis; hobby-specific prostheses

anticipate using their prosthetic devices. Based on this assessment, these individuals are then categorized into five functional levels, as displayed in Table 12.3.

12.8.1 Sockets and Liners

Socket fabrication is similar to that in upper limb amputation. The most common forms of socket designs for lower limbs are:

Transtibial designs

Patellar-tendon-bearing (PTB) socket: This is a total-contact socket that distributes weight over the patellar tendon, pretibial muscles, calf muscles, lateral fibular shaft, and medial tibial flare while reducing pressure over the crest of the tibia, fibular head, distal tibia, distal fibula, and hamstring tendons.

Total-surface-bearing (TSB) socket: This allows distribution of pressure throughout all anatomical structures of the residual limb, and is usually used in combination with a gel sleeve interface.

Hydrostatic socket: A subtype of TSB design, this socket uses principles of fluid mechanics and a compression chamber to distribute pressure equally throughout the residual limb. It uses a silicon suction suspension system.

Transfemoral designs

Quadrilateral socket: This is narrow in the anterior and posterior dimensions and wider in the medial and lateral dimensions. Historically, the quadrilateral socket had a flat shelf on which the ischial tuberosity and gluteal muscles sat, and where most of the weight was transferred. Today these sockets are designed for total contact, where weight-bearing is taken by all pressure-tolerant areas. This usually requires use of an external hip joint and suspension belt. It is less stable and often uncomfortable.

Ischial containment socket: This design is wider in the anterior and posterior dimensions and narrow in the medial and lateral dimensions so as to align the residual limb more anatomically inside the socket. The ischium is contained within the socket, and weight bearing is over the whole limb and ishium, thus providing a total contact socket of fit. Although more expensive and difficult to fabricate, this design is more stable and more comfortable.

Liners typically used for the lower limb are locking liners, as in upper limb prostheses, or sometimes suction liners that create a negative pressure with an expulsion valve and sealing mechanism.

12.8.2 Suspension

In addition to the suspension provided by the liners, varying types of belts can be used for additional suspension in transfemoral limb loss. Silesian belts, Neoprene total elastic suspension belts, and pelvic band and belt with hip joints are examples. In addition, sockets for transtibial amputations can extend higher above the knee joint, as in supracondylar brim suspension, or have additional supracondylar cuffs with waist belts or thigh corsets.

12.8.3 Hip and Pelvic Components

Hip disarticulations generally require an exoskeletal prosthesis that extends to the intact side. The Canadian prosthesis is the one typically used. Sometimes transfemoral limb loss requires the use of quadrilateral sockets and hip joints. The prosthetic hip joint is directly connected to the socket and located anterior to the acetabulum of the user's hip joint. The option of manually locking the hip joint is typically available, which provides more stability to users while standing.

12.8.4 Upper Pylon

In endoskeletal prostheses for transfemoral limb loss, an upper pylon, or shaft, is sometimes needed between the socket and knee joint.

12.8.5 Knee Joints

Categorizing prosthetic knee componentry is based on function, or how easily they extend during the swing phase of gait and how stable they are in the stance phase (Obert et al., 1993). Allowing the knee to bend freely, the *single-axis hinge* is the simplest articulated knee. However, this simple knee does not provide control during the swing phase, and thus is rarely used outside of developing countries. An adjustable friction cell, which presses a fixed force against the knee axle to provide swing-phase damping, can be added to the single-axis hinge. The *manual locking knee* has a single-axis hinge that generally remains locked in extension throughout gait. This is used for elderly or debilitated amputees. The *single-axis, constant-friction knee* is stable only when the net ground reaction force passes anterior to the knee. Such a device requires a consistent gait speed. *Stance control* or "safety knees" produce friction when weight is placed on them in stance. They allow stability for up to 20° of flexion. These are most appropriate for amputees with weak hip extensors or some geriatric individuals. In *four-bar polycentric knees*, the position of the center of rotation changes as the knee is flexed, providing a biomechanical advantage of continued knee flexion. In addition, they offer greater toe clearance at midswing. They are usually prescribed to individuals who walk at one speed only; however, a fluid-controlled swing-phase system added to the basic polycentric framework allows wearers to vary their walking cadence. Hydraulic or pneumatic knees use fluid or air to dampen the swing phase and permit ambulation with varying cadences. *Pneumatic knees* are prescribed for slow- to moderate-cadence walkers. *Hydraulic knees* allow smooth ambulation at any speed from slow to a race-walking pace, and they can tolerate heavier loads. Whereas fluid-controlled knees control the knee's velocity during the swing phase, some do have stance control that gradually decreases resistance to knee flexion just prior to the swing phase.

12.8.6 Electromechanical Knees

Intelligent prostheses are those comprised of sensors and actuators controlled by a microprocessor. For example, the Rheo Knee system by Ossurs (Figure 12.9) uses a magnetorheologic (MR) fluid actuator and advanced sensor technology, guided by a microprocessor to control the stance and swing phases during ambulation. The Rheo Knee is typically partnered with an energy-storing foot, such as the Flex-Foot (Figure 12.10b). This intelligent knee system can automatically adapt to different walking styles and walking conditions experienced by the user. Over time, the microprocessor optimizes the settings of the knee system to optimize the gait of the user.

The C-Leg by Otto Bock (Figure 12.11) is another example of an intelligent prosthesis. A unique microprocessor controls swing and stance phases, allowing for customized setting and real-time adjustments. The C-Leg allows varying gait speeds and ambulation on uneven terrain.

By adjusting hydraulic resistance as demands change, microprocessor controls permit ambulating with a broad range of speeds. The current literature demonstrates that these microprocessor-controlled knees are more energy efficient for the wearer during ambulation (Otto Bock, 2004).

12.8.7 Lower Pylons

Endoskeletal prostheses have another pylon between the knee and ankle joints. Recent use of shock- and/or torque-absorbing pylons has proved to be an effective strategy in improving tasks such as descending stairs and walking with more comfort. In a

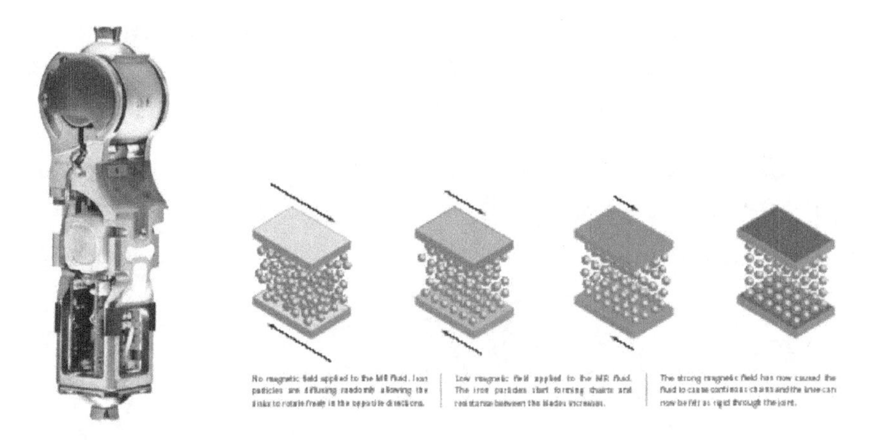

FIGURE 12.9 Rheo Knee system by Ossurs. (*Source*: Ossurs, www.ossur.com/shared/filegallery/sharedfiles/ossurcom/bionics/brochures/rheokneebrochlr.pdf.)

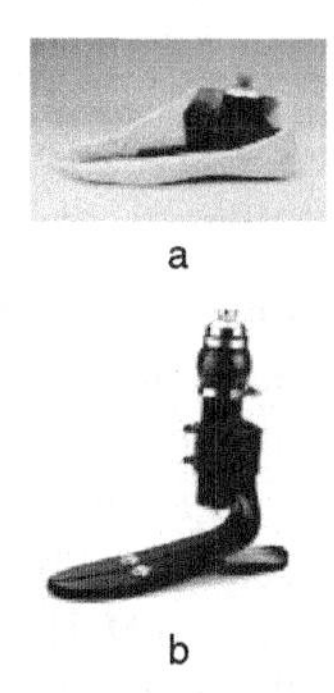

FIGURE 12.10 (a) SACH foot with molded pyramid. (*Source*: http://www.ossur.com/template110.asp?pageid=24); (b) Flex-Foot from Ossurs. (*Source*: http://www/owwco.com/CategoryDetail.aspx?Key=63.)

study conducted on amputees from trauma and vascular disease who used a shock-absorbing pylon for a month, improvements in walking performance, speed, and quality of motion were reported, along with self-reported improvement in comfort (Grad and Konz).

12.8.8 Ankles and Feet

The *single-axis ankle* is effective in providing passive knee stability to users during ambulation by allowing for dorsiflexion and plantarflexion, which is controlled by adjustable internal bumpers. However, this foot has a rigid keel, is heavy, has minimal shock absorption, and may require more service because the bumpers wear. The *solid-ankle cushioned heel* (SACH) was developed in the 1950s and is constructed of

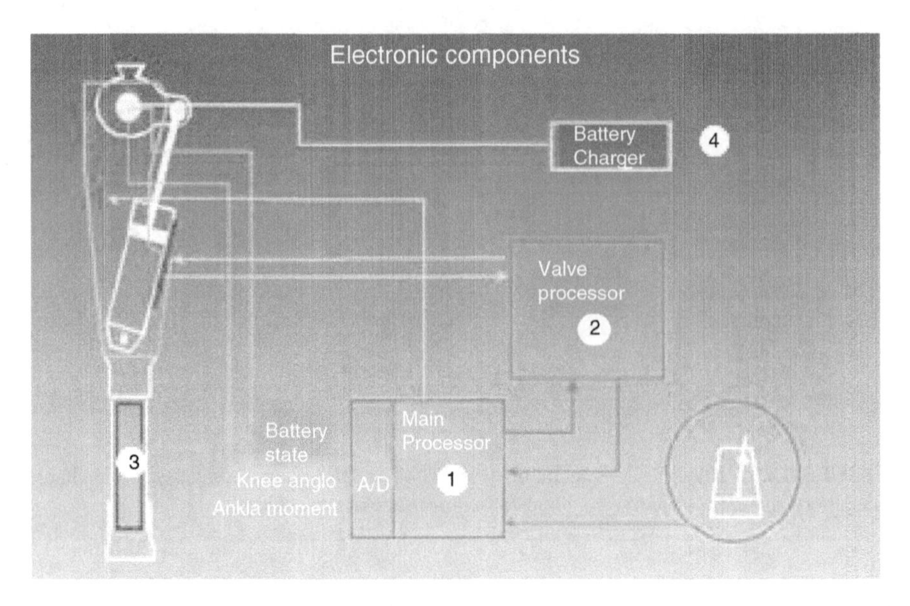

FIGURE 12.11 Model of C-Leg components. (*Source*: http://www.healthcare.ottobock.com/info_download/pdf/646D129_GB_C_Leg.pdf.)

urethane materials with a less dense material incorporated at the heel. This difference in material densities simulates ankle motion, but the ankle joint is actually solid. With ambulation, the softer heel moves the center of gravity forward. The SACH foot (Figure 12.10a) contains a keel made of a hardwood or composite material covered by belting material to prevent the keel from breaking through the urethane on the bottom of the foot. During ambulation, the SACH heel simulates eccentric lengthening of the anterior shin muscles (dorsiflexors) that let the toes come down slowly to the ground immediately following heel-strike. The keel simulates the posterior skin muscle groups (plantarflexors) and also prevents the loss of anterior support during the push-off of the toe. Because gait and weight vary, different heel densities are available for the SACH. The *multiaxial ankle* allows movements of inversion, eversion, and rotation in addition to plantar and dorsiflexion. This type of foot is ideal for use over uneven surfaces; however, the disadvantages of complexity, frequent servicing, and added weight persist with this foot. The extended flexibility to the entire forefoot provided in the *flexible-keel foot* reduces weight, making walking more efficient and easy. Absence of a rigid keel in this foot, however, limits effective push-off during ambulation. Finally, the *dynamic response foot*, which is also referred to as an "energy storing" foot, is useful for amputees who are active ambulators. The action of this foot mimics the action of a diving board, with the storing of energy from foot flat to heel-off and returning of energy from heel-off to toe-off in the later phase of stance. It can also be used with a split-toe design or different densities of urethane that allow for multi-axial movements and can be paired with shock and torque absorbers. This results in more efficient gait with less energy expenditure. Research has shown that energy-storing feet result in comfortable ambulation speeds and stride lengths about 7 to 13% longer than a conventional (SACH) foot in both traumatic and vascular transtibial amputations (Barr, 1992; Hsu, et al., 2006).

Other commonly used components in lower limb prostheses are torque absorbers and positional rotators. Torque absorbers function to absorb the forces generated by the ground and transferred to the prosthetic component. This reduces stress on the residual limb. A positional rotator, which is a locking turntable, allows passive rotation of the prosthesis for activities such as sitting or squatting.

12.9 REJECTION OF UPPER AND LOWER LIMB PROSTHESES

Long-term acceptance of prostheses depends on a number of factors, including cosmesis, ability to perform ADLs without a prosthesis, chronic pain, and weight of the prosthesis. When individuals with amputations begin to perform ADLs without the use of prosthesis or use their preserved limb only, they typically find prostheses cumbersome and are less receptive to fitting. Generally, lower limb prostheses are accepted more commonly than upper limb prostheses. For rates of rejection of various upper extremity prostheses, see Table 12.4. Prostheses are usually fitted within 30 days of surgery to promote acceptance by the amputee (Malone et al., 1984). In several large studies (Millstein et al., 1986 Crandall and Tomhave, 2002; Wright et al., 1995) that followed-up subjects to 49 years, 38% of individuals with unilateral upper limb amputations reject their prostheses, whereas all individuals with bilateral amputations continued to use their prostheses.

The abandonment or change rate of lower limb prostheses is approximately 15% within 1 to 5 year of discharge from a rehabilitation program (Gauthier-Gagnon et al., 1998). No rejection statistics by type of lower limb prostheses are available. Of those who did use their lower limb prostheses, approximately 64% used their prostheses for outdoor mobility and 53% used them for ADLs in their homes (Gauthier-Gagnon et al., 1998).

TABLE 12.4
Rate of Prostheses Rejection

Type	Rate
Upper Limb	
Wrist disarticulation	94%
Transradial amputation	7–11%
Transhumeral amputation	43%
Higher than transhumeral (shoulder disarticulations)	40%

12.10 INTERNATIONAL STANDARDS

To eliminate technical barriers to trade and establish a consensus of technical cohesion, the International Standards Organization (ISO) is a legal association that facilitates the exchange of goods and services. Two groups of ISO standards have been adopted for prosthetic devices. One ISO standard provides a system of nomenclature and related terminology to allow all parties involved in lower limb and upper limb prosthetic and orthotic treatments to apply a standard terminology (Millstein et al., 1986; Crandall and Tomhave, 2002; Wright et al., 1995). The second outlines a system of test methods for the verification of essential requirements on prosthetic and orthotic devices related to the safety of the users. The American Board for Certification in Orthotics and Prosthetics (http://www.abcop.org/) and the International Society for Prosthetics and Orthotics (http://www.ispo.ws/) are good sources for finding the ISO standards.

12.11 SUMMARY

The naming of congenital, acquired and traumatic amputations is based on the level at which the amputation occurs. For congenital amputations, a system advocated by the International Society of Prosthetics and Orthotics (ISPO), called the International Terminology for the Classification of Congenital Limb Deficiencies, is the most widely used. In both adults and children, the first prosthesis may require multiple revisions before fabrication of a definitive prosthesis. After initial fitting, rehabilitation begins and may last for weeks. Close follow-up is needed after this initial fitting, to ensure that functional goals are being met and medical complications such as skin breakdown are avoided. Many people with limb loss benefit from using more than one type of prosthesis, depending on their functional needs.

12.12 STUDY QUESTIONS

1. How do prosthetic devices differ from orthotic devices?
2. What is the difference between transverse amputations and disarticulations, and terminal and intercalary amputations?
3. What is the importance of having an extended lever arm in the residual limb?
4. Why are the epiphyses preserved in surgical pediatric amputations?
5. Which factors contribute to the adoption of prostheses?
6. How does the solid-ankle cushioned heel (SACH) differ from energy-storing feet such as the Flex-Foot™ prosthetic?
7. What types of suspension systems are available for upper and lower limb prostheses, and how do these systems differ?
8. Who mandates prosthetic standards, and what are the two areas that these standards address?
9. How does the C-Leg differ from the hinge-joint prosthetic knee?
10. For which task has the recent use of shock- and/or torque-absorbing pylons provent to be an effective strategy for improving consumer comfort?

BIBLIOGRAPHY

Crandall RC, Tomhave W. Pediatric unilateral below-elbow amputees: retrospective analysis of 34 patients given multiple prosthetic options. *J Pediatr Orthop*, 2002;22:380–3.

Datta D, Selvarajah K, Davey N. Functional outcome of patients with proximal upper limb deficiency — acquired and congenital. *Clin Rehabil*, 2004;18:172–7.

DeLisa JA, Gans BM, Bockinek WL et al. *Rehabilitation Medicine: Principles and Practice.* 3rd ed. Philadelphia: Lippincott-Raven; 1998; p 260.

DeLisa JA, Gans BM, Walsh NE et al. *Rehabilitation Medicine: Principles and Practice.* 4th ed. Philadelphia: Lippincott Williams and Wilkins; 2005.

Edmonds L, James L. Temporal trends in the prevalence of congenital malformations at birth based on the birth defects monitoring program: United States, 1979–1987. *CDC Surveil Summ Morbid Mortality Weekly Rep*, 1990;39:19–23.

Ertl J. Über Amputationsstümpfe. *Chirurgie*, 1949;20:218–24.

Esquenazi A, Meier RH, Rehabilitation in limb deficiency. Limb amputation. *Arch Phys Med Rehabil*, 1996;77(3 Suppl):S18–28.

Esquenazi A. Amputation rehabilitation and prosthetic restoration. From surgery to community reintegration. *Disabil Rehabil*, 2004;26:831–6.

Fraser CM. An evaluation of the use made of cosmetic and functional prostheses by unilateral upper limb amputees. *Prosthet Orthot Int*, 1998;22:216–23.

Gauthier-Gagnon C, Grise, MC, Potvin D. Predisposing factors related to prosthetic use by people with a transtibial and transfemoral amputation. *JP&O*, 1998;10:99–109.

Houston VL, Mason CP, Beattie AC, et al. The VA-Cyberware lower limb prosthetics-orthotics optical laser digitizer. *J Rehabil Res Dev*, 1995;32:55–73.

Kay HW, Day HJ, Henkel HL, et al. The proposed international terminology for the classification of congenital limb deficiencies. *Dev Med Child Neurol Suppl*, 1975;34:1–12.

Malone JM, Fleming LL, Roberson J, et al. Immediate, early, and late postsurgical management of upper-limb amputation. *J Rehabil Res Dev*, 1984;21:33–41.

Millstein SG, Heger H, Hunter GA. Prosthetic use in adult upper limb amputees: a comparison of the body powered and electrically powered prostheses. *Prosthet Orthot Int*, 1986;10:27–34.

Oberg T, Lilja M, Johansson T, Karsznia A. Clinical evaluation of trans-tibial prosthesis sockets: a comparison between CAD CAM and conventionally produced sockets. *Prosthet Orthot Int*, 1993;17:164–171.

Wright TW, Hagen AD, Wood MB. Prosthetic usage in major upper extremity amputations. *J Hand Surg [Am]*, 1995;20:619–22.

Zheng YP, Mak AF, Leung AK. State-of-the-art methods for geometric and biomechanical assessments of residual limbs: a review. *J Rehabil Res Dev*, 2001;38:487–504.

13 Orthotic Devices

Michael Dvorznak, Kevin Fitzpatrick, Amol Karmarkar, Annmarie Kelleher, and Thane McCann

CONTENTS

13.1 LEARNING OBJECTIVES OF THIS CHAPTER

Upon completion of this chapter, the reader will be able to:

Learn the basic principles of the biomechanics used in orthoses
Know the common types of materials used in orthoses
Learn about the basic methods of orthoses design
Understand the problems for which orthoses are used
Gain knowledge about the principles used in orthoses

13.2 INTRODUCTION

The intent of this chapter is to provide a general overview of the principles, design, fabrication, and function of spinal, lower extremity, and upper extremity orthoses. It is acknowledged that entire books and atlases have been dedicated to this subject. Orthotics is considered an art and skill, requiring creativity and knowledge in anatomy, physiology, biomechanics, pathology, and healing. Evolving into a competent clinician, who is comfortable with designing and fabricating orthoses to meet the unique needs of each individual client, requires practice of these skills and knowledge.

The word *orthosis* derives from the Greek orthos, meaning "to straighten." An orthosis is an orthopedic device that provides functional stability to a joint or prevents, corrects, or compensates for a deformity or weakness. Typically, this is accomplished through external bracing (although the term *brace* is somewhat deprecated because of institutionalized stereotypes and implications of static fixation), whereas orthoses also provide dynamic joint control. An orthosis can be as simple as an off-the-shelf, prefabricated shoe insert that one could purchase at a pharmacy or department store, or something more complex, such as a reciprocating gait orthosis usually consisting of custom-molded plastic solid ankle and thigh shells connected via bilateral uprights with locking knee joints, dual cables for controlling alternating gait, and gas-filled struts or springs to aid in knee and hip flexion.

The term *splint* is synonymous with orthosis and is commonly referred to in the literature, especially with regard to upper extremity orthoses. For consistency, this chapter will use the term *orthosis* throughout.

13.2.1 Classification

Standard nomenclature for orthoses varies, depending on the type, i.e., upper extremity vs. spinal vs. lower extremity. Names for orthoses originate from the joint or joints they encompass (e.g., wrist orthosis), the function provided (e.g., reciprocating gait orthosis), condition treated (e.g., tennis elbow splint), appearance (e.g., airplane splint and halo brace), or person who designed it or place where it was designed (e.g., Milwaukee brace and Jewett orthosis). In addition, organizations such as the American Society of Hand Therapists (ASHT) and International Standards Organization (ISO) have developed standard nomenclature for orthoses.

The current method of naming an orthosis is by creating of an acronym from the English words for the joints that the orthosis crosses, in sequence from proximal to distal. The letter "O" is appended to signify orthosis. For instance, an orthosis that covers the foot and attaches to the leg to compensate for weakened ankle dorsiflexors (drop foot) would be called an ankle–foot orthosis, or AFO. Often, additional letters are added, describing the device, such as AFO–SA (solid ankle) or AFO–PLS (posterior leaf spring).

13.2.2 Function

Orthoses are classified as static or dynamic. Static orthoses are designed to prevent or limit motion and have no moveable parts. Dynamic orthoses are designed to facilitate movement and have one or more movable parts. The three general functions of orthoses are categorized as (1) immobilizing, (2) restrictive, and (3) mobilizing. Immobilizing and restrictive orthoses, both provide static support. Immobilizing orthoses prevent any movement in the joints involved, whereas restrictive orthoses limit movement in a specific aspect of joint range of motion.

The static support helps to reduce stress and maintain joint alignment; prevent deformities and soft tissue contractures; scar reduction; provide rest to reduce inflammation and pain; positioning to facilitate proper healing; and protection against further injury.

Mobilization or dynamic orthoses are designed to increase range of motion (stretch soft tissue contractures) and assist muscle weakness or spasticity to improve function. These orthoses are also used for exercise to improve the range of motion and strength.

13.3 BIOMECHANICAL PRINCIPLES

13.3.1 Three-Point Pressure System

Orthoses act to restrict joint motion via a three-point pressure system. Although other pressure systems exist, this is the most basic one. The terminology *three-point force* is also used to describe this system; however, pressure is preferred as it reminds us that it is an applied force distributed over an area. The three-point pressure system generally consists of a principal force acting at or near the affected joint, opposed by two forces, one proximal and the other distal to the joint, to stabilize the joint.

13.3.2 Leverage

Leverage goes hand in hand with the three-point pressure system. Leverage is the mechanical advantage of a force applied at a distance from a fulcrum. The moment of this force causes the body to tend to rotate about the fulcrum. The larger the perpendicular distance of the force to the fulcrum, the greater the moment generated. To stabilize the joint, the moments from all applied forces (including gravity and inertia) must sum to zero. By increasing the lever arm length, the force necessary to stabilize the joint can be reduced, thus increasing comfort.

13.3.3 Ground Reaction Force

The ground reaction force, or force generated by the floor on the patient due to gravity and body accelerations, is also important in controlling joint motion. The ground reaction force can be utilized to stabilize a joint more proximal to the orthosis. When the foot contacts the ground during gait, it produces a moment about each of the joints of the lower extremity. Depending on the line of action of the force, the moment could be in flexion, extension, or through the joint center, causing no moment. A heel cutoff on a shoe can compensate for weak quadriceps muscles by shifting the contact point forward and the direction of the ground reaction force closer to the knee center, thus reducing the flexion moment about the knee. Alternatively, an AFO–SA with an anterior band provides a posteriorly directed force at the tibia that limits knee flexion by resisting forward motion of the tibia.

13.3.4 Axial Forces

Axial force is a force directed along the long axis of the bone. Reduction of axial forces in lower limb fractures is typically via crutches or in unweighting AFOs that unload the ankle or foot by transferring the weight through the orthoses.

13.3.5 Pressure

In general, it is beneficial to distribute the forces the orthosis applies to the body over a large area. This minimizes the stress on soft tissue. An AFO with a custom-molded plastic calf shell would be more comfortable than a leather–metal AFO; however, this would be contraindicated in cases of edema and dermatitis or hot climates.

13.3.6 Shear Stress

It is important to minimize shear stress at the interface of the patient and orthosis. Shear stress is typically caused by tangential forces applied to the load-bearing surface of the body. These shear forces can cause motion between the underlying skin, muscle, and fascia and the orthosis. The deformation may restrict blood and lymph flow, causing ulcers and tissue damage. Shear stresses can be reduced by optimal design, proper fit and alignment of the orthosis, and use of slippery elastic padding at the patient–orthoses interface.

13.3.7 Creep

Creep is the time-dependent strain, or change in shape, of a material due to exposure to stresses and loading. This creep occurs in a matter of seconds or weeks, depending on the stiffness and viscoelastic properties of the material. Creep occurs in both muscles and soft tissues bearing load due to the orthosis, and is critical in the design and fit of orthoses.

13.4 DESIGN CONSIDERATIONS

13.4.1 Materials Selection

Plastics are the most common material in modern orthoses. Plastics are divided into two major categories, thermoplastics and thermosets. Thermoplasts become pliable when heated and rigid when cooled. Thermoplasts can be reheated and reformed. Thermosets can be worked into shape when heated; however, they develop a permanent shape once cooled. Thermoplasts that become workable under temperatures of 80°C are considered low-temperature thermoplasts and, with care, can be molded directly to a body segment. Because of limited strength and fatigue resistance, low-temperature thermoplasts are found primarily in fracture splints and upper extremity orthotics. Thermoplastics workable at temperatures over 80°C are called high-temperature thermoplasts and must be shaped over a mold or cast. They are stronger and more resistant to creep than their low-temperature counterparts, and are therefore used in more permanent and higher-load-bearing orthoses. High-temperature thermoplasts can be injected with air or other gases to form soft foam padding and liners in many orthoses.

Thermosetting plastics are made by curing a polyester or epoxy resin. The resin is poured over fabric such as nylon, fiberglass, or carbon fiber, and layers are laminated over a mold. The resulting material can have strength-to-weight ratios comparable to steels and must be cut, machined, ground, or sanded to modify its shape. Thermosets are typically used where strength is of importance; however, the use of thermoplastics with carbon fiber has limited their use.

Understanding the properties of commonly used materials is critical in custom designing orthoses and modifying prefabricated orthosis. Prefabricated orthoses are mass produced and, with minor adjustments, are ready to wear and fit the needs of many users. When available, these easy-to-use and time-saving options are often considered first.

13.5 SPINAL ORTHOSES

The appropriate orthotic prescription to address pathology of the spine (i.e., back pain, spine deformity, or injury) requires an understanding of the biomechanics and kinetics of the spine. The spine is composed of 7 cervical, 12 thoracic, 5 lumbar, and 5 fused sacral vertebrae. The articulation of the vertebrae with supporting ligaments, muscles, and other skeletal elements is responsible for the unique characteristics of each of the four regions of the spine. Movement occurs between sequential vertebrae and is

TABLE 13.1
Spinal Orthosis Nomenclature

Cervical	CO: Philadelphia, Miami
Cervicothoracic	CTO: Halo, SOMI, Minerva
Thoracolumbosacral	TLSO: Jewett, CASH
Lumbosacral	LSO: Chairback, Knight
Sacroiliac	SO: Sacral corset

described as having six degrees of freedom, with rotation about the three orthogonal axes and translation through the same axes. Muscles attached to the vertebrae act to extend, flex, rotate, and laterally bend the spine. The sum of intervertebral movements provides a gross functional range of motion (Table 13.1). This flexibility makes the spine vulnerable to injury and degeneration. A spinal orthosis is prescribed to protect an injured segment, limit intervertebral movement, limit gross movements of the spine, and apply forces to correct or inhibit the progression of some deformity. The unique anatomy and kinesiology of the cervical, thoracic, lumbar, and sacral vertebrae necessitates that orthosis be designed to address the specific requirements to treat pathology of each of those segments.

There are four general curves along the spine — the cervical and lumbar spines are lordotic (concave posteriorly) and the thoracic and sacral spines are kyphotic (convex posteriorly). The intervertebral discs make up 25 to 30% of the total spine length. Load bearing of the spine consists of transferring the compressive force through the individual vertebrae and between adjacent vertebrae through the intervertebral disc. The disc, composed of a fibrous annulus and gelatinous inner layer (nucleus pulposus), has specific viscoelastic properties that allow it to dampen the transmitted forces. Injury to either the skeletal components or to the surrounding musculature may create instability, resulting in pain, degeneration, and deformity.

The spinal orthosis is designed to address this instability. The optimal spinal orthosis design is made of material that is rigid enough to control motion at the particular vertebral segments while minimizing discomfort. Three-point systems are used in alignment of the spine, with the corrective force midway between the opposing forces. The efficacy of spinal orthosis is inversely related to the distance between the inner surface of the orthosis and the spine. The angles between the axis of rotation of the vertebral segment of interest, the edge of the brace, and the long axis of the spine are important factors in designing a spinal orthosis. This angle is affected by both the thickness of the tissue and the length of the brace.

13.5.1 Cervical and Cervical Thoracic Orthoses

The cervical spine segment is very flexible, based upon the shape of the individual vertebrae and their articulation with adjacent vertebrae. Flexion, extension, lateral bending, and axial rotation are the movements of this segment. Cervical orthoses (CO) are divided into soft and hard orthoses. The most common soft cervical collar is made

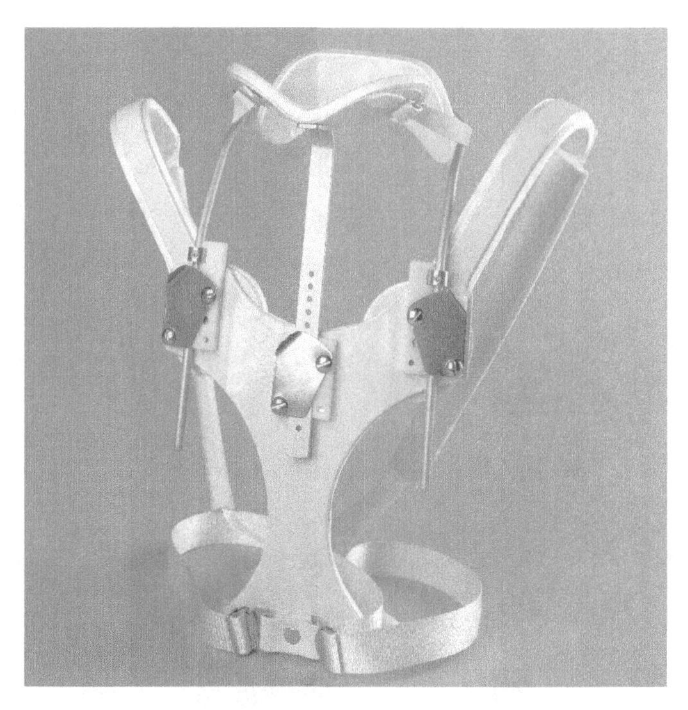

FIGURE 13.1 Sternal occipital mandibular immobilizer (SOMI). (Courtesy of De LaTorre Orthotics and Prosthetics, Inc.)

of foam and has a stocking sleeve. The soft orthosis does not provide control of the cervical segment; however, it does serve as a proprioceptive reminder to the patient to avoid excessive motion. The hard cervical collar is available in a variety of prefabricated models. The hard orthoses control cervical motion in the sagittal plane via the mandibular and occipital pads. These collars have poor control of lateral flexion and axial rotation.

Superior control of cervical segment motion is possible by attachment of metal posts to the hard collar, which is then attached to a pad that rests on the shoulders. Greater vertebral segment control can be provided by extending the orthosis to involve the thorax, thereby providing additional leverage to control cervical spine motion. This extension results in a cervical thoracic orthosis (CTO), the most commonly prescribed one being the SOMI (sternal occipital mandibular immobilizer) (Figure 13.1). This orthosis consists of a sternal plate, shoulder pads, mandibular and occipital pads, and bars. This wearer maintains lumbar flexion.

The next CTO, and considered the most invasive, is the HALO orthosis (Figure 13.2). This orthosis provides motion control in all three planes, and consists of a ring fixed to the skull with a series of pins. The ring is connected via bars to a rigid jacket. This degree of control makes this orthosis the preferred one to allow for fracture healing; however, complications include pin site infections, slippage of pins, and pain and discomfort interfering with the activities of daily living.

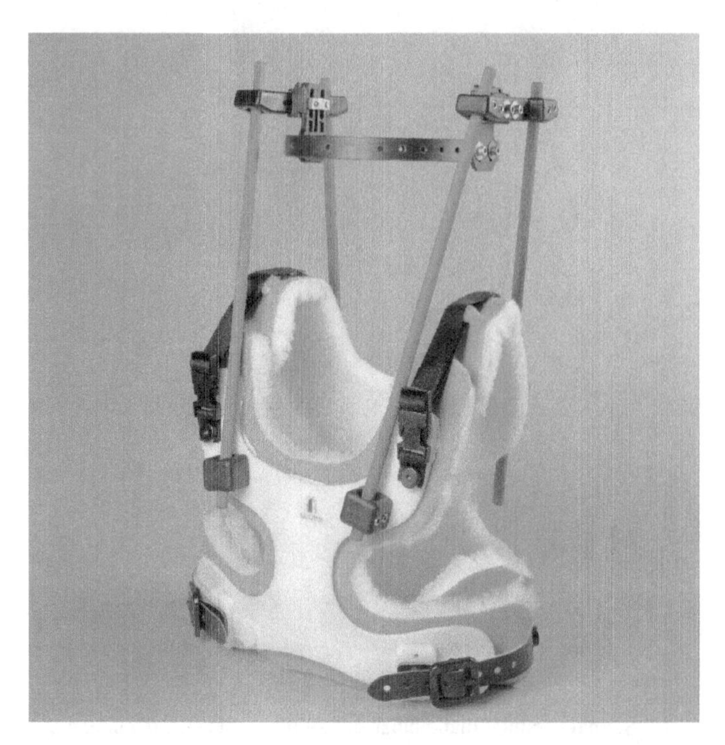

FIGURE 13.2 HALO orthosis. (Courtesy of De LaTorre Orthotics and Prosthetics, Inc.)

13.5.2 Thoracolumbosacral and Lumbosacral Orthoses

The thoracolumbosacral orthosis (TLSO) and the lumbosacral orthosis (LSO) are typically prescribed to address deformity, fractures, and musculoskeletal back pain. Back pain is second only to cough as the presenting chief complaint to the primary care physician. Over 80% of people will experience low-back pain, and fortunately, more than 85% of these will resolve within 1 year. A variety of orthoses have been prescribed, ranging from the lumbosacral corset to the custom-molded thermoplastic TLSO. In addition to the orthosis prescription, an exercise program to strengthen the supporting spinal musculature, as well as proper lifting techniques, have proven effective. The corsets serve as both a proprioceptive reminder to limit motion and provide warmth to the injured area, but they provide limited motion control (Figure 13.3). Some corsets incorporate metal stays to provide increased rigidity, as well as thermoplastic inserts or air bladders. LSOs incorporate thoracic and pelvic bands that encompass the torso at each respective level. Paraspinal bars are shaped to follow paraspinal musculature, and lateral bars are along the midaxillary trochanteric line. Using these basic principles, a variety of orthoses are created and adapted according to the plane of motion that needs to be controlled.

Sagittal control is provided by an LSO (also known as chairback) consisting of a thoracic band, pelvic band, and two paraspinal bars. Two three-point systems are utilized. To control motion in the sagittal and coronal plane, lateral bars are added to

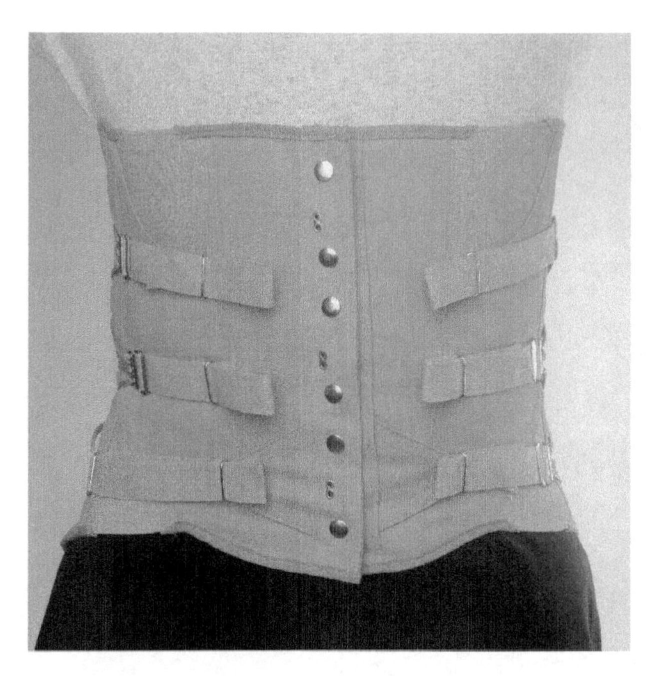

FIGURE 13.3 Lumbosacral corset. (Courtesy of De LaTorre Orthotics and Prosthetics, Inc.)

the chairback orthosis, in effect, adding an additional three-point pressure system in the coronal plane. To provide control in the extension coronal plane, oblique bars are added, resulting in free flexion with limited spinal extension.

The TLSO incorporates the thorax, with the same principles; the basic design may be modified based on the level and motion control needed. For flexion control, the orthosis is designed with a metal frame, with pads at the pubic bone, sternum, and at the lateral midline of the trunk. The Jewett orthosis is the archetype, with a three-point pressure system, two directed posteriorly and one directed anteriorly (Figure 13.4). The result is reduced lumbar flexion with full extension. Motion in the sagittal plane is controlled by modifying the TLSO by incorporating paraspinal bars, pelvic and interscapular bands, and axillary straps. Two three-point systems are demonstrated, with the interscapular strap providing anterior force and the axillary straps providing posterior-directed force. The sagittal coronal movement may be controlled with the combination of thoracic, pelvic and interscapular bands, axillary straps, and both paraspinal and lateral bars.

13.5.3 Cervicothoracolumbosacral Orthosis

No discussion of spinal deformity would be complete without discussing scoliosis, control of this progressive spinal deformity being the ultimate test of design, and clearly demonstrating the three-point model of orthoses. Scoliosis is a disorder resulting in progressive spinal deformity, the most common cause of which is idiopathic. The biomechanics of progression of the deformity is described by Euler's theory of

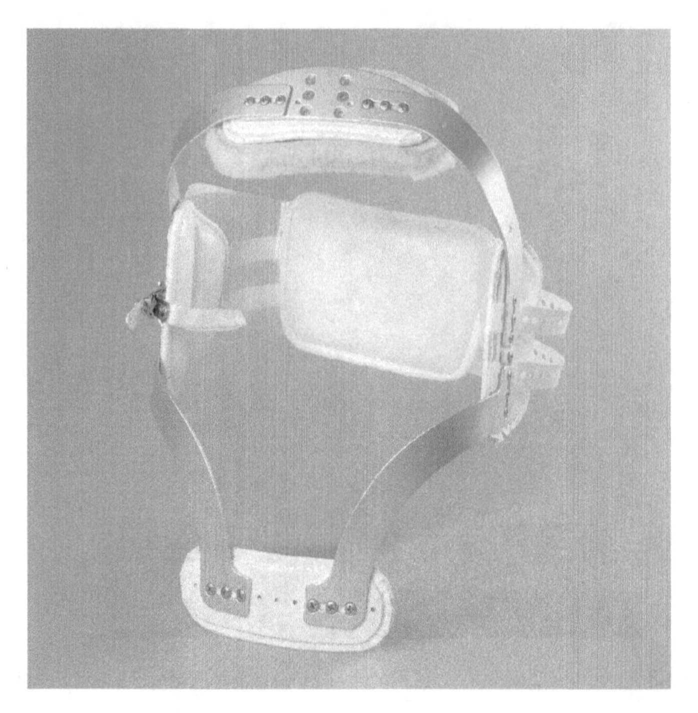

FIGURE 13.4 Jewett orthosis. (Courtesy of De LaTorre Orthotics and Prosthetics, Inc.)

elastic buckling of a slender column. In Euler's theory, a straight flexible column, fixed at the base is placed under an axial load. There is some threshold value of the force beyond which the column buckles. In much the same way, as the child ages, the spinal column lengthens, and with associated weight gain in the upper extremities and trunk, the aforementioned threshold is achieved, and the spinal curvature progresses. The comparison is an imperfect one, as unlike in the theory, in clinical practice the critical load (threshold axial force) is correlated to the degree of spinal curvature, large curves having lower critical loads; and the deformation is permanent even with removal of the axial load.

The orthosis designed for treatment of scoliosis has three functions: (1) end-point control, (2) transverse support, and (3) correction of the curve. The standard orthosis that demonstrates this is the Milwaukee brace (cervicothoracolumbosacral orthosis, CTLSO) (Figure 13.5). End-point control is provided by pelvic attachment to stabilize the base of the spine and a neck ring to keep the neck in line with the pelvis. A mandibular and occipital pad provides longitudinal distraction, while a lateral pad is placed on the convex side of the segmental deformity. Orthotic management of scoliosis is indicated in the management of those with spinal segment curves ranging from 20 to 30°. Surgical correction of the spinal deformity is recommended for curvatures greater than 40°, as the amount of corrective force required by the orthosis would be intolerable. The Milwaukee brace is the only CTLSO used in the treatment of spinal deformity; lower-profile orthoses, such as the Miami TLSO, are more

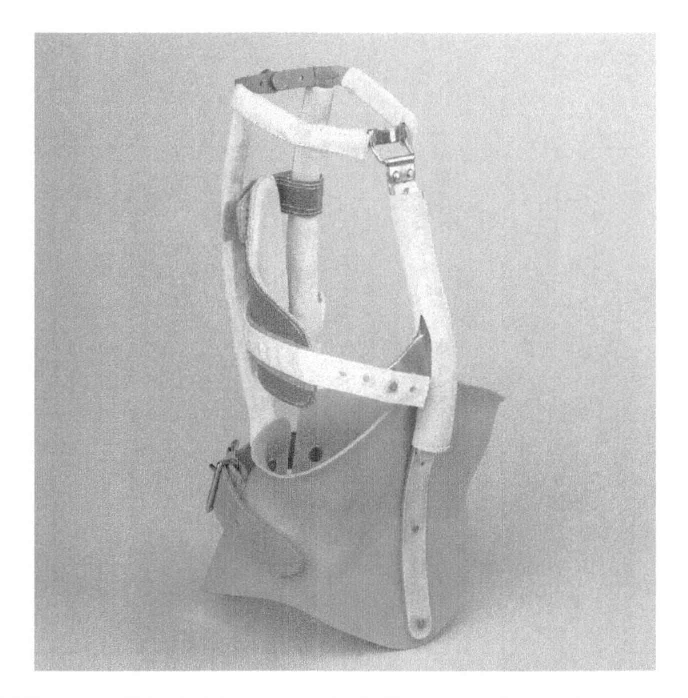

FIGURE 13.5 CTLSO. (Courtesy of De LaTorre Orthotics and Prosthetics, Inc.)

desirable by patients because of comfort and cosmesis. The TLSO does not offer the same segmental control as the CTLSO. The basic design of three points of contact is demonstrated by the spinal orthosis, which is an excellent nonoperative intervention to reduce pain, protect a vertebral segment, and reduce and prevent progression of deformity.

13.6 LOWER EXTREMITY ORTHOSES

13.6.1 Foot Orthoses (FOs)

Foot orthoses (FOs) are the most common type of lower limb orthosis. FOs are used to realign the foot, change the distribution of pressure in the foot, and reduce pain. In addition, they can correct for problems at more proximal joints (e.g., leg length discrepancy, weak quadriceps, and osteoarthritis of the knee). An FO can consist of an insert that fits inside the shoe, an internal shoe modification, an external modification to the heel or sole of the shoe, or combinations of the previous (Figure 13.6). Shoes may be considered FOs when prescribed for therapeutic reasons. A shoe with an enlarged toe box can be used to relieve pressure on claw, hammer toes, and bunions. Inserts inside the shoe are used to reduce pressure on load-bearing areas. Internal shoe modifications can provide the same benefits as inserts but cannot be transferred to another shoe. An important anatomical consideration in the use of foot orthoses is the goal of keeping the subtalar joint in a neutral position. The subtalar joint is the joint

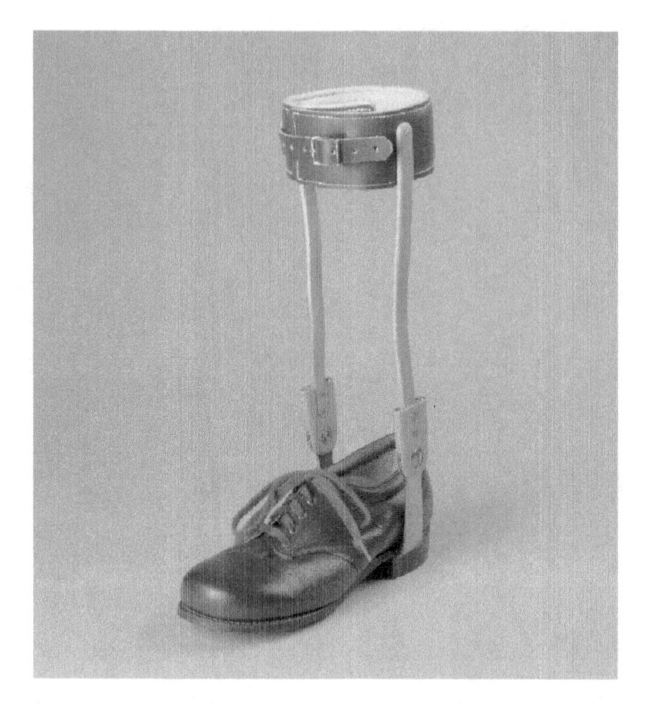

FIGURE 13.6 Metal AFO. (Courtesy of De LaTorre Orthotics and Prosthetics, Inc.)

between the calcaneus and talus. It allows for the motions of inversion and eversion as well as supination and pronation at the ankle. Abnormalities in the alignment of this joint can limit proper functioning of the foot, including gait abnormalities, and lead to pain and degeneration in all joints of the lower limb. Maintaining neutral subtalar joint alignment is the primary goal of FOs.

13.6.1.1 Common Indications for Foot Orthoses

Degeneration of the joints within the foot, or osteoarthritis (OA) of the foot, can lead to pain and difficulty with ambulation. By limiting motion at these joints, as well as limiting the amount of stress placed on specific joints, FOs can be used in the treatment of this condition. The University of California Biomechanics Laboratory (UCBL) orthosis is a device that can be inserted into a shoe to prevent excessive pronation. It is a custom-molded polypropylene plastic device molded to the shape and size of the patient's foot. It has a flat surface that contacts the bottom of the hind foot and prevents motion at the joints of the hind foot. Devices to treat midfoot OA can be as simple as a standard shoe with a very stiff sole. Steel shanks can be customized to fit inside a patient's shoes to limit midfoot motion in the sagittal plane. Finally, rocker-bottom shoes limit bending at the midfoot during ambulation by providing a transition from heel strike to push-off that does not require motion at the midfoot.

In OA of the knee, even though the site of pathology is at the knee, orthotic treatment can be accomplished with FOs. In typical knee OA, the medial knee joint is more significantly affected. When the medial knee joint decreases in size, the knee tends to deform such that the lower leg and foot are medially displaced (a condition known as genu varum, or "bowlegged"). A lateral heel wedge can be used to compensate for this condition and keep the leg in a normal alignment, thereby alleviating pain and preventing further degeneration.

Plantar fasciitis is a condition of very small tears in the fibers of the plantar fascia at its site of insertion on the bottom of the calcaneus. This creates inflammation at this site, which results in heel pain. The condition can be caused by excessive mechanical forces placed on the plantar fascia, for example, in an excessively pronated foot, tight Achilles tendon, or high arches. This condition can be treated with an orthotic device that distributes the weight to minimize pressure on the painful area. A device as simple as a heel pad (or heel wedge) can be used for this purpose. The UCBL orthosis can be used to maintain neutrality at the subtalar joint if excessive pronation is the underlying cause. An elevated arch support may be useful to decrease stress on the longitudinal arch of the foot, and therefore decrease stress on the plantar fascia.

Pes planus is a term used to describe a flat foot, or a foot with a reduced longitudinal arch. This condition is most frequently identified by excessive pronation, or "inrolling" of the foot, where the medial side of foot is too close to the ground. This deformity is most commonly caused by dysfunction of the posterior tibialis tendon, which plays an important role in the dynamic support of the longitudinal arch. When the tendon is weakened or ruptured, pes planus occurs. Correction of this deformity can be accomplished with an upward force on the medial side of the calcaneus. In order to provide the necessary forces, the distal part of the orthosis should extend beyond the metatarsophalangeal joints so that the lever arm is of sufficient length. In order to prevent further complications and adequately treat this condition, the subtalar joint must be in a neutral position when the foot is in the orthosis. Off-the-shelf shoe inserts are available for these purposes. In addition, a UCBL orthosis is a custom made device that can provide the necessary forces.

13.6.2 Ankle–Foot Orthoses

AFOs contain a component that crosses the ankle joint. Their primary function is to control the amount of dorsiflexion and plantarflexion that the ankle can move through. However, all AFOs have a component that connects to the foot, and can therefore be useful in controlling movement at the subtalar joint as well as providing medial–lateral ankle support. As with FOs, AFOs can be useful not only in conditions that affect the foot and ankle but in conditions of more proximal joints as well. For example, an AFO can be useful in the treatment of quadriceps weakness, a condition of instability at the knee.

Unlike FOs, AFOs cross a joint and have attachments both above and below the joint. Therefore, more components are required and the device as a whole is somewhat more complex. It is useful in the discussion of AFOs to divide them into two broad categories: metal and plastic. Metal AFOs have largely been replaced by plastic AFOs. However, a discussion of metal AFOs is important because many of

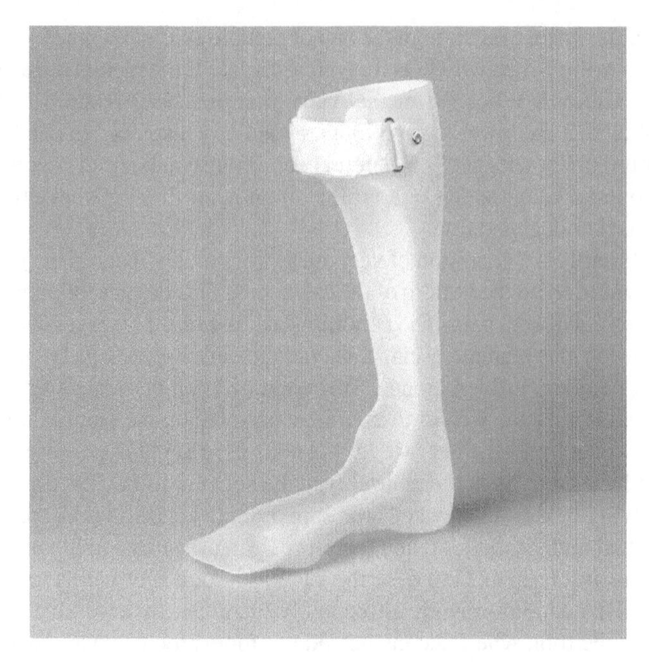

FIGURE 13.7 Plastic AFO solid ankle. (Courtesy of De LaTorre Orthotics and Prosthetics, Inc.)

the components are the same as those used in plastic AFOs, and the principles that guide the design and prescription are the same in both groups.

The metal AFO is made up of at least four components: (1) the calf band, which attaches the device to the leg, (2) the metal uprights, (3) the ankle joint, and (4) the attachment to the shoe (Figure 13.6). The calf band is made of metal only in its posterior half. The anterior half is a Velcro band. The function is to attach the AFO to the leg and to prevent posterior movement of leg relative to the ankle. The metal uprights connect the calf band to the ankle joint and shoe attachment, and can provide some medial–lateral support. The ankle joint is made up of several components that will be discussed further in the following text. The attachment to the shoe may be a solid or a split stirrup. A solid stirrup is permanently attached to the shoe, whereas a split stirrup (or calipers) allows the uprights to be removed from the sole plate that attaches to the bottom of the shoe.

The metal ankle joint is made up of two channels, one anterior and one posterior. Depending on the clinical application, pins or springs can be inserted into each channel. A pin is placed in a channel to prevent motion. A pin in the anterior channel will prevent dorsiflexion of the ankle, and it is known as an "anterior stop" or "dorsiflexion stop." A pin in the posterior channel will prevent plantarflexion of the ankle, and it is known as a "posterior stop" or "plantar stop." A spring placed in a channel will assist with motion at the ankle. A spring in the posterior channel will help the ankle dorsiflex, for example, in a patient with weakness of their dorsiflexion muscles that results in foot drop during ambulation, causing frequent falling. Springs are not typically used in the anterior channel because the amount of force needed to provide for functional plantarflexion would be too great to be accomplished with a spring.

As mentioned earlier, plastic AFOs have virtually replaced metal AFOs in clinical use. AFOs made from plastic are less expensive, more cosmetic, lighter, and provide better foot support when compared to metal AFOs. Cosmesis is improved because the plastic AFO can be worn inside the patient's shoes. They can be prefabricated (off-the-shelf) or custom molded to the specific dimensions of the patients and the specifications of the physician. Plastic AFOs can be divided into two categories: solid or hinged. A solid AFO is made from a single piece of plastic, and it has no extra ankle component (Figure 13.7). Even though there is no ankle joint, some motion is allowed at the ankle due to the inherent flexibility of the material. The angle between the foot and the leg will be determined by the clinical needs of the patient. For example, in a patient with dorsiflexion weakness, the angle should be set to 90° to allow for toe clearance during ambulation. A hinged or articulated plastic AFO has two plastic components connected by a metal ankle joint. They are used when some ankle motion is wanted, and complete restriction of movement is not necessary. The metal ankle joints are identical to those used in metal AFOs, and have the same components and functions.

The stability of the ankle joint in a plastic AFO is determined by three factors. (1) the trim line, which describes the line of the most anterior extension of the part of the AFO that is posterior to the ankle (the posterior leaf); if the trim line is moved anteriorly, the ankle becomes more stable; (2) the thickness of the plastic; a thicker plastic will provide more stability; and (3) corrugations in the posterior leaf, whose presence makes the ankle more stable.

13.6.2.1 Common Indications for AFOs

An equinovarus deformity is frequently seen in the lower limbs of children with cerebral palsy (CP). An equinovarus deformity describes a foot that is inverted, plantarflexed, and adducted. In CP, this is usually due to spasticity in the calf muscles. An AFO can be used to prevent or treat these deformities, which can make ambulation more feasible in patients affected. A posterior rod in the ankle joint would be useful in this case to prevent dorsiflexion.

Foot drop, or weakness of ankle dorsiflexors, is a condition frequently caused by an injury to the peroneal nerve. Depending on the severity of the weakness, this condition can lead to gait abnormalities including foot drop (failure to clear the toes during the swing phase of gait) and foot slap (failure to control the speed with which the foot returns to the ground after heel strike). An AFO can be used to substitute for the weak muscles and return the patient to a normal gait pattern. The thickness of the posterior leaf can be varied to give the desired amount of resistance to plantar flexion as well as the amount of passive dorsiflexion. If an ankle joint is used, a posterior channel spring can substitute for dorsiflexors that have no strength at all.

As discussed in the case of FOs, when a joint is affected by OA, limiting the movement of that joint can help treat the pain associated with OA. In ankle osteoarthritis, AFOs can be used to limit the motion at the ankle joints. In this case, the AFO should be custom molded to the patient's size and shape and should provide a great deal of ankle stability to limit the motion of the ankle. Simply by providing a well-fit AFO that effectively limits ankle motion during ambulation, a patient with significant ankle OA can have dramatic reductions in pain.

Knee instability can be a tremendous barrier to ambulation. In particular, quadriceps weakness due to a spinal cord injury can be devastating in terms of prognosis for ambulation. However, AFOs can help to overcome these barriers. When the quadriceps are weak, there is nothing to prevent "knee buckling," which would occur during the stance phase of gait any time there was flexion at the knee. This can be overcome by using an AFO to create a knee extension moment during stance phase. By limiting the amount of dorsiflexion that can occur (e.g., with an anterior channel pin), a knee extension moment can be created. This is because during stance phase, dorsiflexion creates a knee flexion moment and plantarflexion creates a knee extension moment. This can be visualized if one imagines a foot flat on the ground during the stance phase of gait. If the ankle is dorsiflexed, the tibia moves forward in space. In order to prevent the whole body from moving forward and falling over, the knee must flex to restore balance. Thus, dorsiflexion has created a knee flexion moment. In the case of quadriceps weakness, the aim is to prevent knee flexion, and therefore prevent "knee buckling." Therefore, by limiting the amount of dorsiflexion that can occur, the knee is made more stable, and the prognosis for ambulation has increased.

13.6.3 Knee Orthoses

A knee orthosis (KO) is an external device that crosses the knee in order to provide support, correct deformity, or prevent injury to the knee itself. Because the knee is the only joint involved, their use is limited to conditions that affect only the knee. Although they are used for a wide variety of applications, the use of KOs is a subject of some controversy. Although there are a few indications for KOS that are well supported by scientific evidence, the amount of knee orthoses prescribed far outnumbers those for which there is consensus regarding their utility. It is widely accepted that KOs are useful patients with genu recurvatum (hyperextension of the knee). By applying the three-point system of orthosis design, a Swedish knee cage can keep the knee from hyperextending (Figure 13.8). One brace is placed posterior to the knee joint at the level of the knee. The other two bands are placed anterior to the knee joint, one above and one below the level of the knee. With the use of a knee joint that allows full knee flexion while limiting extension beyond neutral, the patient can be protected from genu recurvatum.

In contrast to the Swedish knee cage, KOs can be used in patients with knee OA. Typically, in knee OA, the medial compartment of the knee joint is the most severely affected, and therefore has lost the most joint space. There is an asymmetry within the knee where the lateral joint space is larger than the medial joint space. This results in a valgus deformity at the knee (the foot is displaced laterally compared with a normal leg). By applying three-point systems in a different plane, this deformity can be corrected with a knee orthosis. This orthosis would have one strap on the lateral side of the leg at the level of the knee. The other two straps would be on the medial side of the leg, one above and one below the level of the knee. This system allows the pressure on the medial compartment of the knee joint to be minimized.

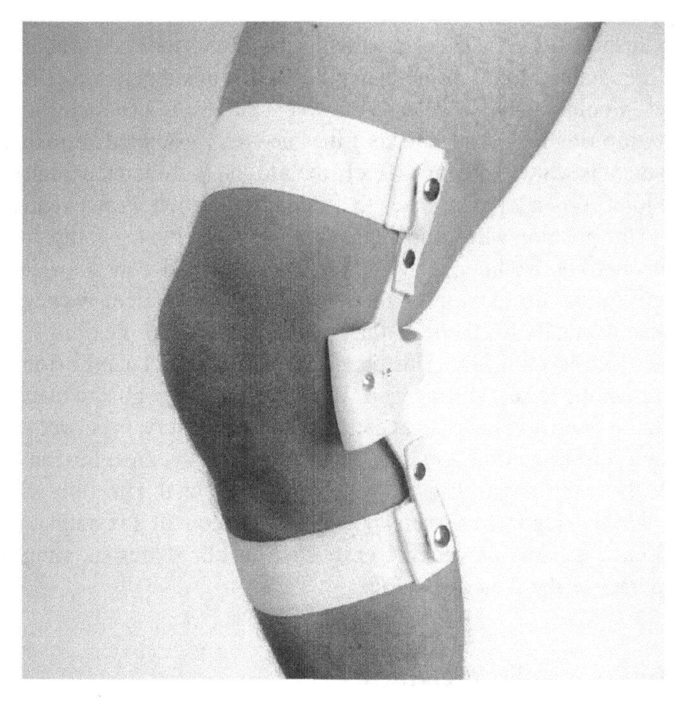

FIGURE 13.8 Swedish knee cage. (Courtesy of De LaTorre Orthotics and Prosthetics, Inc.)

Many providers and patients have advocated the use of KOs for prophylaxis against knee injury or in the rehabilitation from a knee injury. These braces are often prescribed to be worn during athletic activity. Despite their widespread use, there is little scientific evidence to support their efficacy, and the use of these devices is controversial. In some cases, the devices may be detrimental to the patient by increasing the amount of energy required for walking or running, or by providing the wearer with a false sense of security.

Regardless of the indication for its use, a KO must have some form of knee joint. A knee joint may be either single axis or polycentric. A single-axis knee joint rotates about a single axis, whereas a polycentric knee joint has an axis of rotation that varies with the position of the knee. Even though the polycentric knee joint more closely resembles the rotation of a human knee joint, it has not been found to improve ambulation. With few exceptions, the polycentric knee joint is limited to applications involving sports KOs.

A single-axis knee joint may be placed in line with the natural knee joint or may be offset in the anterior–posterior plane. A knee joint that is offset posteriorly can help stabilize the knee. If the knee is offset posteriorly, the ground reaction force that is applied to the lower limb from the ground will be directed in front of the knee. This will have the effect of promoting knee extension. By promoting knee extension, knee buckling can be prevented, thus making the knee more stable.

A knee lock is a component of an orthotic knee joint that locks the knee in a given position. For example, some patients may require full knee extension for safe ambulation. In this case, a knee lock would be used to keep the knee in a fully extended position during ambulation. For the patient to sit down and perform other activities, the knee lock would need to be released. Two examples of knee locks help illustrate their use. A drop-ring lock is used to keep the knee fully extended during ambulation. When the patient is seated, the ring is freely mobile and is around the thigh component of the knee joint. When a patient rises to a standing position, gravity causes the ring to descend to the position where the thigh component and the leg component overlap. The ring surrounds both components and prevents any relative motion between them, therefore preventing any flexion of the knee. When the patient wishes to flex the knee, they can manually lift the ring, thus freeing the joint for flexion.

A ratchet lock has a different mechanism that prevents knee flexion. When the patient is seated, the lock is disengaged. When the patient begins to extend the knee (for example, to stand up), the ratchet catches at various intervals between full flexion and full extension. Each time the lock catches, it prevents knee flexion beyond the point where the catch is set. Therefore, if a patient is unable to fully support himself/herself while rising from a seated position, the knee will flex only to the point of the most recent catch and prevent the knee from buckling, thus stopping the patient from falling back to the seated position.

13.6.4 Knee–Ankle–Foot Orthoses

Knee–ankle–foot orthoses (KAFOs) differ from KOs in that they cross the ankle joint and contain a component that contacts the foot (Figure 13.9). The knee joints available

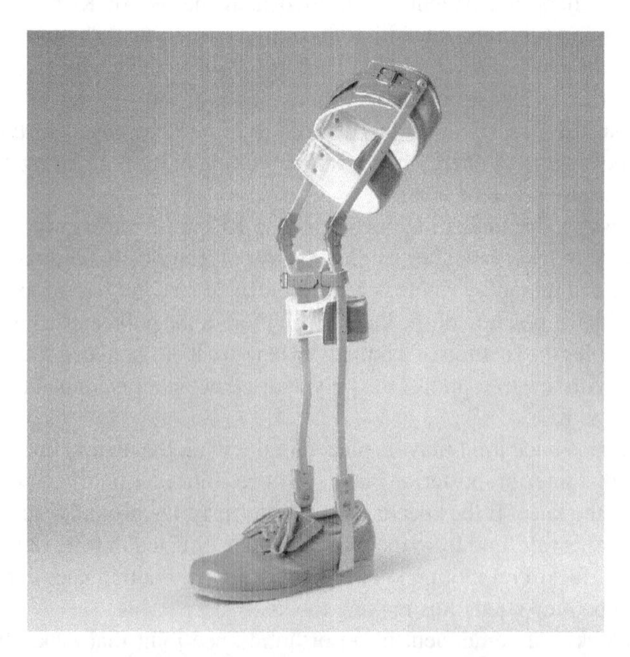

FIGURE 13.9 KAFO metal. (Courtesy of De LaTorre Orthotics and Prosthetics, Inc.)

for KAFOs are no different from those used in KOs, just as the ankle components are identical to those used in AFOs. The only difference being that, in a KAFO, all components are present in the same device. KAFOs are generally used in patients with severe knee extensor and hamstring weakness, knee instability, and spasticity of hamstring muscles, most commonly in patients with paraplegia due to spinal cord injury. Many patients with paraplegia are able to achieve ambulation with bilateral KAFOs used in combination with walkers or crutches. Therefore, it is crucial that users have good upper body and trunk control to be considered candidates for the use of KAFOs. Because of the weight and bulk of these devices, ambulation with KAFOs requires a great deal of energy expenditure and is rather inefficient. As a result, most patients with paraplegia who are able to ambulate with KAFOs will still rely on a wheelchair as their primary means of ambulation.

13.6.5 Hip–Knee–Ankle–Foot Orthoses and Reciprocating Gait Orthoses

Hip–knee–ankle–foot orthoses (HKAFOs) are very similar to KAFOs, but contain a component that attaches to the patient's trunk. The most common type of HKAFO is a reciprocating gait orthosis (RGO). An RGO is a bilateral HKAFO with both limbs connected. It is composed of a series of cables and pulleys. The purpose of the equipment is to provide for unilateral hip flexion simultaneously with contralateral hip extension. When a patient wearing a RGO lifts one limb off the ground, the cables and pulleys provide hip flexion of that limb, thereby advancing the limb, or accomplishing a step. Next, the other limb is lifted from the ground, and that limb is advanced to complete one full stride. RGOs are used in conjunction with upper limb crutches to assist in ambulation for patients who may otherwise not be able to ambulate. Much like KAFOs, RGOs are typically used in paraplegic patients. RGOs are usually prescribed to children aged 3 to 6 years. Similar to ambulation with KAFOs, ambulation with RGOs is slow, inefficient, and energy consuming. Almost all patients will choose to use a wheelchair as a primary means of ambulation, but RGOs can provide an excellent resource for exercise and may allow the patient to experience the world "at eye level."

13.7 UPPER EXTREMITY ORTHOSES

Upper extremity orthoses are mainly prescribed in order to provide support, prevent deformities; maintain function; and/or restore function of upper limb. Depending on the intended purpose, these orthoses are either static, providing complete immobilization, or dynamic, allowing only functional motion while restricting others. Typical conditions under which upper extremity orthoses are prescribed are fractures, nerve injuries (median, ulnar, or radial), brachial plexus injuries, burns, inflammatory joint conditions (rheumatoid arthritis), degenerative joint conditions (osteoarthritis), postsurgical management of tendon injuries, spasticity (postcerebrovascular accident or cerebral palsy), and repetitive stress/strain disorders (carpal tunnel syndrome). The prescription guidelines for all orthoses are very specific, which depends on each patient's medical condition and functional requirements. Also, orthosis wearing schedules and follow-up are crucial factors in maximizing the effectiveness of orthosis and potentially eliminating or reducing adverse effects that may result from its inappropriate use.

13.7.1 Finger Orthoses

The first category in finger orthosis comprises the immobilization orthoses, which are typically used to restrict motion at the metacarpophalangeal joint (MP), proximal interphalangeal joint (PIP), and distal interphalangeal joint (DIP) individually. The dynamic finger orthoses are used for mobilization of PIP and/or DIP joints. Two types of orthoses are prescribed for individuals with rheumatoid arthritis for prevention of deformities of boutonniere and swan-neck. Boutonniere deformity results in flexion of PIP and hyperextension of DIP, whereas swan-neck results in hyperextension of PIP with flexion of DIP joints. The orthosis prescribed for boutonniere deformity prevent flexion at the PIP joint and extension at DIP. The orthosis prescribed for correction of swan-neck deformity prevents hyperextension at PIP and flexion at DIP joints. The two aforementioned orthoses are important as they prevent occurrence and/or aggravation of deformities while not restricting functional use of the hand in other activities.

13.7.2 Hand Orthoses

The immobilization orthoses in this category limit motion at the MP, PIP, and DIP all together. These orthoses are also prescribed for prevention of MP joints ulnar deviation, which is commonly seen in individuals with arthritis. Two types of mobilization hand orthoses mainly used are: PIP flexion mobilization orthoses and PIP extension mobilization orthoses. These dynamic orthoses are effectively used in flexor tendon contracture and postsurgical management of extensor tendon.

13.7.3 Wrist–Hand Orthoses

Wrist immobilization orthoses (WHO) that are typically used to prevent motions of wrist extension and wrist flexion are the dorsal wrist immobilization orthosis and volar wrist immobilization orthosis. An ulnar wrist immobilization orthosis is used to prevent excessive ulnar deviation and the radial wrist, and thumb immobilization orthosis is used to prevent radial deviation while performing activities. The aforementioned orthoses are useful in conditions such as rheumatoid arthritis and deQuervain's condition, which result because of acute inflammation of the abductor pollicis longus tendon. Emphasis, while immobilizing the wrist joint, is placed on maintaining the hand in a functional position of 25 to 30° wrist extension. WHO that allow mobilization of the wrist and other joints in the hand are the wrist extension MP flexion mobilization orthosis and the wrist flexion MP extension mobilization orthosis (Figure 13.10). The purpose of this orthosis is to develop natural grasp and release pattern after both central nervous dysfunction such as a stroke and peripheral nerve injuries such as radial nerve palsy.

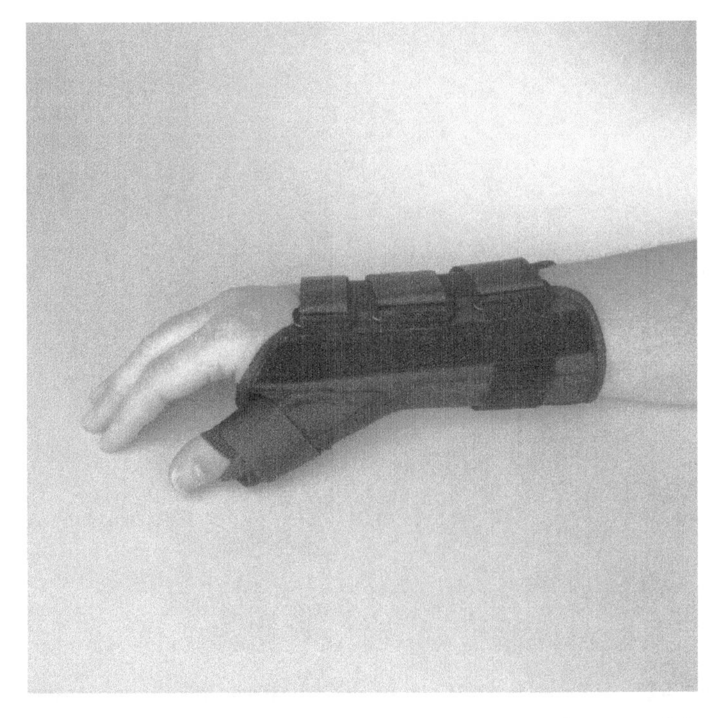

FIGURE 13.10 Wrist–hand orthosis with thumb extension. (Courtesy of De LaTorre Orthotics and Prosthetics, Inc.)

Other forms of mobilization WHO consist of MP flexion and MP extension mobilization orthoses, which are commonly used in postsurgical rehabilitation after finger flexor and extensor tendon injuries. WHO are used in decreasing muscle tone in the hand after a stroke or other types of disorders involving upper motor neurons. Antispasticity WHO commonly prescribed are: hard cone and thumb/fingers spreader. Despite extensive clinical prescription, the evidence for the effectiveness of these orthoses in controlling spasticity is limited.

13.7.4 Elbow Orthoses

Two types of immobilization elbow orthoses (Eos) are available: posterior elbow immobilization orthosis and anterior elbow immobilization orthosis. The anterior elbow immobilization orthosis is commonly prescribed after burns in order to prevent elbow flexion contractures. A nonarticular circumferential elbow orthosis is used frequently in acute-stage intervention for a condition known as lateral epicondylitis or tennis elbow. A mobilization orthosis for elbow joint is prescribed after elbow extensors contracture to provide slow, progressive flexion motion for lengthening elbow flexors.

13.7.5 Shoulder Orthoses

Shoulder abduction orthosis is used for immobilization of the shoulder joint in an abducted position and for prevention of an adduction/internal rotation deformity,

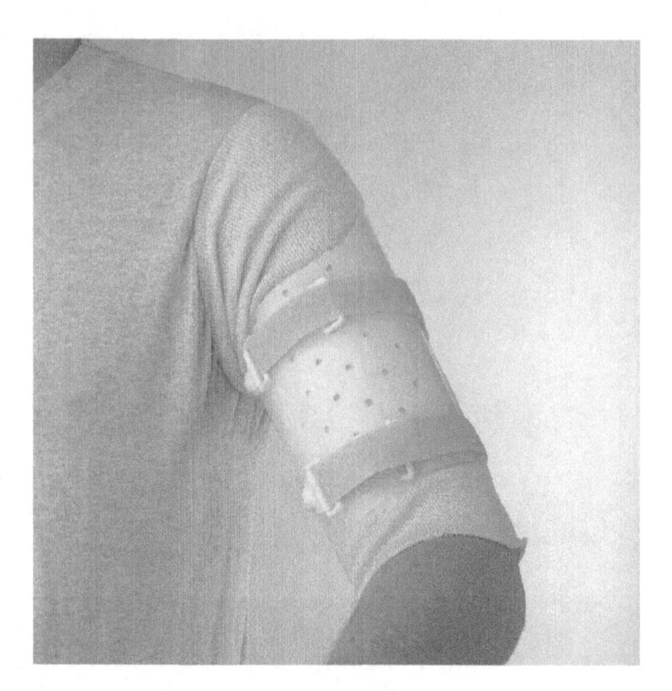

FIGURE 13.11 Humeral brace. (Courtesy of De LaTorre Orthotics and Prosthetics, Inc.)

which is very common after brachial plexus injury or after burn. Humeral orthosis, which is a nonarticular orthosis and is also referred to as a functional brace, refutes the old concept of strict immobilization after fracture, using casting or other techniques (Figure 13.11). This orthosis, which is circumferential, is prescribed after a fracture of the shaft of the humerus and is effective in enhancing the healing process while simultaneously reducing complications due to prolonged immobilization. The concept used in designing this orthosis is that partial loading on the bone after fracture enhances activities of osteoblasts, which are the primary factor in bone healing. After hemiplegia/hemiparesis resulting from a stroke, use of shoulder orthoses, also referred to as *hemiplegic sling*, has been indicated as an effective intervention for further prevention of shoulder joint dislocation or subluxation. This orthosis passively holds the shaft of the humerus in the glenoid fossa, preventing dislocation by compensating for lack of voluntary control of muscles and laxity of ligaments around the shoulder joint. However, evidence for their effectiveness as an intervention after stroke is lacking. Also, use of this form of orthosis is recommended to combine with other rehabilitation techniques, such as therapeutic exercises and activities, in order to prevent future nonuse or neglect of the affected extremity.

13.7.6 Shoulder–Elbow–Wrist–Hand Orthoses

The most common form of shoulder–elbow–wrist–hand orthoses (SEWHO) is one given to patients after brachial plexus injury. This orthosis keeps the upper extremity in abduction, externally rotated at the shoulder joint, extended and pronated at the elbow, and extended at the wrist joint, thus preventing occurrence of further deformity. This "airplane orthosis" is also useful after burns, especially to the axillary region for prevention of contractures.

13.7.7 Advances in Upper Extremity Orthoses

Advances in upper extremity orthoses has been indicated in several domains. Use of functional electrical stimulation (FES) adjunct with an upper extremity orthosis is one such emerging intervention. Use of an FES system combined with a hand orthosis was reported by Weingarden et al. (1998) in reducing muscle tone of hand muscles along with significant improvement in the functional abilities of individuals in chronic stages of hemiparesis following stoke and traumatic brain injuries. Other forms of upper extremity orthoses are those that utilize external power sources for either providing continuous motion, such as continuous passive motion (CPM) orthosis, or to provide assistance during upper limb motion. One such form of powered upper extremity orthosis was designed and tested for providing continuous assisted or resisted motion of shoulder, elbow, and forearm joints altogether or individually. This externally powered orthosis called a motorized upper limb orthotic system (MULOS) was reported to be effective in prevention of upper extremity joint contractures, assisting in daily functional tasks, and improving strength of upper extremity muscles.

13.8 SUMMARY

Successful outcome with the use of an orthosis is dependent on many factors. The therapists' or orthotists' skill levels can have a direct effect on the design and fit of the device. If the device interferes with participation in home, work, school, or social activities, poor compliance is eminent. It is also important that the wearer is able to don/doff the device independently or have adequate help to follow the wearing schedule. In addition to proper fit and function, comfort and perceived cosmesis must also be considered as important aspects in the design of orthoses. Even if a device has been meticulously designed to provide the ideal biomechanical properties, if that device is uncomfortable or not cosmetically pleasing to the wearer, it may be discarded or seldom worn. Of course, the benefits of an orthosis cannot be realized if it is not used routinely. For these reasons, it is important to take a client-centered approach to the design and prescription of orthoses so that all issues can be addressed with the unique situation of the patient in mind.

13.9 STUDY QUESTIONS

1. Provide a definition for orthosis.
2. Describe the difference between static and dynamic orthoses.
3. Describe the four general curves along the spine.
4. Provide a definition of scoliosis and its most common route cause.

5. A single-axis knee joint may be placed in line with the natural knee joint or may be offset in the anterior–posterior plane. Explain why the offset knee joint would be used.
6. Explain why most individuals with paraplegia are able to ambulate with KAFOs but nearly all still rely on a wheelchair as their primary means of ambulation.
7. Name seven typical conditions under which upper extremity orthoses are prescribed.
8. What is an RGO?
9. In addition to proper fit and function, what other two factors must also be considered as important aspects in the design of orthoses.
10. Define creep as it applies to orthoses.

BIBLIOGRAPHY

Ada L, Foongchomcheay A, Canning C. Supportive devices for preventing and treating subluxation of the shoulder after stroke. Cochrane Database of Systematic Reviews 2005:CD003863.

American Society of Hand Therapists. Splint classifications system. Garner, NC: The American Society of Hand Therapists, 1992.

Andriacchi T. Milwaukee brace corrections of idiopathic scoliosis. *J Bone Joint Surg* 58A:806,1976.

Aoyagi Y, Tsubahara A. Therapeutic orthosis and electrical stimulation for upper extremity hemiplegia after stroke: a review of effectiveness based on evidence. *Top Stroke Rehabil* 11:9–15,2004.

Axelsson P. Lumbar orthosis with unilateral hip immobilization, effect on intervertebral mobility determined by roentgen sterophotogrammaetric analysis. *Spine* 18:876–879,1993.

Bonnett C, Tosoonian R. Results of Milwaukee brace treatment in seventy patients. *Orthop Rev* 7:79,1978.

Bono CM, Berberian WS. Orthotic devices: degenerative disorders of the foot and ankle. *Foot Ankle Clin* 6:329–340,2001.

Buonomo LJ, Klein JS, Keiper TL. Orthotic Devices: Custom-made, Prefabricated, and Material Selection. *Foot Ankle Clin* 6(2):249–252,2001.

Cailliet R. Biomechanics of the Spine. *Phys Med Rehabil Clinics North Am* 3:1–28,1992.

Coppard BM, Lohman, H. (2001). *Introduction to Splinting: A Clinical-Reasoning & Problem-Solving Approach*, 2nd ed. Mosby: St. Louis.

DeLisa J, Gans B, Walsh N (Eds.), *Physical Medicine and Rehabilitation, Principles and Practice*, Fourth Edition, Lippincott Williams & Wilkins, 2005.

Deyo R, Tsui-Wu Y. Descriptive epidemiology of low-back pain and its related medical care in the United States. *Spine* 1987 12:264–268,1987.

Fast A. Low back disorders: Conservative Management. *Arch Phys Med Rehabil* 69:880–891,1988.

Fishman S. Spinal orthoses. *From American Academy of Orthopedic Surgeons: Atlas of Orthotics, Biomechanical Principles and Applications*, 2nd ed., St Louis, 1985, CV Mosby.

Frontera W, Silver J, Eds. *Essentials of Physical Medicine and Rehabilitation Review and Self-Assessment*. Philadelphia: Hanley and Belfus; 2003.

Garfin S. Complications in the use of the halo fixation device. *J Bone Joint Surg* 68:320–325,1986.

Gavin T. Preliminary results of orthotic treatment for chronic low back pain, *J Pros Orthos* 5:5/25–9/29,1993.

Goldberg B, Hsu JD. *Atlas of Orthoses and Assistive Devices.* St. Louis, MO: Mosby, Inc., 1997.

Hartman J, Palumbo F, Hill B. Cineradiography of the braced normal spine: A comparative study of five commonly used cervical orthoses. *Clin Orthop* 107:97–102,1975.

Jacobs MA, Austin NM. *Splinting the Hand and Upper Extremity: Principles and Process.* Lipincott Williams & Wilkins: Baltimore, 2003.

Johnson R. Cervical orthoses: a study comparing their effectiveness in restrictin cervical motion in normal subjects. *J Bone Joint Surg* 59A:332–339,1977.

Johnson GR, Carus DA, Parrini G, Scattareggia Marchese S, Valeggi R. The design of a five-degree-of-freedom powered orthosis for the upper limb. *Proceedings of the Institution of Mechanical Engineers Part H — Journal of Engineering in Medicine* 215:275–284,2001.

Lannin NA, Herbert RD. Is hand splinting effective for adults following stroke? A systematic review and methodologic critique of published research. *Clin Rehabil* 17:807–816,2003.

Lewin P. Cotton Collar: a physical therapeutic agent. *JAMA* 155:1155–1156,1954.

Loke M. New Concepts in Lower Limb Orthotics. *Phys Med Rehabil Clin North Am* 11:477–496,2000.

Lonstein J, Carlson M. The prediction of curve progression in untreated idiopathic scoliosis. *J Bone Joint Surg* 66A:1061,1984.

Lonstein J, Winter R. Milwaukee brace treatment of adolescent idiopathic scoliosis — review of 1020 patients. *J Bone Joint Surg* 76A:1207,1994.

McKee P, Morgan L. *Orthotics in Rehabilitation: Splinting the Hand and Body*. F.A. Davis Company: Philadelphia, 1998.

Nachemson A, Peterson L. Effectiveness of brace treatment of moderate adolescent idiopathic scoliosis. *J Bone Joint Surg* 77A:815,1995.

Neeman RL, Neeman HJ, Neeman M. Application of orthokinetic orthoses in habilitation of a person with upper extremity incoordination secondary to spastic quadriplegia due to cerebral palsy. *Can J Rehabil* 1:145–154,1988.

Noll KH. The Use of Orthotic Devices in Adult Acquired Flatfoot Deformity. *Foot Ankle Clin* 6(1):25–36,2001.

Quigley M. Evaluation of two experimental spinal orthoses. *Orthot Prosthet* 28:23–41,1974.

Rothstein JM, Roy SH, Wolf SL. *The Rehabilitation Specialist's Handbook*, 3rd ed. FA Davis Company: Philadelphia, 2005.

Sarmiento A, Burkhalter WE, Latta LL. Functional bracing in the treatment of delayed union and nonunion of the tibia. *Int Orthop* 27:26–29,2003.

Schultz AB, Ashton-Miller J. Biomechanics of the human spine. In Mow V, Hayes W (Eds.): *Basic Orthopaedic Biomechanics*, New York, Raven Press, 1991.

Stillo J, Stein A, Ragnarsson K. Low-back orthoses. *Phys Med Rehabil Clin North Am* 2:57–94,1992.

Weingarden HP, Zeilig G, Heruti R. et al. Hybrid functional electrical stimulation orthosis system for the upper limb: effects on spasticity in chronic stable hemiplegia. *Am J Phys Med Rehabil* 77(4):276–281,1998.

Whitehill R, Richman J, Glaser J. Failure of immobilization of the cervical spine by the Halo vest: A report of five cases. *J Bone Join Surg* 68A:326–332,1986.

14 Aids for People Who Are Blind or Visually Impaired

John Brabyn, Katherine D. Seelman, and Sailesh Panchang

CONTENTS

14.1 LEARNING OBJECTIVES OF THIS CHAPTER

Upon completion of this chapter, the reader will be able to:

Learn the approximate size of the visually impaired population
Understand the different types and dimensions of visual impairment and their different impacts on visual task performance
Learn about the basic methods of vision assessment
Understand the problems caused by dual sensory loss

Gain knowledge about the history of what has been done in this field
Gain knowledge of the different approaches to medical and surgical function restoration
Learn how universal design concepts apply to blind and visually impaired persons
Understand the different approaches to assistive technology for way-finding
Understand the problems and options for reading and graphics access
Understand the barriers (and potential solutions) to computers, Internet, and telecommunications access faced by this population
Understand the employment and daily living problems and solutions for blind and visually impaired persons

14.2 TARGET POPULATION

The causes and demographics of blindness vary greatly in different parts of the world. In most industrialized countries, approximately 0.4% of the population is called "legally blind" for the purposes of qualifying for government assistance. With variations by country, this definition means best corrected acuity of 20/200 or worse (i.e., one-tenth of "normal" visual resolution) or a visual field restricted to 20° or less. In the US, this definition includes approximately 1 million people; 3.5 to 5 million have low vision, usually defined as best corrected visual acuity less than 20/70; and almost 14 million have some visual impairment that hampers performance and enjoyment of everyday activities. About 200,000 to 300,000 have no useful vision at all. For our purposes, the term "blind" is restricted to this group, and "low vision" or "visually impaired" is used for those with partial vision. Although most people with low vision are in the older age groups, the totally blind population is more evenly distributed.

In the younger age groups, visual impairment is on the rise due to perinatal birth defects associated with drugs and the ability to save premature babies who are at risk of many medical conditions. About 35% of visually impaired infants have cortical visual impairments, that is, the source of the impairment is in the brain rather than in the eye. This problem is not yet well understood or addressed.

School and working-age blind and visually impaired persons face major barriers in daily living, information access, way finding, and other tasks, and have very high unemployment rates. The prevalence of multiple sensory (e.g., deaf–blind), physical, or cognitive impairments is also growing.

The vast majority of visually impaired people are aged 65 years or over, a group growing twice as fast as the overall population. There is a per-decade increase of 1 to 2 million persons over 65 years with functional limitations in seeing, defined as difficulty in seeing ordinary newsprint. Even less well addressed is the over-85 age group, which has grown ten times as fast as the overall population during the past century. Approximately 10% of this group is legally blind and over 20% has low vision. Vision deficits are one of the first impairments seen in aging, and often go along with hearing impairments.

In other parts of the world, the incidence of blindness is much higher, and is mostly from "preventable" causes. About 1% of Africans are blind, and it is estimated that cataracts alone (which can easily be removed) account for about 20 million cases of blindness in Asia and Africa, which is about half of the world's total blind population. Trachoma, an infection treatable with antibiotics, is endemic in areas with poor hygiene and contaminated water. Similarly, onchocerciasis, or "river blindness," spread by black flies, is treatable but is still a major cause of blindness in Africa. Scarring of the cornea due to vitamin A deficiency, measles, and other causes also results in countless cases of blindness in some developing countries.

14.3 DIMENSIONS OF VISUAL IMPAIRMENT AND THEIR IMPACT ON TASK PERFORMANCE

14.3.1 Definitions of Visual Acuity

Traditionally, visual acuity is expressed as a Snellen fraction such as 20/200, meaning the ability to identify at 20 ft a letter that a "normal" viewer can identify at 200 ft. Thus, it is a measure of angular spatial resolution, and visual scientists often express it as a "LogMAR" number, that is, the logarithm (to the base 10) of the minimum angle of resolution in minutes of arc. Conveniently, 20/20 represents a resolution of about 1', or a LogMAR of 0, and 20/200 represents a LogMAR of 1.0. Most people can achieve 20/15 acuity (LogMAR 0.12) with best optical correction.

14.3.2 Nature of Different Visual Impairments

Different "blinding diseases," accidents, wounds, or ocular pathologies give rise to widely differing forms of functional visual impairments, most of which are not measurable simply as a loss of acuity or resolution. For example, cataracts and other ocular media do degrade acuity but also cause an even bigger loss in the contrast of the image on the retina, known as contrast sensitivity. Scattering of light in the ocular media causes an even more severe reduction in contrast in the presence of glare. Age-related maculopathy (ARM), the most common cause of statutory blindness in the United States, causes loss of resolution and ultimately blind spots ("scotomas") in the center of the visual field, accompanied by a reduction in contrast sensitivity and inability to adapt rapidly to different light levels. Advanced glaucoma and retinopathy of prematurity (ROP) cause a degradation of the peripheral field. Diabetes can cause losses in seemingly random parts of the visual field. Color blindness affects about 8% of the male population and can affect tasks such as interpretation of computer graphics, although it is not considered disabling.

14.3.3 Impact on Task Performance

Naturally, these different functional deficits have widely differing impacts on the performance of different visual tasks. Standard acuity testing only evaluates one's ability to read very-high-contrast letters in very good lighting — a situation that does not correspond to most real-world viewing conditions. Most eye diseases degrade performance even more under nonideal viewing conditions; hence, effective impairment is often far greater than that suggested by an individual's acuity score. Only in recent years has there been greater appreciation of these increased deficits in real-world conditions. Because of this complex interplay between the dimensions of visual deficits, the viewing conditions, the varying visual demands of different tasks, and the impact of an individual's visual impairment on performance of a particular task can be hard to predict. In general, there is increasing evidence that in many cases, visual task performance is more closely related to variables such as contrast sensitivity than to resolution or acuity. For these reasons, in rehabilitation practice, the provision of optimal lighting and contrast, and the elimination of glare, are just as important as providing adequate magnification.

14.3.4 Disability Rating Scales and the International Classification of Function (ICF)

Because of the difficulty of characterizing the interaction between the many dimensions of visual function, real-world abilities, and task performance, eligibility for disability benefits has historically been based on the simplest medical measures of visual impairment, namely visual acuity and visual field extent. ICD-9-CM (1978) (the official U.S. clinical classification of diseases) introduced a gradual scale of mild/moderate/severe/profound/total vision loss to replace the old black-and-white dichotomy of legally sighted vs. legally blind. The 5th edition (2001) of the *AMA Guides to the Evaluation of Permanent Impairment* added a numerical scale to this, with 20/20 rated as 100 points, 20/200 as 50 and 20/2000 as 0 (similar calculations factor in losses of visual field). Although any such scale is somewhat arbitrary, a recent study (Fuhr et al., 2003) showed that this scale, developed mostly by Colenbrander (Colenbrander, 1977), correlated better than other scales with self-reported quality of life.

In a recent report on disability determination for individuals who are visually impaired, the National Research Council (NRC, 2002) took a similar approach, emphasizing the continuity of function from normally sighted to totally blind. It recommended the use of a very similar scale to that outlined here, using the visual acuity rating (VAR), which can be calculated as follows: VAR $= 100 - 50$ LogMAR (NRC, 2002). This also gives the current 20/200 "statutory blindness" level a score of 50%.

The aforementioned scales adhere to what has been called the "medical model of disability." In recent years, another framework for classifying functional deficits has emerged in the form of the International Classification of Functioning (ICF) of the WHO. This classification adheres to the "social model of disability," based on the difficulties people experience in their participation in societal activities. To fully describe a disability, both aspects are needed, because the difficulties a person experiences result from the interaction between individual abilities and societal demands. For example, the AMA scale measures visual acuity and relies on the fact that visual reading ability is related to visual acuity. The ICF lists "reading" and includes visual reading as well as Braille reading, because both contribute to literacy, which in turn contributes to participation. This emphasis on ability, function, and participation is an increasing trend in rehabilitation theory and terminology.

14.3.5 Visual Function Assessment

Measurement of standard visual acuity is performed with the familiar letter chart, either at distance (usually 20 ft or 6 m) or near (usually 16 in. or 40 cm). It is important that the subject wear the appropriate refractive correction for the test distance and that the measurement is made in good lighting. For psychophysical rigor, it is also important that the test chart is one with equal numbers of letters on each line; the old fashioned Snellen charts, still often used, are quite unreliable for measuring severely impaired vision as the largest (usually 20/200) line has only one letter, invariably "E," which everyone has memorized.

Visual fields are normally measured using an automated perimeter, an expensive piece of equipment that requires the subject to fixate on a central target light while spots of light are presented randomly at different positions in the periphery. The subject is asked to press a switch when a peripheral target is seen.

The most commonly used measure of low-contrast vision is the Peli–Robson contrast sensitivity chart, calibrated in log units of contrast sensitivity, with large letters on a white background, all the same size but with decreasing contrast from top to bottom.

Measurements of other aspects of visual function such as acuity in the presence of glare, glare recovery, and vision in poor lighting are as yet far less standardized and are seldom performed routinely, even though it is increasingly recognized that they are extremely important in relation to real-world task performance.

14.4 CONCEPTUAL FRAMEWORK FOR REHABILITATION

14.4.1 Enhancement vs. Substitution

In engineering terms, sensory loss can be regarded as a reduction of information input, reducing access to the information needed to perform any given task or to interact with the environment. Rehabilitation engineering for this population takes two main approaches to replacing the lost information and thereby improving abilities and function:

1. Enhance the effectiveness of, and information provided by, remaining vision (mainly but not exclusively through various techniques for improving magnification and contrast).
2. Substitute for lost visual information by providing auditory and/or tactile information.

In visual enhancement, the challenge is complicated by the fact that vision loss has to be regarded as multidimensional, with several different (not necessarily independent) vision variables caused by different disease processes. In visual substitution, the challenge is to present the necessary information effectively under the constraints imposed by the much lower effective "bandwidth" of the other senses compared to the vision channel.

In low vision, the usual goal of assistive technology is to try and enhance and make best use of the residual vision, whereas in the case of blindness the approach is generally to substitute input from other senses (primarily hearing and touch) to help replace the lost information. These distinctions can often be blurred; for example, a person with low vision might use both magnification and synthetic speech for computer access.

14.4.2 General vs. Task-Specific Solutions

Another dichotomy found in the field of sensory aids for both visual and hearing impairment is the division between general-purpose devices, which are intended for a multitude of tasks, and devices or adaptations intended to deal with one specific task. A head-mounted display, for example, may be intended by its inventor to provide visual enhancement that can be used in any situation or visual task, from reading to driving a car. Text enlargement software is an example of the opposite approach, where the adaptive technology is intended to focus solely on improving computer access for a visually impaired person.

There have been numerous research projects designed to produce a general-purpose substitution for the lost visual sense, the latest examples being the retinal implant projects, which aim to restore vision in suitable candidates to a level that would obviate the need for other adaptive technology regardless of the visual task. Until this can be achieved, however, the more common rehabilitation engineering approach for the totally blind population is to address tasks one at a time. This has led to different families of assistive technology for computer access, travel, reading, and a large array of very specific adaptive devices for specific tasks such as operating a lathe, setting an oven dial, or reading an electrical quantity on a meter.

The concept of universal design can be regarded as another general-purpose approach, making the environment and products easier to use for people with a wider range of visual and other abilities. In the case of blindness, this concept often offers only a partial solution, but it can be extremely effective for milder visual impairments.

14.5 HISTORICAL OVERVIEW OF TECHNOLOGY FOR BLIND AND VISUALLY IMPAIRED PERSONS

Early developments in rehabilitation technology for blind persons included the development of the six-dot tactile Braille code in the 19th century and the advent of formal long cane training methods after World War II to assist blind persons with safe travel. For persons with low vision, early technological aids came in the form of magnifiers for reading and telescopes for finding and identifying objects in the distance (such as street signs and information on blackboards). Since the 1960s and 1970s, rapid advances in electronics combined with the expansion of rehabilitation programs in general have led to a much wider array of assistive technologies for this population.

The most common approach has been to develop task-specific solutions for activities such as reading, mobility, or computer access. However, there was also early experimentation with more general-purpose approaches to replacing the lost visual input, such as the tactile vision substitution system, which converted the camera images to tactile inputs on the skin using vibrotactile or electrocutaneous stimulator arrays. Another example was the early work by Brindley and Lewin (1968) on direct stimulation of the visual cortex to reproduce visual images. In both cases, very simple image elements could be interpreted, but real-world images were far too complex, and task-specific solutions became predominant.

During this period, early steps towards universal design and access were also made, but were driven primarily by accessibility for wheelchair users, which resulted in relatively few benefits to blind persons. The following sections summarize the developments in assistive technology that followed these early experiments, including the practical solutions that are available today.

14.6 GENERAL-PURPOSE ASSISTIVE TECHNOLOGY SOLUTIONS

14.6.1 Medical or Surgical Approaches to Restoring Function

Blurring the line between medicine and rehabilitation are the many current research projects aimed at restoration of partial visual function by implanting electrodes or bioelectric materials in various parts of the visual system. Most researchers feel that these technologies have a considerable way to go before becoming practical, and for the foreseeable future, eligible candidates for the proposed implants would be a limited subset of the overall blind and visually impaired population. However, considerable progress is being made, and there is an interesting race developing between these approaches and new medical treatments that would prevent or delay the progression of the major diseases that cause blindness or visual impairment.

14.6.2 Cortical Implants

The pioneering work in this field was the investigation in the 1960s of direct stimulation of the visual cortex using implanted electrodes to create "phosphenes" or perceptions of patches of light in the visual field. A more recent system by William Dobelle uses two 64-element electrode arrays, one for each hemisphere. A percutaneous arrangement delivers power to the electrodes, with the image signal derived from a spectacle-mounted camera. At least eight blind patients have been implanted overseas. Results varied, but some patients were able to detect simple objects. On one patient an equivalent visual acuity (using tumbling Es and Landolt Cs) in the region of 20/1000 was reported, and the visual field covered by the resulting phosphenes was reported to be about $22 \times 14\,\text{in.}^2$ at arm's length.

A similar approach has been under development by a team at the University of Utah led by Richard Norman. Rather than surface electrodes, this team has used a penetrating electrode array to communicate directly with large numbers of individual nerve cells at low current levels (1 to 10 μA). Research has included investigation of methods to unscramble the means by which the visual pathway maps images onto cortical structures in a complex and seemingly unpredictable way. Recent versions have used 100-element (10 × 10) arrays with electrodes 1.5 mm long and approximately 0.4 mm apart. The eventual aim is to provide a total of about 625 electrodes.

Compared to the retinal implant approach, the cortical approach is likely to be more promising for persons with nonfunctioning retinas or optic nerves. Difficulties to be overcome include the somewhat irregular mapping of retinal geometry on the visual cortex and disruption of the images due to uncontrolled eye movements.

14.6.3 Retinal Implants

Larger numbers of researchers have been pursuing approaches involving implanting electrodes in the retina in order to stimulate the visual sense. Active groups include Harvard University, MIT, Johns Hopkins, the Doheney Eye Institute, Optobionics, and groups in Germany, Japan, and Australia. Initial implant experiments have been performed in human volunteers. It has been shown that transretinal electrical currents can evoke visual percepts and simple patterned percepts in individuals with retinitis pigmentosa by using implants with four electrodes.

As of this writing, the Doheney group in collaboration with Second Sight, Inc. (part of the Mann group that was involved in the development of cochlear implants), implanted six totally blind retinitis pigmentosa patients with 16-element (4 × 4) epiretinal arrays attached to the retinal surface by tacks, covering about 12° of visual angle. The electrodes stimulate the ganglion cells and their axons traveling across the surface of the retina so that the perceived points of light or phosphenes do not necessarily correspond exactly to the geometrical pattern of the electrodes and some remapping will be required. The next implants planned will have 60 elements.

The Optobionics group inserted a suspension of nonpowered microphotodiodes under the retina to stimulate the remaining layers of retinal neurons. Improvements in vision have been reported by test subjects, but it is not clear that the photodiodes were producing direct retinal nerve stimulation. It is hypothesized that the benefits may be due to upregulation of growth or neurotrophic factors. There is some debate as to whether passive electrodes can produce sufficient current for direct neural stimulation, and most research groups are focusing on implants that are externally powered.

Some researchers are working on methods for more accurate stimulation of individual cells rather than attaching electrodes that may affect a region of cells. The Stanford University group is exploring an array of micromachined apertures in soft materials as a spatially controlled fluidic delivery device that would stimulate the retinal cells by supplying neurotransmitters. Nerve regeneration would be used to grow dendrites to these apertures so that single cells can be addressed individually.

Apart from the technical issues remaining to be addressed in retinal implants, questions remain about how to deal with eye movements, spatial remapping, interactions between electrodes, and the ability of degenerated and deprived retinal and cortical cells to process the stimuli. Overall, the retinal implant approach avoids most (but not all) of the problems of geometric mapping of images on the visual cortex and preserves more of the visual system's processing. However, it presents greater challenges of technical and surgical practicality and requires the patient to have a functional optic nerve pathway.

14.6.4 Optic Nerve Stimulation

One research group has investigated the possibility of simulating the optic nerve using a cuff electrode wrapped around it. Experiments have been conducted in a blind human volunteer using four surface electrodes. A difficulty with this approach is to present a geometrically meaningful map of visual space because of the bundled arrangement of the optic nerve, so it may be necessary to use large numbers of penetrating electrodes and sophisticated computation to drive such an implant.

14.6.5 Auditory and Tactile Information Display

Researchers have long worked on tactile and auditory display technologies that could have application to multiple tasks. Tactile information display work by Bach-y-Rita et al. (1969) presented images on the skin of the stomach and back using vibrotactile and electrocutaneous arrays as large as 1000 points (32×32). In general, it was found that very simple images could be interpreted readily, and the commercial success of the Optacon, which presented one letter or part of a letter at a time, confirmed this finding. More complex images such as video images of real-world scenes were nearly impossible to interpret, and so means of simplifying and processing the images were investigated, including the use of scanning sonar as input instead of a camera. Continuing pursuit of this line of work by Bach-y-Rita and collaborators (Sampaio et al., 2001) has led to displays that can present tactile images to the tongue using a flexible 12×12 point electrode array.

In the area of auditory information display, "sonification" methods have been proposed and explored for converting video images into auditory images using various transformations. The family of ultrasonic spatial sensors developed by Kay can also be used as a general-purpose environmental probe for a range of different tasks.

14.6.6 Computer Vision

Computer vision is the application of computing power to analyze visual images and extract information of interest. In concept, this technology could be adapted to any visual task, and applications are already being developed in several areas including display reading, finding and reading signs and text, analyzing terrain for curbs and drop-offs, and analyzing intersections to advise the blind traveler how and when to cross. Eventually, it is possible that all these and other functions could be combined in a single device, making a general-purpose aid with broad application.

14.6.7 Head-Mounted Displays and Image Enhancement for Low Vision

There have been efforts to develop general-purpose head-mounted electronic visual aids to enhance the image to match the individual's remaining vision. A pioneering effort under the leadership of Robert Massof led in 1992 to a head-mounted system using binocular cathode ray tube (CRT) displays for maximum brightness and contrast (Massof and Rickman, 1992). A wide range of image manipulations were investigated, such as optimized threshold and contrast functions, adjustment of gamma function, and "remapping" of the image to place sections of the scene that were obscured by a central scotoma on good functioning areas of the retina. As the technology of miniature video cameras and head-mounted displays continues to evolve, other systems combining head-mounted cameras and displays such as the Jordy and Keeler Nu Vision are slowly emerging with more cosmetically acceptable configurations. Simultaneously, research is being conducted on new algorithms for enhancing images to optimize them for low-vision users, regardless of subject matter or display modality.

14.7 ENVIRONMENTAL ADAPTATIONS AND UNIVERSAL DESIGN

The barrier-free movement initially focused mainly on physical impairments but has steadily been spreading to benefit those with sensory disabilities. Although regulations mandate the placement of Braille signs, these are of limited use in many places as they have to be found before they can be read. Some of the assistive technology discussed in other sections of this chapter also amounts to environmental adaptation, for example, the installation of Talking Signs® or audible traffic signals to assist visually impaired travelers. These and their more advanced derivatives can also provide useful information for the general population. Tactile warning strips have been installed on many railway platforms and wheelchair curb cuts. Legal action by blind activists has resulted in manufacturers of public terminals such as ATMs making their machines accessible. The first accessible (talking) building entry systems and voting machines have also emerged.

For people with low vision, many simple modifications to the home or work environment can be of assistance and enhance safety. For example, painting white or yellow stripes on the edges of steps considerably enhances their visibility. Use of contrasting black or white counter surfaces helps in performing many everyday tasks such as cooking, pouring liquids, or eating. Boosting light levels benefits almost all persons with low vision. Methods to reduce glare, and elimination of the latest high-intensity discharge headlights on cars (with their high short-wavelength spectral content) would reduce the glare problem for elderly drivers with poor vision. These and other modifications can greatly benefit not only those with severe visual impairments but the wider population of aging individuals with less than perfect vision.

14.8 TASK-SPECIFIC ASSISTIVE TECHNOLOGIES

14.8.1 Blind Mobility Aids

The task of safe, independent travel through the environment is challenging with little or no visual input. A family of mobility aids has been developed to detect nearby objects and obstacles. For example, the Mowat Sensor, a handheld ultrasonic device, uses a vibratory code to warn of the presence and range of an object in its beam. The Sonic Pathfinder is a head-worn device with ultrasonic beams controlled by a microcomputer. The Nurion Industries Laser Cane uses laser beams to detect objects, and incorporates the ability to warn of drop-offs. Taking a different approach is the family of wide-bandwidth frequency-modulated sonars developed by Kay, beginning with the Sonic Torch. These are sometimes termed environmental sensors, due to their ability to provide information about the nature of the surface being sensed in addition to its distance and direction. The ultrasonic swept FM transmissions are multiplied by the received signals to produce an audible difference signal whose pitch is proportional to range and whose timbre indicates the nature of the target. Variants have included the head-worn Sonicguide, in which two wide-beam receivers are splayed apart so that the interaural amplitude difference of the received signal gives a directional cue. The Trisensor or KASPA system combined a narrow central beam superimposed on the wide peripheral beams to mimic the manner in which central and peripheral signals are processed in the visual system. The current version of this technology is the BAT "K" sonar cane, which is a single-channel narrow-beam version of this sonar system that can be handheld or clipped to a long cane.

14.8.2 Orientation and Navigation Aids

Beyond the difficulties of steering a safe path through the immediate environment, the broader aspect of the travel problem is variously known as navigation, "orientation," or "way finding." Navigational difficulties are considerable without access to the usual cues (such as signs and landmarks) used by sighted persons. Technology to address this aspect of the travel problem has a shorter history, and devices in this category have only recently entered commercial production.

14.8.3 Remotely Readable Infrared Signage

The infrared Talking Signs® system was developed as an environmental labeling system to allow blind travelers to locate and identify landmarks, signs, and facilities of interest in the environment. It uses coded infrared transmitters as labels, and the user's handheld receiver converts the transmissions into speech. The infrared beam pattern provides control of range and coverage, and the directional nature of infrared light allows the user to accurately locate each sign. Since this concept was prototyped in 1979, a number of alternative systems have been proposed. An infrared system ("Pathfinder," modeled on Talking Signs) was evaluated in a London subway station. Similar approaches have been taken in the European OPEN (Orientation by Personal Electronic Navigation) project, the SEAL Pilot-Light system, the Tele-Sensory Marco system, the RNIB Infra InfraVoice, and the AudioSigns infrared orientation system. Systems using speech labels triggered by a user carried device include the REACT system (1987) and The Open University device (1991). Verbal Landmark® demonstrated a radio-based system in 1993 (see evaluation by Bentzen and Mitchell, 1995) in which a portable receiver detects messages transmitted from an electromagnetic loop. Numerous proposals have been made to use Radio Frequency

Identification (RFID) tags for this function, but as with other radio-based systems they lack directional information. While these and other proposals have come and gone, the original Talking Signs system has been in continuous production and spreading steadily. It has also spawned many variants including incorporation into bus arrival announcing systems, and museum tour guide systems with different messages for different audiences (children, sighted adults, blind persons, etc.). Transmitters now incorporate a digital code that identifies their position, and future versions will use this information to access (possibly via the Internet) information of interest in the vicinity, such as nearby restaurant locations and menus, train time tables, etc.

14.8.4 GPS Technology

GPS technology can be of some assistance in orienting blind persons to open outdoor environments. Loomis et al. (1994, 2001) have systematically studied this possibility combined with externalized sounds for locating environmental features. Another version was developed by Arkenstone, Inc., using a notebook computer packaged with the GPS and synthetic speech in a backpack. A commercial version of this approach, GPS-Talk, is now available through the Sendero Group LLC. Another example now available is the Trekker from Pulse Data.

The European consortium project named MoBIC (Mobility of Blind and Elderly People Interacting with Computers) (1994 to 1996) proposed using GPS technology and a protocol, based upon ISO's Open Systems Interconnection architecture (1978), to interface other technologies that could be used for orientation and navigation. The MoBIC Project and Brabyn et al. (2002) found that the accuracy of GPS is severely degraded on the sidewalks in city areas next to multistory buildings, and needs to be supplemented by other forms of information. One possibility is dead-reckoning systems using inertial navigation sensors.

14.8.5 Computer Vision

"Computer vision" technology, mentioned under general-purpose solutions, is now being explored for several specific tasks encountered in travel. The concept is to use portable computing power to analyze images from a digital camera and extract features of interest, such as street signs, intersection crossing signals, etc., and convey this information to the traveler via synthetic speech and/or enhanced image presentations. The development of this technology involves not only difficult problems in algorithm design to find and extract the desired information from images but also a variety of human factors problems including capturing usable images and effective information presentation without causing sensory overload. As of this writing, prototype systems have been developed with head-mounted cameras and with computing power supplied by a notebook computer carried in a backpack.

14.8.6 Audible Pedestrian Signals

For the specific problem of street crossing safety, at least 11 accessible traffic signal systems are available to cities. Most provide a simple auditory signal when the "Walk" signal is on. Some indicate other information such as the position of the crossing light activation control by an auditory beeper or ticking sound. However, these systems are implemented only in a few locations in relatively few cities. In some areas they are turned off at night (arguably when they are most needed) to avoid disturbing local residents.

14.8.7 The Future

Eventually, some combination of these and other technologies should give blind and visually impaired persons more options for effective navigation information. GPS technology is ideal for informing users of their approximate location in outdoor areas away from tall buildings. Infrared transceivers are ideal for providing immediate, real-time identification and labeling of key landmarks, facilities, and entrances, with exact directional information and extensive message information about a specific situation. These systems, in conjunction with digitally stored maps, will be able to provide directions to desired destinations. Computer vision approaches using text recognition will allow more generalized access to environmental signs and other printed labeling information where other solutions are not available.

14.9 TRAVEL WITH LOW VISION

Less attention has been paid to travel problems of persons with low vision than those who are blind. Relatively little vision is needed to maintain a straight path along a sidewalk and avoiding large obstacles and traffic, but many visually impaired persons suffer from acute glare problems and field and contrast losses, which make travel difficult. Even "normally sighted" older persons, let alone the visually impaired, have trouble reading signs positioned against the sky, which causes glare. Most people with low vision are elderly, and these extra problems of travel, coupled with the possibly life-threatening consequences should they suffer a fall, discourage many from independent outdoor travel. Among those who do travel, use of a cane is common both as an aid to avoid tripping on low visibility obstacles or surface irregularities as well as an alert to traffic and other pedestrians. When traveling in unfamiliar areas, a spotting telescope is sometimes used to read street signs and similar information in the distance.

Remotely readable signage technology such as Talking Signs may well be useful to a large number of elderly and visually impaired persons. As mentioned earlier, computer vision technology to locate points of interest such as signs in an outdoor scene and enhance the image (magnify, increase contrast, and highlight) of the object of interest is under development. For persons with missing sections of visual field, optical devices that "multiplex" or superimpose selected parts of the scene in the blind hemifield onto the scene in the sighted hemifield are being investigated. Use of miniature cameras and head-mounted displays for mobility is problematic, partly due to the need to obtain an exact 1:1 magnification if optical orientation is to be maintained and nausea avoided. However, new see-through displays may offer innovative applications in this area.

14.10 TECHNOLOGY FOR READING, WRITING, AND GRAPHICS ACCESS

14.10.1 BRAILLE

Braille is a code in which each character is represented by an arrangement of six raised dots arranged in two columns of three side by side, with an interdot spacing of slightly less than 0.1 in. (2.5 mm). Presence or absence of particular dots identifies the character. Advanced readers use "contractions," in which one character may represent several letters. While individuals who become blind at a young age pick up Braille easily, those who lose vision as adults often find it difficult to learn or they have reduced motivation.

After a period in the late 20th century, when its use was considered superfluous by many, Braille has undergone a revival as appreciation of its importance as a means of literacy has spread. Technology for producing Braille has also improved steadily in its sophistication, availability, and affordability. A manual slate and stylus is still commonly used for making the dots in paper for small Braille notes, but electronic notetakers with six-key Braille keyboards are becoming common. For output, these usually include a 20-character display of mechanical Braille dots that are raised and lowered electronically to form a line of Braille characters.

Personal Braille embossers, equivalent to printers for sighted persons, are becoming more attainable. Other means of producing Braille from computer outputs include heat-sensitive capsule or swell paper, which can be printed with Braille or other tactile dot patterns on a regular ink printer and then run through an infrared heating machine. The black dots absorb more heat than the lighter surround, and so the paper swells under the dots, producing Braille. The technology for producing this type of output is inexpensive, but the paper itself is expensive if large quantities are required. Irrespective of the method of Braille production from computer output, software is needed to produce the Braille code. Companies such as Duxbury systems supply software packages to translate regular text (in the form of ASCII code, PDF files, and other computer formats) into Braille and format it for printing.

14.10.2 BOOKS ON TAPE AND DIGITAL FORMATS

Talking books have long been a mainstay of reading material for blind persons. A long history of special tape and vinyl disk formats and players for this purpose has included special indexing codes for finding chapters and sections rapidly. Organizations such as the American Printing House for the Blind produce books and magazines on tape.

Digital Talking Books are now available to accomplish the same ends, using the new "Daisy" standard developed by an international consortium. These are designed to benefit not only blind persons but anyone who has trouble reading normal print for any reason such as a learning disability or inability to hold a book. They offer improved quality and navigation features over the older talking books, and can be produced on any available digital media including CD and DVD. Special players are available, but these can also be played on computers using suitable software. Increasing amounts of reading material are becoming available in this format from such organizations as the American Printing House, Bookshare.org, and the Library of Congress National Library Service.

14.10.3 Braille Note Takers

Over the past 10 to 15 years, numerous Braille Note takers have evolved, serving much of the purpose of a notebook computer. They commonly use a Braille keyboard, which is convenient as it is smaller than a conventional keyboard, requiring only six keys (one for each dot in the Braille cell) and a space bar. (Often eight keys are provided to accommodate the eight-dot computer Braille code). Output is via a refreshable Braille display, usually of about 20 cells. These small computers (most will fit in a large pocket) can be interfaced to a variety of peripherals and perform other functions such as a calculator, calendar, etc. Most use proprietary word processing software. The pioneer in this field was the Blazie Engineering "Braille 'n Speak," now produced by Freedom Scientific.

14.10.4 Optical-to-Tactile and Optical-to-Auditory Conversions

An early reading aid for blind persons was the "Stereotoner," which a user could scan across a line of print to convert it into a combination of tones. A more widely used device, the Optacon, developed by John Linville and James Bliss, uses a handheld camera that the user scans across a line of print to produce a tactile image on a 144-pin array of vibrating piezoelectric pins placed under a finger of the other (stationary) hand. With some training and practice, a blind user could read any type of print including handwriting — an ability still not available on automated reading machines. Proficient users could reach up to 80 words per minute. Production ceased in 1996. A few years later, a new device called the Video TIM, from German manufacturer ABTIM, emerged with a 256-pin (nonvibrating) $4 \times 4\ \text{cm}^2$ display.

14.10.5 Reading Machines

Early research on reading machines for blind persons (designed to "read" print and convert it into the spoken word) produced technologies (document scanners and synthetic speech) that have subsequently become commonplace in the general consumer market. Accordingly, most stand-alone reading machines have been replaced by software packages designed for use with personal computers and scanners. However, stand-alone reading machines still exist, such as the Galileo and the Portset Reader, which are more portable than PC-based systems. The problem of reading handwriting has still not been cracked, and current machines are still nowhere near the capability of a sighted reader (still used by most blind persons for intelligent analysis and scanning over documents), but they have come a long way since the early days and are eminently capable of dealing with straightforward text-reading tasks.

14.10.6 Access to Graphics and Maps

Access to graphical, pictorial, and map information is very problematic for blind persons. Tactile maps, graphs, and pictures can be made by a variety of manual methods producing raised line drawings on various materials, but these are laborious. The result is that few graphical educational materials are available to blind children, and so they do not become well exposed to tactile spatial representations.

14.10.7 Graphics Access and Production

In recent years, increased efforts have been exerted to make the production of graphics from computer Braille embossers more feasible. There are numerous problems involved: generating a perceptible tactile equivalent of a visual picture, and drawing or map is not straightforward, because the density of information needs to be much lower and the tactile sense operates different from that of vision. One recent development in this field is the advent of the View Plus "Tiger" embosser, which can emboss dots of variable heights as well as closer together than standard Braille dot spacing.

The "holy grail" of graphics access for blind persons is the full-page volatile Braille/graphics tablet, with an array of dots that can be raised or lowered using computer control. Many efforts have been devoted to this problem, but the technological problems are formidable and to date the largest arrays available are the Dotview units of 30 × 40 dots. Present volatile Braille and tactile graphics displays use piezoelectric or electromagnetic technology, and current research efforts are devoted to harnessing other technologies such as smart polymers, electrorheological fluids, and micromachining.

Meanwhile, other methods of graphics access have become available, including touch tablet technology in combination with speech output, allowing a user to trace out spatial information and receive spoken feedback. However, raised overlays for touch pads usually have to be specially produced for each diagram or picture being explored. Examples include the Nomad, Concept Keyboard, and Tag Pad. The "Vertouch," a computer mouse with two small tactile arrays on it, had software available for tactile games and exploration of computer screen information. Force feedback is another technology that can be used for tactile and haptic exploration. Examples include a force-feedback mouse and the PHANTOM system for exploration of virtual three-dimensional objects. For displaying graphs, SKDATA Tools makes graphical output from MATLAB accessible through auditory and Braille outputs.

14.10.8 Maps

Traditionally, street maps suitable for pedestrian navigation have been extremely scarce due to the difficulty and expense of making them. For some areas, such as the Washington, D.C., Metro, tactile maps are available as a general guide to the system. Even these, however, cannot include the level of detail of a printed street map. Auditory maps and travel directions can be a substitute, and are available, for example, using the Atlas Strider system. Another approach combines a touch tablet with a tactile map or graphic overlay, interfaced to a computer to produce a talking map. Examples include the Talking Tactile Tablet from Touch Graphics and the Nomad system by Quantum Technology.

Advances in computer, Internet, and Braille embossing technologies are about to revolutionize tactile map production. For example, the TMAP project will make it possible for a blind user to access a Web site on which he/she can specify any address or set of cross streets of which a street map is desired. The computer system will then generate a street map centered on that address for download and printing on the user's own Braille embosser. If the user does not have one, the map can be sent to a Braille embossing service for production and then mailed to the user.

14.10.9 Low-Vision Reading Problems and Solutions

Reading difficulties in low vision depend heavily on the type of visual impairment. In the most common type, due to age-related maculopathy, the resolution of the central vision is reduced drastically, and training is often given to help the individual use the remaining (lower-resolution) areas of the surrounding retina for reading (known as "eccentric viewing"). Considerable magnification is often needed. Large print, where available, is a partial solution to this problem. Some periodicals and books are produced in large print (for example, by the American Printing House for the Blind).

Often neglected is the fact that providing proper lighting can make an enormous difference in the ability to read. A good, bright, glare-free light, coming from over the reader's shoulder, can often compensate for minor visual impairments without any magnification requirements.

The traditional low-vision reading aid is an optical magnifier. These come in many forms, most commonly handheld or mounted on stands. The practical degree of magnification obtainable with these is modest, about two to four times, but is often enough to make a difference. For people with limited dexterity, stand magnifiers can make positioning of the magnifier relative to the page easier, with the page staying in focus. Both types can have internal illumination to ensure adequate lighting on the reading material (which has a major effect for most low-vision readers). To gain higher magnifications, high-powered reading lenses (e.g., 10 or 20 dpt) on spectacle frames are sometimes used. Special short-range telescopes mounted on spectacle frames can be designed to preserve normal reading distance while providing the desired degree of magnification.

Higher magnifications still are possible with CCTV, or closed circuit television, magnifiers, usually consisting of a camera and CRT monitor mounted on a stand with a table underneath upon which the reading material is placed. The viewing table is commonly mounted on sliders or rollers so that the material can be moved easily for scanning. All systems allow adjustment of magnification and contrast (including contrast reversal to produce white print on a black background, thus reducing glare), and most also enable the user to choose the background color. When text or graphical images are magnified, the viewer can only see a small portion of the overall page or picture, often making interpretation more difficult. The ability to see an overall view on one part of the screen and the magnified view on another part may provide an advantage in this respect, and many CCTV systems provide a split-screen feature.

In recent years, much smaller versions of these CCTV magnifiers have become available using solid state cameras and displays. There are now several handheld, pocket-sized electronic magnifiers (such as the Pocket Reader, Quicklook, PICO, and Pulse Data Pocketviewer) that can simply be placed on the page surface to magnify it by 5 to 7 times. These devices conveniently enable the user to look at details and price labels in shops, sign credit card payments, read restaurant menus, theatre programs, and timetables, check lottery results, check television and radio programs, browse through magazines, and read books while traveling. Another variant is the head-mounted camera and display combination described earlier and in the following section; this concept can be used for a variety of reading and writing tasks.

14.11 COMPUTER AND INTERNET ACCESS

14.11.1 Computer Output and Input

A third-party "screen reader" software (such as Jaws for Windows and Window-Eyes) provides synthetic speech output and a command structure for exploring computer screen contents. (Interestingly, synthetic speech was first developed as assistive technology, and later spread to many other applications). Whereas in augmentative communication the most "natural sounding" speech is the goal, in the screen reader naturalness per se takes second place to speed and intelligibility, and most blind computer users run their synthesizers far above "natural" speed. Access systems using volatile Braille displays for information output (for example, the ALVA system) are preferred by most Braille users if given a choice. Unfortunately, they involve expensive electromechanical technology, and so they are more popular in countries where rehabilitation agencies will pay for them.

For persons with dual sensory loss (vision and hearing), Braille output systems are currently the only practical computer access method. However, this is problematic because most individuals with dual sensory loss lose their vision after deafness, and the need to learn Braille is a significant additional hurdle.

For low-vision computer users, standards in some software applications enable on-screen magnification. For more magnification and features, commercial software packages such as ZoomText are designed to ease screen reading by visually impaired users. Some commercial CCTV systems can alternate between camera and computer input, with some providing a split-screen capability on which both can be displayed simultaneously.

The mouse is not an effective input tool for a blind person; so keyboard commands for all functions are necessary. Voice input, if truly perfected, would be equally useful to blind and sighted persons. If voice interaction with computers becomes more common in the future, it may reduce the dependence on graphics-based interactions (mouse, touch screen, etc.) and such a trend should benefit blind computer users.

14.11.2 Portable Computing Devices

With the trend away from all-purpose computers and towards portable computing devices in all their various forms, it will be extremely important to provide access to such systems for blind users. As of this writing, there are very few PDAs (personal digital assistants) that are accessible to blind users.

14.11.3 Internet and Web Access

While access technology lagged considerably behind when computers were first introduced, efforts to make the Internet accessible have been much more proactive. The public/private Web Access Initiative (WAI) works in concert with the World Wide Web Consortium (W3C) through several small working groups, including an Authoring Tool Guidelines Working Group, which develops guidelines to ensure that Web-page-authoring software automatically produces accessible pages. Accessibility guidelines have also been produced by WAI and other groups and user participation is active in e-mail lists to enhance accessibility. A second component of Web access is the development of accessible Web browser software. Two main approaches have been taken to address this problem. One is to use any commercial screen reader software in conjunction with standard Web browsers (e.g., Internet Explorer). The other is to use a special accessible browser, such as the IBM Home Page Reader. These mechanisms taken together have been fairly successful at overcoming the worst problems of Web access for blind users, but heavily graphics-intensive Web sites are still problematic.

14.12 COMMUNICATION TECHNOLOGY

14.12.1 Visually Impaired Access to Telephones and Cell Phones

Modern communication technology is posing its own accessibility problems. The old-fashioned telephone was always accessible by blind users (except that it had to be located before it could be used). Cell phones have eliminated the problem of finding a phone, but as they have become more like computers, their accessibility has suffered. Indeed, all modern phones and answering machines are steadily becoming more like other consumer appliances, with low-contrast LCD displays and operation via complex command structures and nested menus. As of this writing, there are at least two models of cell phone that have been designed to be usable by a blind person — using "screen readers" with synthetic speech to describe all menus and commands.

As the cell phone becomes ubiquitous and encompasses more and more functions (GPS, camera, browser, etc.), maintaining accessibility will become an ongoing challenge. On the other hand, as technology matures and expands, it is likely that the computing power inherent in the cell phone will be harnessed for more and more functions useful to a blind person, assisting in everything from note taking to way finding. Already, it is possible to take a picture of the scene in front of you and transmit it to a sighted friend for analysis. It is vital that these and other functions are implemented with the broadest possible user population, including blind, hearing impaired, and visually impaired persons in mind.

14.12.2 Communication and Telecommunication with Dual Sensory Loss

For persons with combined vision and hearing impairments, communication becomes a much more major problem. For example, deaf persons used to communicating by sign language and lip reading experience increasing difficulties when they become visually impaired. A common situation as people age is the onset of vision loss in the center of the visual field, in turn affecting lip reading. As yet, there is little or no technology specifically designed to help with this problem, although it is possible that magnification with optical aids or head-worn electronic magnifiers would be of some assistance.

When both vision and hearing losses are absolute, communication is extremely problematic. Deaf–blind people can communicate among themselves using a finger-spelling code in which each individual letter is spelled out by the fingers, and sensed tactually by the recipient by holding the sender's hand. An alternative but less common communication code, finger Braille, is based on the Braille code and uses four fingers on each hand to represent the dots in eight-dot Braille, which can be transmitted to the recipient by hand contact. Research to produce and receive deaf–blind finger-spelling codes on robotic finger-spelling hands and gloves has produced prototype devices capable of sending and receiving this information and translating it to and from electronic text, but no commercial device of this sort yet exists. The finger Braille code is much simpler to convert to electronic format, and prototypes for this purpose have been developed, but again, these are not commercially available.

For those who begin life in the world of blindness and subsequently lose their hearing, special versions of the telecommunication devices (TDD) for the deaf incorporating refreshable Braille displays were developed through collaboration between Smith-Kettlewell and Telesensory Systems Incorporated. This allowed deaf–blind persons to gain access to the telephone. A current version of this technology, the Com Lite by Freedom Scientific, interfaces with a Braille note taker to provide similar functionality.

14.12.3 Television Access

Access for deaf people to television has outpaced that for blind persons as the adoption of captioning has spread. For blind television viewers, the equivalent solution is Descriptive Video, a process wherein the action on the screen is described in a verbal narration that does not interfere with the normal program sound track. Such a narration has to be specially produced and recorded and then inserted into the broadcast signal. Various means of providing video description have been investigated, and it now uses the Second Audio Program (SAP) channel on televisions or VCRs. Early investigations into the feasibility of spreading video description throughout the broadcast system were undertaken by William Gerrey of Smith-Kettlewell, and pioneering implementation of described video was carried out by Larry Goldberg at the WGBH Media Access Group. In July 2000, the Federal Communications Commission (FCC) adopted rules to mandate that a certain amount of programming contain video description. These rules took effect in April 2002, but in November 2002 a federal court struck down the rules, leaving only a requirement that broadcasters provide emergency information in a form that is accessible. Fortunately, some broadcasters and other program distributors (notably the Public Broadcasting Service) continue to provide video description in their programming.

The Media Access Group also developed the Mopix service for making movies in theaters accessible to both deaf and blind patrons. This is available for selected showings in a limited number of theaters in most major urban areas.

14.13 VOCATIONAL AND DAILY LIVING AIDS

14.13.1 Vocational Instruments

There are many vocational and daily living tasks that are difficult for blind and visually impaired persons. In many cases, assistive devices can be designed to help ease the problem. A simple example is the light probe, of which several versions exist, designed to provide an auditory output, which varies according to the intensity of incident or reflected light. This tool can be used for many tasks from determining the pattern of stripes on clothing to detecting the lights required for operating telephone consoles.

Taking readings of physical or electrical quantities is a common problem in employment and daily living. A few talking instruments now exist, such as a tape measure, multimeter, and blood pressure monitor. Talking watches and alarm clocks are now readily available. Speech outputs are effective for measuring relatively static quantities, but for information that varies, auditory tones with pitches corresponding to signal levels are far more satisfactory than speech outputs and more closely simulate an analog visual display.

A simple example is a meter or continuity tester display, which gives a rapid, approximate indication of voltage, resistance, or other quantity purely through auditory pitch. This type of display has found application in many other instruments, including more complex ones such as the Smith-Kettlewell Auditory Oscilloscope, in which the horizontal position of a cursor superimposed on the trace is controlled by a knob with a Braille scale, while vertical trace deflection at the cursor location is coded by the pitch of a tone. By "scanning" across the display with the knob, the user obtains a remarkably accurate picture of the displayed signal. Another example is the Smith-Kettlewell Dynamic Meter Reader. When a dial on the device is rotated to the point where chopping begins, the signal level can be read from the Braille scale. Variations in the signal level are readily evident from changes in the pitch of the output tone, and the knob can be adjusted in advance for any desired "set point."

Aside from solutions involving electronics or other sophisticated technology, a multitude of other engineering solutions can be applied to vocational tasks including the use of jigs, guides, and fixtures to make the task at hand easier to accomplish without vision.

14.13.2 Low-Vision Aids for Vocational and Daily Living Tasks

For many near- and intermediate-distance tasks such as those involved in many work and home activities, low-vision aids different from those employed purely for reading come into play. For example, monocular and binocular telescopes and microscopes designed for different viewing distances are sometimes employed in a variety of manual tasks. Watching television, recognizing faces or facial expressions, or seeing the blackboard may require a telescopic aid. In the latter task, portable CCTV units are available in which the camera can be aimed at distant objects.

A relatively recent development is the use of head-mounted displays for a variety of vocational, educational, and daily living tasks. The most well-known example of this technology is the Jordy system (named after the *Star Trek* character) by Enhanced Vision Systems. It consists of a head-worn camera and binocular display that looks like a large pair of goggles. The camera incorporates digital zoom and autofocus. A separate control and battery box can be clipped on to the user's belt. Working distances from near to infinity with magnifications of up to 30× are obtainable. The head-worn unit can also be placed in a docking unit that operates more like a conventional CCTV reading aid with a movable reading table, allowing any standard monitor to be used for viewing.

14.14 ACCESS TO CONSUMER APPLIANCES

An ever-expanding range of appliances for both the home and job site are inaccessible to blind or visually impaired users. Many items such as televisions, VCRs, stereos, etc., have control panels with black buttons on black backgrounds, digital displays, and menus that are difficult or impossible to access. The use of low-contrast LCDs has spread to almost every electrical appliance in the home and office, making operation difficult for a person with low vision and impossible for a blind person. Similarly, difficult to operate are touch-pad controls, keypads with no tactile feel, and complex control menus. There is a need for appliance designers to be more aware of the range of abilities of their consumer populations, as very simple design modifications would obviate many of these problems.

On the few appliances that can still be found with knobs and simple buttons for controls, Braille markings and labels can go a long way towards making them accessible. For devices with digital displays, handheld digital display readers are being developed using computer vision to find the display, analyze it, and read it to the blind user in synthetic speech. Researchers are also working on possible future accessibility standards that would require such appliances to make the information on the display available in some standard format (e.g., using the IrDA infrared, Bluetooth, or similar standards) so that third-party manufacturers could make access devices that would enable a blind or other disabled user to access them. These efforts are still ongoing.

14.15 INDUSTRIAL AND SERVICE DELIVERY CONTEXT

14.15.1 Costs and Benefits

As in other areas of rehabilitation, it has been demonstrated that, apart from the valuable benefits in human terms, expenditures on rehabilitation technology are repaid to the economy many times over in increased economic output and tax revenues (Sensory Aids Foundation, 1985–1988). Unfortunately, the costs and benefits are seldom accounted for together, and so rehabilitation tends to be seen as a net expense. The expenses tend to come out of a different compartment of the public purse from the one the benefits flow into (e.g., increased tax receipts from employed persons enjoying the fruits of rehabilitation). Thus, society as a whole is often slow to recognize the full benefits and returns from efforts spent on developing and applying rehabilitation technology.

14.15.2 Blindness and Low-Vision Assistive Technology Delivery System

The medical rehabilitation system impinges only peripherally on blindness and low vision. Historically, these disabilities have fallen outside the scope of "physical medicine and rehabilitation." Whereas orthotics and wheelchairs can be prescribed and paid for within the normal medical model, this is not the case for sensory aids.

In low vision, patients are initially served by their regular eye care practitioners — ophthalmologists and optometrists with their respective expertise in addressing the underlying eye disease and prescribing optimal refractive correction (glasses). Ideally, they are then referred to a low-vision clinic or service with specialized knowledge of rehabilitation and devices. These clinics, often in major medical centers, are staffed by professionals from optometry, ophthalmology, and occupational therapy. They conduct special testing and needs assessment, and prescribe (and sometimes dispense) optical and electronic visual aids. They may also provide eccentric viewing training to enhance reading proficiency and training in the use of the various assistive technologies as well as daily living skills. Some also provide support groups and follow-up services.

For people who are functionally blind, rehabilitation service and assistive technology provision is completely outside the medical system. In the US, blind persons of employable age are usually clients of state rehabilitation departments, which may help pay for assistive devices and the training needed for a job. For school age children, special education services are provided and may also help pay for accommodative technology.

Considerable support is available from the U.S. Veterans Health Administration (VHA) for blind and visually impaired veterans of all ages. The VHA operates a network of blind rehabilitation centers with residential training programs, and a network of special outpatient services for blind and low-vision veterans. They pay for an array of assistive technologies for veterans.

Over the past 40 years, the industry that manufactures and sells technology for blind and visually impaired persons has matured from virtual nonexistence to one with a number of large and small firms worldwide. Because large parts of the service delivery system are outside the medical arena, these firms have a hybrid marketing system, often dealing directly with the customer as well as the low-vision clinic or the state department of rehabilitation.

14.16 STUDY QUESTIONS

1. What is visual acuity?
2. Describe the three dimensions of vision loss other than acuity.
3. What is statutory blindness?
4. What is cortical visual impairment?
5. How many totally blind persons are there in the United States?
6. What are the differences between cortical and retinal implants?
7. What is computer vision and how could it be applied to this population?
8. How do the principles of universal design apply to those with visual impairments?
9. What technologies exist for blind travelers? Discuss their relative merits.
10. How can deaf–blind persons access computers?
11. How can a person with a central scotoma read?
12. What is descriptive video?
13. How can measuring instruments be adapted for use without vision?
14. What technology is available to allow a deaf–blind person to use a telephone?
15. What features make consumer appliances less accessible for blind users?

BIBLIOGRAPHY

Bach-y-Rita, P., Collins, C.C., Saunders, F.A., White, B., Scadden L., Vision substitution by tactile image projection, *Nature*, 1969;221:963–4.

Bentzen, B.L., Mitchell, P.A., Audible signage as a wayfinding aid: comparison of verbal Landmark® with Talking Signs®. *J Vis Impairm Blind*, 1995;89:494–505.

Bentzen, B.L., Tabor, L.S., Accessible pedestrian signals, U.S. Access Board, Washington, D.C., 1998.

Bliss, J., Optical-to-tactile image conversion aids for the blind, *Arch Phys Med Rehabil*, 1967;48:352–3.

Brabyn, J.A., New developments in mobility and orientation aids for the blind, *IEEE Trans Biomed Eng*, 1982;29:285–9.

Brabyn, J., Schneck, M., Haegerstrom-Portnoy, G., Lott, L., The Smith–Kettlewell Institute (SKI) longitudinal study of vision function and its impact among the elderly: an overview. *Optom Vision Sci*, 2001;78:264–9.

Brabyn, J., Alden A., Use of GPS in urban settings, *Proceedings of the RESNA Conference*, June 2002.

Brabyn, J.A., Coughlan, J., Shen, H., Some applications of computer vision technology to assist blind and visually impaired persons, *Vision 2005*, London, July 2005.

Brindley, G.S., Lewin. W.S., The sensations produced by electrical stimulation of the visual cortex, *J Physiol (Lond)*, 1968;196:479–93.

Burton, D., 2004, Now they're talking!: A review of two cell-phone based screen readers, *Access World, American Foundation for the Blind*, 5, November 2004.

Clark, J., Building accessible websites. *New Riders*, 2002.

Cocchiarella, L., Andersson, G.B.J. (Eds), *Guides to the Evaluation of Permanent Impairment*, 5th ed. American Medical Association, 2000.

Colenbrander, A., Dimensions of visual performance — low vision symposium, *American Academy of Ophthalmology, Transactions AAOO*, 1977;83:332–7.

Denham, J., Leventhal, J., McComas, H., Getting from point A to point B: a review of two GPS systems, *Access World, American Foundation for the Blind*, 2004;4.

Fuhr, P.S.W., Holmes L.D., Fletcher D.C., Swanson M.N., Kuyk, T.K., The AMA guides functional vision score is a better predictor of vision-targeted quality of life than traditional measures of visual acuity or visual field extent, *Visu Impairm Res*, 2003;5:137–46.

Crandall, W., Brabyn, J., Bentzen, B.L., Remote infrared signage evaluation for transit stations and intersections, *J Rehabil Res Dev*, 1999;36(4):341–55.

Delbeke, J., Pins, D., Michaux, G., Wanet-Defalque, M., Parrini, S., Veraart, C., Electrical stimulation of anterior visual pathways in retinitis pigmentosa, *Investig Ophthalmol Vis Sci*, 2001;42:291–7.

Dobelle, W.H., Artificial vision for blind by connecting a television camera to a visual cortex, *ASAIO J*, 2000;46:3–9.

Fowle, T., The Fowle Gimmique, Smith–Kettlewell Technical File, Summer 1982.

Fructerman, J., Talking maps and GPS systems, Paper presented at the Rank Prize Funds Symposium on Technology to Assist the Blind and Visually Impaired, Grasmere, Cumbria, England, March 25–28, 1996.

Gilden, D., Smallridge, B., Deaf–blind users grasp the idea with dexter, a prototype robotic finger-spelling hand, *Proceedings of the RESNA Conference*, 1996.

Groves, N., Subretinal device improves visual function in RP patients, *Ophthalmology Times*, front page, August, 2003.

Guides to the Evaluation of Permanent Impairment, American Medical Association, Chicago. 5th ed., 2001.

Heyes, D., Anthony, The sonic pathfinder: a new electronic travel aid, *J Visu Impairm Blind*, 1984;77:200–2.

Hoiuchi, H., Ichikawa, A., Teletext receiver by finger Braille for deaf-blind, *Proceedings of the CSUN Conference*, Northridge, CA, 2001.

Humayun, M.S., de Juan Jr., E., Weiland, J.D., Dagnelie, G., Katona, S., Greenberg, R.J., and Suzuki, S., Pattern electrical stimulation of the human retina, *Vision Res*, 1999;39:2569–76.

Humayun, M.S., Weiland, J.D., Fujii, G.Y., Greenberg, R., Williamson, R., Little, J., Mech, B., Cimmarusti, V., Van Boemel, G., Dagnelie, G., deJuan, E., Visual perception in a blind subject with a chronic microelectronic retinal prosthesis. *Vision Res*, 2003;43: 2573–81.

Kaper, H.G., Tipei, S., Wiebel, E., Data sonification and sound visualization, *Comput Sci Eng*, 1999;1:48–58.

International Classification of Functioning, Disability and Health (ICF), World Health Organization, Geneva, 2001.

International Classification of Diseases, 9th Revision — Clinical Modification (ICD-9-CM). First edition: Commission on Professional and Hospital Activities, Ann Arbor, 1978.

Kay, L., An ultrasonic sensing probe as a mobility aid for the blind. *Ultrasonics*, 1964;2:53.

Kay, L., Acoustic coupling to the ears in binaural sensory aids, *J Vis Impairm Blind*, 1984;77:12–16.

Kay, L., Auditory perception of objects by blind persons using bioacoustic high resolution air sonar, *JASA*, 2000;107:3266–75.

Kirchner, C., Schmeidler, E., Prevalence and employment of people in the United States who are blind or visually impaired, *J Vis Impairm Blind*, 1997;91:508–11.

Koch, P.P., Accessibility and Usability, *Digital Web Magazine*, February 18, 2004, available at http://www.digital-web.com/articles/accessibility_and_usability/.

La Grow, S., The Use of the Sonic Pathfinder as a Secondary Mobility Aid for Travel in Business Environments: A Single Subject Design.

Lewallen, S., Courtright P., Blindness in Africa: present situation and future needs. *Br J Ophthalmol*, 2001;85:897–903.

Loomis, J.M., Golledge, R.L., Klatzky, R.L., Speigle, J., Tietz, J., *Proceedings of the First Annual International ACM/SIGCAPH Conference on Assistive Technologies*, Marina Del Rey, California, October 31–November 1, 1994, New York: Association for Computer Machinery, pp. 85–90.

Loomis, J.M., Klattzky, R.L., Golledge, R.G., Navigating without vision: basic and applied research. *Optom Vision Sci*, 2001;78:282–9.

Loughborough, W., Talking lights, *J Vis Impairm Blind*, 1979;243.

Massof, R.W., Rickman, D.L., Obstacles encountered in the development of the low vision enhancement system, *Optom Vision Sci*, 1992;69:32–41.

Meijer, P.B.L., Seeing with Sound: Wearable Computing for the Blind, Invited presentation at NIC2001 (Nordic Interactive Conference), Copenhagen, Denmark, Thursday, November 1, 2001.

Miele, J.A., Smith-Kettlewell, Display tools: A project description. Proceedings of the Ninth Annual International Conference on auditory displays, 2003, 288–91, Details available at http://www.ski.org/Rehab/SKDtools/index.html.

Miele, J., Marston, J., Tactile Map Production: Project Update and Research Summary, CSUN Conference, March 2005, available at http://www.csun.edu/cod/conf/2005/proceedings/2239.htm.

National Eye Institute, A National Plan for Eye and Vision Research. National Eye Institute, Bethesda, M.D., 2004, available at http://www.nei.nih.gov/strategicplanning/np_index.asp.

National Research Council (Lennie P, Van Hemel S, Eds), Visual Impairments: Determining Eligibility for Social Security Benefits. National Academy Press, Washington, DC, 2002.

National Society to Prevent Blindness, Vision problems in the US: a report on blindness & visual impairment in adults age 40 and older, *Prevent Blindness America*, Schaumburg, IL, 1994.

Peli, E., Vision multiplexing: an engineering approach to vision rehabilitation device development, *Optom Vision Sci*, 2001;78:304–15.

Tang, J., Kim, J., Peli, E., Image enhancement in the JPEG domain for people with vision impairment, *IEEE Trans Biomed Eng*, 2004;51:2013–23.

Sampaio, E., Maris, S., Bach-y-Rita, P., Brain plasticity: “visual” acuity of blind persons via the tongue, *Brain Res*, 2001;908:204–7.

Schmeidler, E., Halfman, D., Statistics on Visual Impairment in Older Persons, Disability in Children, Life Expectancy, *J Visu Impairm Blind*, 1997;91:602–6.

Sensory Aids Foundation: Return on Investment Report, Published annually, 1985–8.

Veraart, C., Raftopoulos D. et al., Optic nerve electrical stimulation in a retinitis pigmentosa blind volunteer, Society for Neuroscience, Los Angeles, 1998.

Yuille, A.L., Snow, D., Nitzberg, M., Signfinder: Using Color to Detect, Localize and Identify Informational Signs, *Proceedings of the 6th International Conference on Computer Vision*. Bombay, India, January 1998.

15 Maximizing Participation for People with Hearing Loss

Elaine Mormer, Amanda Ortmann, Catherine Palmer, and Katherine D. Seelman

CONTENTS

15.1 LEARNING OBJECTIVES OF THIS CHAPTER

Upon completion of this chapter, the reader will be able to:

Differentiate the communication needs of people who are hard of hearing and people who are deaf
Describe instruments used to assess the communication needs of persons with hearing loss
Compare technologies intended to substitute vs. enhance participation for given listening situations
Diagram the major components of a hearing aid
List factors contributing to signal-to-noise ratio (SNR)
Explain the benefits of remote microphone technology
Name the types of wireless transmission modes used in hearing assistance Technology
Explain the limitations of hearing aids used for speech understanding in Noise
Understand the barriers (and potential solutions) to computer, Internet, and telecommunications access for this population
Define hearing assistance technology

15.2 INTRODUCTION

The World Health Organization (2002) (Facts about Deafness, 2004) estimates that 250 million people in the world function with a disabling hearing loss. The presence of an uncorrected hearing loss serves to reduce an individual's level of participation in work, home, school, family, and social domains. The consequences may result in poorer academic outcomes, decreased vocational opportunities, social isolation, and depression (Keller et al., 1999; Karchmer and Allen, 1999). Assistive technology for this population is referred to as hearing assistance technology (HAT). HAT devices are designed to either enhance the degraded auditory signal caused by the hearing loss or to substitute the degraded auditory signal via alternative modes of information input (mainly visual or tactile).

15.3 POPULATIONS AND DEFINITIONS

The population of individuals with hearing loss can be categorized into two groups, differentiated by the primary mode of communication used during daily activities. Those who are deaf primarily rely on visual or tactile modes of communication for sending and receiving communication or alerting messages. For those in this group, speech messages are generally sent and received via manual communication and/or lip reading, whereas alerting signals (e.g., alarm clocks and smoke detectors) are received via flashing lights or vibratory devices. People who are hard of hearing use oral/auditory modes as their primary means of sending and receiving communication. For this group, speech messages are usually sent via spoken words and received via auditory input. Enhancement of speech signal reception is achieved via amplification of the auditory input or the addition of visual input to the auditory signal. There are those individuals who fall somewhere in between the aforementioned categories and use various combinations of HAT following a continuum of enhancement of the auditory signal to substitution of that signal with alternative sensory input.

The affected population extends across all age groups, with the greatest prevalence among the older adult population. In the United States, 60% of people older than 70 years have some degree of significant hearing loss (Gratton and Vazquez, 2003). In the pediatric population, 83 out of every 1000 children in the US have what is termed an educationally significant hearing loss (Public Health Service, 2000). Approximately 30% of children who are hard of hearing have a disability in addition to a hearing loss, adding further challenges to the successful use of HAT (Wolff and Harkins, 1986). Clearly, the population of persons with hearing impairment is significant.

15.4 TYPES OF HEARING IMPAIRMENT

There are three main categories used to differentiate types of hearing impairment. The first, sensorineural, describes hearing loss resulting from damage to the hearing nerve or inner ear. Hearing loss of this nature is, in most cases, irreversible. The communication impairment associated with sensorineural hearing loss usually includes a reduced sensation of loudness as well as a reduction in the clarity of signals. Persons with sensorineural hearing typically note that the speech can be heard (amplified or not) but is difficult to understand. The second type of hearing loss, conductive hearing loss, results from damage to the outer or middle ear. These structures contain the mechanisms that conduct sound to the hearing nerve. Persons with conductive hearing loss typically experience reduced loudness of the signal; however, clarity is preserved. With amplification, these individuals usually find that conversational speech is heard with reasonable clarity as long as background noise is not excessive. The third type of hearing loss is referred to as mixed hearing loss, meaning that the hearing loss results from damage at both conductive and sensorineural levels. Underlying causes of both conductive and sensorineural hearing loss include congenital anomalies; pre-, peri-, or postnatal infections; disease processes; ototoxic drug reactions; and degeneration associated with the aging process.

The decibel (dB) is the unit of measure used to express a given level of sound intensity. It is a logarithmic unit of sound pressure. Severity of hearing loss is conventionally categorized based on the decibel level necessary for the individual to sense the presence of pure tone signals across a range of frequencies. The hearing assessment instrument used to identify the type and degree of hearing loss is called an audiogram. A mild or moderate degree of hearing loss refers to one whereby the individual is able to sense the presence of pure tone signals that are between 26 and 70 dB hearing levels (HLs). Individuals who require a greater intensity signal, that is, above 70 dB HL are said to have a severe or profound degree of hearing loss. The categories of hearing loss degree are based solely on the perception of pure tone test signals. Such signals are seldom heard in activities of daily life. Thus, the conventional categorization of hearing loss severity fails to describe an individual's degree of function, disability, or level of participation in life activities (Palmer and Mormer, 1999).

15.4.1 Impact on Task Performance and Participation

The functional consequence of impaired hearing is the reduced reception of auditory signals. For humans, the most vital auditory signal encountered on a daily basis is that of speech. The sending and receiving of oral speech signals facilitates a wide array of human interactions for those persons with normal hearing. Additionally, nonspeech signals that are auditory in nature facilitate warning or alerting information. For hearing individuals, auditory signals play a role in human life that ranges from keeping us safe (hearing a police siren) to learning to speak (imitating speech sounds) to building relationships (sharing thoughts). When the auditory system is impaired, these functions are disrupted. For developing infants and young children, the acquisition of speech and language skills can be markedly delayed or disordered in the presence of a hearing loss. For older children, the learning that takes place at home and at school is compromised, as is social developmental, if adequate accommodations are not made. For adults, vocational opportunities can be limited and family and social relationships degraded by hearing loss.

15.4.2 Function and Participation

The World Health Organization International Classification of Functioning (WHO-ICF) provides a framework for describing conditions of health, disability, and function (World Health Organization International Classification of Functioning, Disability, and Health, 2001). According to the WHO-ICF, the objective of a rehabilitation program is to reduce the individuals' activity limitations (difficulties in the execution of a task, or disability) and participation restrictions (problems experienced in involvement in life situations, or handicap) resulting from the hearing impairment. WHO-ICF also acknowledges that exogenous factors found in the individual's environment play a role in either enhancing or hindering the rehabilitation program. Environmental factors are the physical, social, and attitudinal surroundings of the individual with hearing loss. It is apparent that environmental factors will interact with the hearing-impaired individual's activities and participation. For example, a hard-of-hearing student is able to function better in a smaller classroom where the lecturer is clearly seen and heard than in a large lecture hall where there is an increase in background noise and reverberation and a decrease in lighting and visual cues from the speaker. Table 15.1 shows an example of the WHO-ICF framework as might be applied to an otherwise active and healthy 80-year-old individual with hearing loss.

15.4.3 Hearing Function Assessment

The audiogram, used to assess the type and degree of hearing loss, does not specifically assess the individual's communication disability. Measures of hearing function are conducted in the form of interviews, needs assessment tools, questionnaires, or structured communication interactions. These instruments enable the clinician to solicit information regarding the individual's participation levels while engaged in communication activities at home, work/school, community, or travel environments. For example, the Hearing Handicap and Disability Inventory (HHDI) asks the listener to rate his or her performance on 40 items related to communication activities

TABLE 15.1
ICF Example

Health condition	Impairment	Activity limitation	Participation restriction	Environmental factors
Presbycusis (age-related hearing loss)	Loss of hearing sensation for high-frequency sounds	Difficulty understanding speech	Unable to participate in weekly church Bingo	High level of background noise in church hall, highly reverberant room

Note: This table shows an example of World Health Organization International Classification of Function (WHO-ICF) applied to an otherwise healthy 80-year-old woman.

(e.g., my hearing loss discourages me from using the telephone) (Van den Brink, 1995). Instruments such as this are used both for assessment of hearing disability and as tools to plan for which HAT devices may be appropriate to maximize participation. Bentler and Kramer (2000) reported on the psychometric properties of 20 different self-report questionnaires designed to assess communication disability and/or needs (Bentler and Kramer, 2000). Their work highlighted the proliferation of these instruments over the past decade; however, most of these instruments do not specifically target HAT needs. A few instruments are specifically designed to direct clinicians to appropriate HAT solutions for persons with hearing loss. The Hearing Demands and Needs Profile (Table 15.2) is an example of such a tool, creating an inventory of an individual's listening needs at home, when traveling, or at work or school (Palmer and Mormer, 1999). Recently, an assessment tool, called the TELEGRAM, was proposed so that clinicians could easily conduct a comprehensive assessment of communication needs and potential HAT solutions as part of the routine hearing assessment (Thibodeau, 2004). The name of this instrument is based on an acronym that reflects the areas to be considered: telephone, employment, legal issues, entertainment, group communication, recreation, alarms, and members of the family. Information on the listener's disability in each of these areas is investigated and HAT solutions are formulated based on these findings.

15.5 HEARING ASSISTANCE TECHNOLOGY SOLUTIONS

In a given communication environment there are three primary barriers that interfere with the satisfactory reception of a signal or message. The first of these is background noise, some level of which is always present during daily activities. People with hearing loss, whether wearing hearing aids or not, have difficulty understanding speech when background noise is present. The second important barrier relates to room acoustics. For example, a high level of room reverberation is known to contribute to poorer speech understanding. The third barrier relates to the reduction in sound energy that

TABLE 15.2
Hearing Demand and Need Profile

Name:

		Communication Problem Is Present. . . with Hearing Aid:						The Problem Is Due to. . .				Current Compensation (describe)
		Home		School/Work		Travel						
Age	Description of Communication Milestone/Activity	on	off	on	off	on	off	Hearing	Noise	Distance	Visibility	
	Alerting											
	Telephone bell											
	Doorbell											
	Door knock											
	Alarm clock											
	Smoke alarm											
	Siren											
	Turn signal											
	Personal pager											
	Personal Communication											
	Telephone											
	TV/stereo/radio											
	One-to-one(planned)											
	One-to-one(unplanned)											
	Group											
	Large room											
	Other Activities											
	Clubs/games:											
	Lessons:											
	Sports:											
Further information (e.g., status of hearing aids, telecoil, DAI, communication environment):												
Recommendations (assistive technology, communication strategies, environmental manipulation):												

Instrument used to assess a listener's individual communication needs at home, work/school, and when traveling and when hearing aids are on or off.

Source: Healey, J. (1992) and Palmer, C. (1992).

occurs with increased distance from a sound source. All three factors (background noise, room acoustics, and distance from sound source) contribute to the overall SNR at the listener's ear. Persons without hearing impairment require an SNR of +6 dB to perceive speech adequately. Those with hearing loss require an optimal SNR of up to +20 dB to maximize speech reception (Flexer, 2004). A wide variety of technological solutions are used to address the problem of reduced audibility and speech perception. The goals for each of these solutions are dependent on the individual's listening activities and needs. For example, one client might need technology to help increase his/her speech-understanding ability at a work meeting. Another might need a smoke detector that will alert him/her to a fire despite his inability to hear. The solutions across the range of listening difficulties use technologies that either enhance the auditory signal (amplification) or substitute for the auditory signal via an alternative mode (tactile or visual). In some cases, the use of multiple modes is combined in the rehabilitation process. Such is the case when a hard-of-hearing television viewer uses amplification to hear sound while reading printed captions of the auditory signal on the screen (Table 15.3). In addition to HATs, engineering solutions can be aimed at reducing background noise in the listening environment. This can be achieved through a variety of noise reduction approaches aimed at the sound environment rather than at the individual listener's hearing loss. These modifications might include reducing the noise near its source via the use of damping principles or installation of highly absorbent materials (e.g., acoustic tiles, carpeting, or draperies) to the surfaces of a reverberant room.

TABLE 15.3
Continuum of HAT from Enhancement of Auditory Signal to Substitution with Alternative Signal

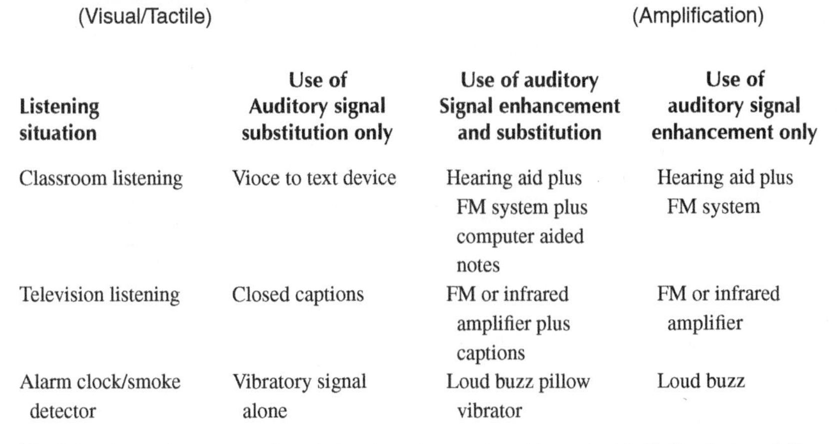

Listening situation	Use of Auditory signal substitution only	Use of auditory Signal enhancement and substitution	Use of auditory signal enhancement only
Classroom listening	Vioce to text device	Hearing aid plus FM system plus computer aided notes	Hearing aid plus FM system
Television listening	Closed captions	FM or infrared amplifier plus captions	FM or infrared amplifier
Alarm clock/smoke detector	Vibratory signal alone	Loud buzz pillow vibrator	Loud buzz

The left column shows examples of situations that can be addressed with HAT solutions falling along the continuum of enhancement to substitution of the auditory signal.

The interface of audiology and engineering occurs when the two professions share the goal of maximizing the deaf or hard-of-hearing person's communication ability. As described earlier, engineering principles allow for the design and production of an array of amplification technologies. Similarly, acoustical engineering addresses problems of noise control and enhancing listening environments. Audiologists and engineers, therefore, work together on a variety of applications across manufacturing and clinical settings.

15.5.1 Historical Overview

One of the earliest forms of HAT was the acoustic horn, which was popular during the 18th century. This device served to funnel sound into the ear canal. Its resonance properties enhanced the acoustic parameters of the sound. Early in the 20th century, a hearing aid employing a body-worn microphone, amplifier box, and receiver earpiece was introduced. Hearing aids have since become the most generally used auditory assistive technology. With miniaturization of the components, including batteries, the devices are no longer body worn. Until the 1990s hearing aids were manufactured with analog circuit technology. Digital circuitry was then introduced and its routine use has continued to grow through the early 2000s.

A hearing aid can be thought of as a head-worn personal amplification system offering a general hearing solution in a variety of listening situations. Currently, hearing aids are available in an assortment of models ranging from a device that is worn behind the ear and coupled to the ear with an ear mold to a custom-made device fitting completely in the ear canal (Figure 15.1). In all cases, the microphone of the hearing aid sits close to the listener's ear and receives the auditory signal and background noise with approximately the same SNR as the ear itself. For most hearing aid users, this results in improved speech reception mainly when background

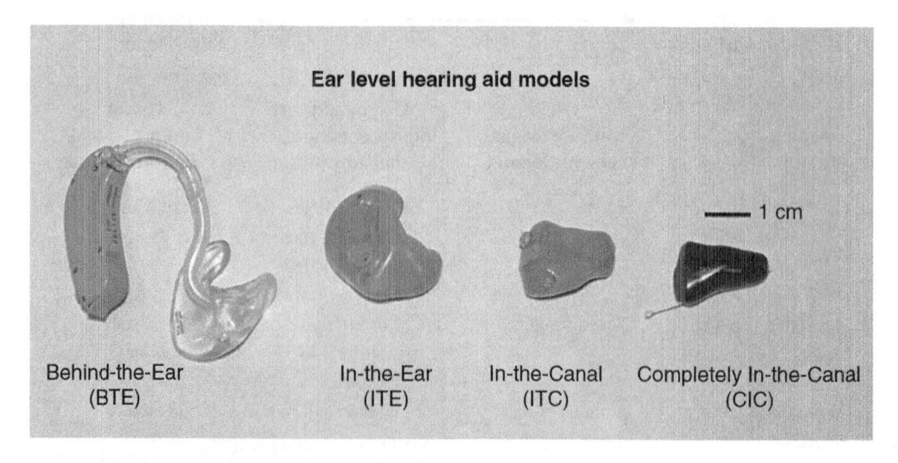

FIGURE 15.1 Ear-level hearing aid types. Shown from left to right are a behind-the-ear hearing aid and custom earmold; in-the-ear hearing aid; in-the-canal hearing aid; and completely-in-the-canal hearing aid.

noise is minimal. In situations of increased background noise, difficulty with speech reception continues or may be exacerbated with hearing aid use. For many decades in the United States, remote microphone technology was used in classrooms for hard-of-hearing and deaf children. These systems used either FM radio signals to transmit the teacher's speech signals as they were spoken into a microphone close to her mouth. Each child wore a receiver that sent the amplified signal to their ears. This configuration of the microphone and receiver helped to overcome the deleterious effects of background noise, poor room acoustics, and distance from the signal source (the teacher) to the listener (the student). This approach resulted in a better SNR than could be achieved with hearing aids alone. In the late 1980s, audiologists began applying this technology to personal listening for children and adults alike. Although current hearing aid technology includes options such as directional microphones and noise suppression features, no hearing aid has been shown to provide as much benefit to speech perception as that provided by a remote microphone system (Nelson et al., 2004). In addition to the remote microphone technology described here, another category of HAT has been incorporated into hearing rehabilitation over the past few decades. These devices create signals that can be used to alert a hard-of-hearing or deaf person to a variety of warning signals such as a smoke detector, doorbell, ringing telephone, or crying baby. Although some of these devices use an enhanced auditory signal, they most often deliver the signal through an alternative mode such as a flashing light or vibrating disk. Throughout the 20th century, the development and use of electronic devices burgeoned. Deaf and hard-of-hearing individuals have been challenged to keep up with society's ubiquitous use of radio, telephone, television, and computers With the passage of the Americans with Disabilities Act in the early 1990s, a variety of devices have been developed and marketed to this population so that their needs can be accommodated (Bengtsson and Brunved, 2000). A description of these devices follows in the next section.

15.6 MEDICAL OR SURGICAL APPROACHES TO RESTORING FUNCTION

Approximately 90% of hearing impairments in adults are sensorineural in nature (Yueh et al., 2003). Sensorineural hearing loss is, in most cases, not treatable by medication or surgery. Recently, developments in the area of inner-ear hair cell regeneration have shown promise for possibly restoring auditory function. More specifically, research has suggested that mammalian cochlear hair cells can be stimulated to regenerate via mitosis or transdifferentiation. Mitosis is a process in the cell cycle that facilitates cell division. Transdifferentiation is the process in which one cell type changes its phenotype to become another, with or without mitosis. Both processes are native to the inner ear of nonmammalian species, such as birds. Interestingly, the mammalian vestibular, or balance, sensory epithelia may also induce both processes. However, the mammalian cochlear, or auditory, sensory epithelium is limited in both. Current research is focused on using gene therapy or stem cells to stimulate mitosis or transdifferentiation in the damaged cochlear sensory epithelium. Success with this type of hearing-loss treatment is certainly far in the future.

Currently, the majority of persons with hearing loss find hearing aids and/or HAT to be the only means available to maximize their communication participation. The technologies available to address this challenge range from surgically implanted devices to body-worn remote microphone technologies.

15.6.1 Surgically Implanted Devices

Recently, a variety of surgically implanted devices have become available for hearing rehabilitation. It is useful to categorize these devices according to the degree of hearing loss for which they are intended. Middle-ear implants (MEIs) and bone-anchored hearing aids (BAHAs) are for those individuals with a useful amount of residual hearing. Cochlear implants are designed for those with severe to profound hearing loss.

MEIs are an adapted hearing aid design in that the receiver of the hearing aid is implanted in the middle-ear space as a mechanical transducer on the ossicular chain (the three bones in the middle ear that convert acoustic energy into mechanical energy). MEI's basic design consists of both external and internal parts. Externally, the user would wear a device that looks very much like a behind-the-ear hearing aid. This device houses the microphone, the digital processing circuit, and the battery. The external device is coupled to the internal device by a magnetic button that is worn on the head, which transmits the signal through to the internal magnetic receiver. A wire then transfers the auditory signal to the transducer attached to the ossicular chain. The MEI provides amplification through the direct enhancement of the middle-ear mechanical vibration. The MEI has several advantages over conventional hearing aids in that acoustic feedback may be reduced and the occlusion effect (resulting from the hearing aid earmold plugging the ear) is diminished. However, it is a surgical procedure, therefore introducing more risks to the patient. Currently researchers are developing a completely implantable MEI so that the microphone of the aid could be embedded in either the canal wall or tympanic membrane.

BAHAs are designed for those individuals with conductive hearing loss. A vibrotactile device is attached to a titanium base that has been surgically implanted in the temporal bone behind the ear. Transduction of sound is highly efficient because the vibrating transducer is directly connected to the skull, where the sense organ of the inner ear is located. The use of BAHAs has shown reduced levels of disability and handicap, and significant patient benefit and satisfaction (McDermott et al., 2002).

Cochlear Implants (CIs) are devices that provide direct electrical stimulation to the auditory nerve. The electrical stimuli result in an auditory sensation. The components worn include the microphone, speech processor, and a magnetic coil that transmits auditory signals to the internal components. The internal components include a magnetic receiver and a multiband electrode that is fed through the shell-shaped turns of the cochlea in the inner ear. The algorithms of CI's signal processing strategies are designed to extract the salient features of speech such as frequency, temporal, and intensity cues and deliver these parameters to the electrode array. The electrode array then electrically stimulates the residual auditory neurons, translating these important speech cues into place, rate, and voltage of stimulation in the inner ear (Tye-Murray, 2004).

15.6.2 Nonsurgical Hearing Aids

Hearing aids are personal amplification devices that improve hearing function through the use of an amplifier circuit. Hearing aids increase the audibility of the acoustic energy entering its microphone. The major advantage of hearing aid use is that audibility is increased in a variety of listening situations. Another advantage of hearing aids is that numerous electronic modifications to the signal processing circuitry can be made. The incoming signal waveform can be divided into numerous bands (up to 32) for frequency shaping and intensity compression. Hearing aid dispensers are able to manipulate the frequency/gain response of the amplified signal via a personal computer to meet the needs of the patient. Although there are many benefits to hearing aids, a major limitation is that the hearing aid is not able to alter the user's environment or to amplify only the wanted signal. Hearing aids come in a variety of sizes and styles (Figure 15.1). Although they may all look different, they contain the same basic components. Sound energy enters the microphone port of the hearing aid and is initially converted to an electrical signal. This signal is amplified via either an analog or a digital circuit. The amplified signal is then delivered to the receiver. Sound exits the hearing aid and is directed to the eardrum via tubing in the shell of a "custom" instrument or the ear mold in a "behind-the-ear" instrument. One optional component is called a telecoil. A hearing aid with a telecoil takes advantage of the stray magnetic field that normally emanates from the receiver in a telephone handset. By placing an induction coil inside the hearing aid, direct transmission of the telephone signal is accomplished, eliminating the need to pass the signal through the hearing aid microphone. Advantages of the telecoil include avoidance of the ambient noise surrounding the listener and elimination of acoustic feedback problems associated with telephone placement near the hearing aid microphone (Yanz and Preves, 2003). Another advantage of the telecoil is its ability to couple with assistive listening technology, thus allowing the hearing aid user to directly receive the signal from a remote microphone system in the hearing aid (Figure 15.2). Alternatively, some hearing aids have direct audio input capability. This feature allows for the direct connection of an auxiliary input signal to electrical contacts on the hearing aid. Audio signals from either a remote microphone system or other audio sources (CD player, telephone, or television) can be delivered directly to the hearing in this way. The direct audio input option eliminates the risk of electromagnetic interference that often occurs with the telecoil.

15.7 ASSISTIVE LISTENING SOLUTIONS

As described earlier, most assistive listening device solutions use a microphone or other input devices placed at the signal source. For example, a teacher in a class of hard-of-hearing students might wear a lapel-type microphone placed within 12 cm of her mouth. Another scenario might involve a hard-of-hearing adult watching television. In this case, a microphone might be positioned within centimeters of the television speaker, or a cable might be plugged into the headphone jack or audio output jack of the television. The microphone or output cable picks up the audio signal from the television, which is then sent to an amplifier. The signal can be transmitted

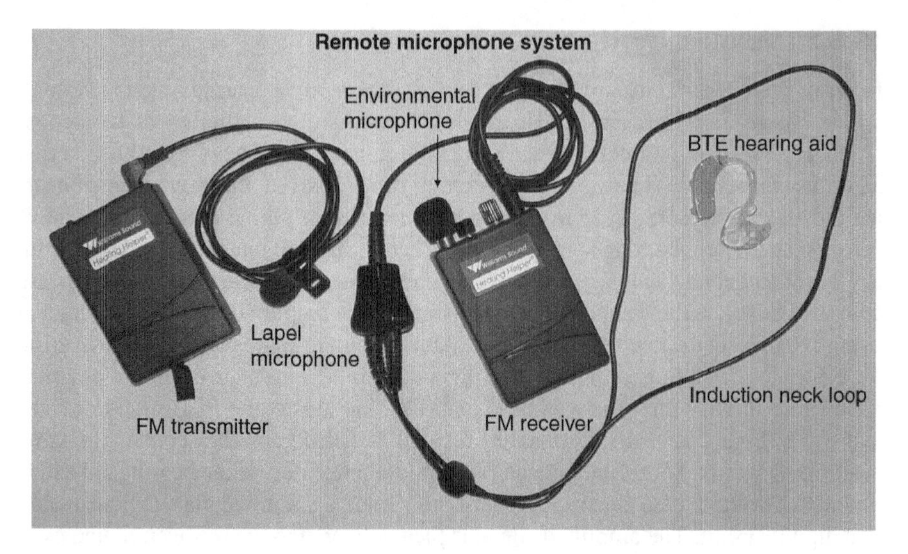

FIGURE 15.2 Photograph of a remote microphone system using FM transmission. The lapel microphone, worn by the speaker, is connected to the FM transmitter. The signal is transmitted to the FM receiver. In this case, the output from the receiver is an inductive signal emanating from a neck loop, worn around the listener's neck. The inductive signal from the neck loop is picked up by the hearing aid telecoil.

TABLE 15.4
Assistive Listening Device Components Options

Input signal	Transmission mode	Ear coupling
Microphone	Hardwired	Headphones/earbuds
Telephone	FM radio waves	Hearing aid via Direct audio input/telecoil
Television	Infrared light waves	Cochlear implant via Direct audio input/telecoil
Portable CD player/radio	Inductive field	Hybrid hearing aid/FM

Note: This table shows some of the options that can be combined to achieve enhancement of the auditory signal. The first column lists some of the input signal options. Column 2 shows the variety of transmission options. Column 3 shows some of the ways that the output signal can be coupled to the listener's ear. Options from each column can be combined with those from the other columns to achieve the desired signal reception.

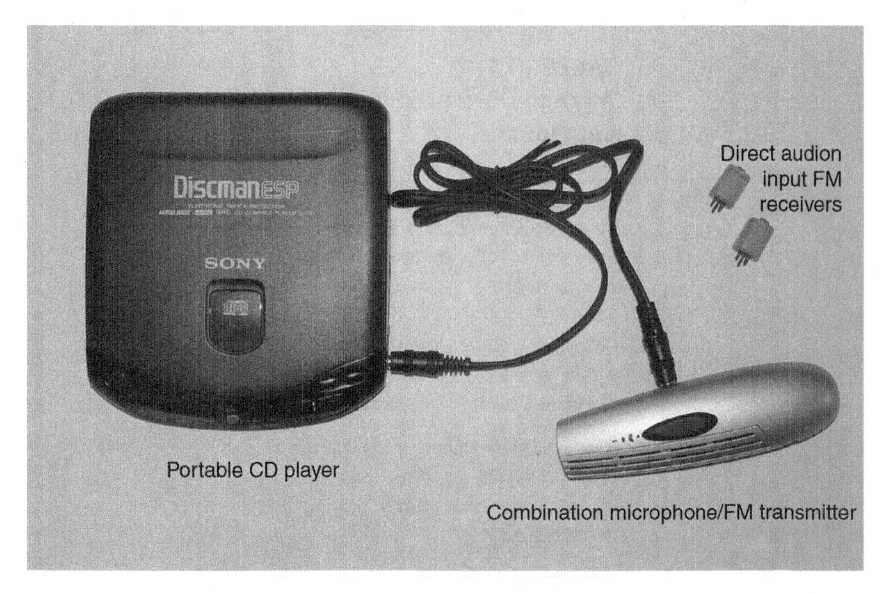

FIGURE 15.3 Photograph showing integration of components in an assistive listening device. The output of a portable CD player is fed to the auxiliary input jack of a combination microphone/FM transmitter. The output is received in two small receivers that can be coupled via DAI to right and left hearing aids.

to a receiver via hardwired, FM radio, infrared light transmission, or induction transmission. The output from this transmission is then coupled to the listener's ears via headphones (when no hearing aid is used) or via the hearing aid telecoil or direct audio input cables and boots (Table 15.4). FM receivers can be built in to behind-the-ear hearing aids or coupled directly via a plug and socket at ear level, affording another option for signal to ear coupling. An example of the integration of these components is shown in Figure 15.3.

15.7.1 Alerting Solutions

Solutions for a variety of alerting needs usually involve substitution of an otherwise auditory signal with a signal that will indeed alert the hard-of-hearing or deaf person. A number of these devices are only necessary because they serve in situations in which the individual would not ordinarily be wearing hearing aids. For example, an older person living alone might be able to hear a household smoke detector when wearing hearing aids during the day. Hearing aids, however, are not meant to be worn during sleeping hours. Thus, an alternative mode of alerting must be used to awaken such an individual if a fire starts in the middle of the night. The most common solution for this situation involves a smoke detector, which, when triggered, sends a signal to either a vibrating disk under the pillow or mattress or to a strobe light. This principle can be applied to a variety of alerting situations such as a ringing telephone, alarm clock, doorbell, or crying baby. A vast assortment of alerting devices is available on

TABLE 15.5
Alerting Device Input and Output Options

Input signal	Output mode
Smoke detector	Shaker/vibrator
Doorbell	Strobe
Baby crying	Flashing light
Siren	Visual text display
Telephone ring	Scent dispenser
Motion sensor	Loud sound
Tea kettle whistle	Dog barking/jumping

This table shows input signals that can be captured by alerting devices and output modes used to alert hard-of-hearing and deaf individuals.

the market or can be created from component parts. Conceptually, each contains a sensor component that responds to the alerting stimulus and transmits a signal to a receiver. The signal is then transduced to whatever form of nonauditory signal is appropriate (Table 15.5). One alerting solution that has gained popularity over the past decade is decidedly low tech in nature. This involves the use of a specially trained dog that alerts its owner to signals across the range of warnings. The response output of the dog includes all sense modalities with signals that include barking, jumping, and licking.

15.8 VISUAL SUBSTITUTIONS TO AUDITORY ACTIVITIES

Visual substitutions of auditory signals can be used in a number of different formats for a variety of listening challenges. Since 1992, all televisions manufactured in the United States have been required to have a closed-caption decoder incorporated into the circuitry. This decoder can be turned on by the user, providing access to a text display of the spoken words and sound effects contained in the broadcasted show. Not all television shows or television broadcasted movies contain the captions, thus this technology is limited by its availability. Open, as opposed to closed-captions, refers to subtitling that is always present on a viewing screen as in viewing the language translation in a foreign made film. Movie theatres can make this technology available to its audience on a limited basis, offering occasional open-captioned viewings of popular films. An alternative system of captioning was developed in the United States in the 1990s. This system, called Rear Window® Captioning System, displays reversed captions on a light-emitting diode (LED) text display that is mounted in the rear of a theater. Deaf and hard-of-hearing patrons use transparent acrylic panels attached to their seats to reflect the captions so that they appear superimposed on the movie screen. The reflective panels are portable and adjustable, enabling the caption user to sit anywhere in the theater (Heppner, 2003).

Computer Aided Realtime Translation (CART) is used regularly for hard-of-hearing and deaf persons in classrooms, meetings, courtrooms, and other venues. This technology grew out of courtroom use of stenotype machines and the need for detailed printed text records of legal proceedings. In the past decade, as this service became integrated with computers, court reporters began to provide real-time translation services in settings outside of the courtroom. The CART provider types into a computer converting spoken words to text in real time. This text can be viewed by the hard-of-hearing or deaf person directly on the computer screen near his or her seat or on a large screen visible to the whole audience. An advantage of this system is that a written transcript can be produced and made available to those in attendance. This solution is often used in college classrooms, where spoken lecture material and student comments are recorded.

Although speech recognition software is still in its early development, it has been used as a component of HAT designed to convert speech to text. This is well demonstrated in a product called the iCommunicator™. This device allows for real-time transcription of spoken signals through speech recognition software. There is no need for speech sounds to be translated into keyboard strokes with this device. This system uses a remote microphone placed at the speaker's mouth. Input is converted to an electrical signal that is delivered to a laptop computer and then converted to printed text appearing across the computer screen. In addition to the conversion from speech to text, this technology allows for conversion from speech to video sign language. The user can choose to view both of these modes simultaneously if desired (Rosenberg, 2003).

In addition to alerting devices that use visual stimuli, there are other HAT devices that use visual signals or text to convey the content of auditory communication. Deaf and hard-of-hearing listeners routinely use the visual cues associated with movements of the face, lips, and mouth during speech. This technique, known as speech reading, is a natural component of speech understanding even for listeners with normal hearing. One HAT device augments its remote microphone technology with the addition of a small, body-worn, remote camera aimed at the talker's mouth. The microphone and camera pick up and amplify both the auditory and visual components of the speech signal (Gagne et al., 2002). The amplified signal and the enlarged image of the talker's face is shown on either a small screen in front of the listener, or on a large screen television at the front of the room. This application of HAT has been used mainly in schools but could be incorporated into community or workplace settings to broaden communication accessibility for children and adults.

15.9 ENVIRONMENTAL ADAPTATIONS AND UNIVERSAL DESIGN

There are a number of design considerations and modifications that play a role in maximizing communication ability for people with hearing loss at home and in public venues. Some of the HATs discussed earlier are included in the desired modifications and designs. It might be useful, however, to envision the home and public facility that best accommodates the hard-of-hearing or deaf inhabitant or visitor. The walls and furniture of this home should be arranged to allow for open spaces. Such unobstructed living space facilitates lines of vision to maximize speech-reading potential or sign language for communication. Ambient noise can be minimized with the application of absorbent wall, floor, and ceiling coverings; however, this must be balanced with the need to feel the vibration of movement across the floor. A view panel at the front door is preferable to a standard peephole, as the resident may not be able to hear a voice on the other side. This house would be modified with the installation of a variety of alerting devices. Smoke detectors on each floor would transmit signals to a vibrating pager worn by the resident of the home. This pager might have four-color coded lights, each of which flashes depending on whether there is a fire in the house, someone at the door, telephone ringing, or baby crying. Telephone jacks would be installed close to electrical outlets to provide power for additional signalers, text telephones, or other telephone related devices.

When the hard-of-hearing or deaf person leaves home and enters public venues, his or her access to communication is maximized if some of the following modifications or designs are used. At a movie theater the sound system should allow for an FM transmitter or infrared emitter to send the audio portion of the movie to receivers available to patrons at the entry. Coupling options for those who do and do not wear hearing aids should be available. For universal accessibility, the theater would offer Rear Window® Captioning. When our subject goes to the airport, he or she hopes to find a text paging system to substitute for the audio paging that is used heavily in that environment. Airports are known to present a variety of communication challenges to hard-of-hearing and deaf travelers. Visual paging systems provide environmentally based solutions to many of these challenges. These systems use wall-mounted signs displaying LED-generated text transmitted from a keyboard. In this manner, flight, boarding, and paging information is accessible to all travelers.

15.9.1 Task-Specific Solutions

The solutions can be individualized as per the needs of its users. What follows are examples of technology solutions in categories of need areas. As seen with the assessment tools described earlier, individual listening needs can be categorized into areas of personal listening situations, group listening, telephone listening, and vocational- or school-related needs. There is overlap of these categories.

15.9.1.1 Personal Listening Situations

Optimal listening conditions for a person with hearing loss require one-on-one conversation with minimum background noise. Often, the use of hearing aids or CIs alone will suffice in this situation, particularly when visual cues from speech reading are available. On the other hand, a hard-of-hearing mom driving in a car with her 3-year-old daughter might be unable to converse under these conditions. The car contains a high level of background noise and visual cues are not available as the child sits in a car seat behind the mom. The solution might be to hang a microphone and transmitter on the child's neck in close proximity to her mouth. Her speech can then be transmitted to an FM receiver that couples to the mom's hearing aid through direct audio input contacts. Additionally, mom might have the option to leave the microphone of her hearing aid on so that she can monitor sounds around the car. Simultaneously, she will hear her daughter's speech as it is directly fed from the remote microphone to the hearing aid amplifier.

15.9.1.2 Group Listening Situations

Let us imagine that the mom described earlier plays Mahjong with a group of women once a week. As the women chat across the table, she has difficulty understanding the speech that is exchanged. She can now adjust the microphone of her FM system, switching it from a directional to an omnidirectional mode. In the omnidirectional mode, she will have improved access to the speech from the women positioned around her at the table. When this mom goes to an evening class at the local college, she again uses the FM system. She asks the teacher to wear the microphone/transmitter as she conducts the lecture. Additionally, as a student, she uses a CART system to capture the comments of her classmates and to create a transcript of the evening's class lecture and discussion.

15.9.1.3 Telephone Solutions

The telephone is a ubiquitous instrument across the industrialized world. Telephone communication may be the most difficult listening challenge the person with hearing loss faces. Individuals with mild to moderate degree hearing loss may be able to use a conventional landline telephone along with the use of a hearing aid. As described earlier, the telecoil allows for coupling directly with the telephone. If the hearing aid telecoil fails to provide enough acoustic gain, an assortment of commercially available amplified telephones is available. These phones include features such as tone and volume control. Some allow for the output of the phone to be coupled to an induction loop worn by the listener. In this manner, the telephone signal can be picked up by the telecoils in both hearing aids for binaural reception. As mentioned previously, the output from any hardwired telephone handset can be fed to a personal FM system that is then coupled to the listener's ears.

Hearing aid compatibility with digital wireless phones is more complex than with landline phones. The hearing aid or CI telecoil not only picks up the electromagnetic waves from the telephone receiver, but also the radio frequency (RF) emissions from the wireless antenna. These RF emissions create an interface signal (a buzzing sound) for the hearing aid user, thus masking the output of the telephone receiver (Preves, 2003). Currently in the United States, digital wireless manufactures are required to provide wireless phones that have reduced RF emissions so that interference between the antenna and the hearing aid's telecoil is minimized. As a result of the difficulty encountered when using cell phones and hearing aids or CIs, a variety of accessory products are now available to couple these devices. Most of these products use enhanced inductive strategies to deliver the telephone signal to the hearing aid while minimizing electromagnetic or radio frequency signal interference.

For individuals with severe or profound hearing loss, use of the telephone often requires a conversion of the auditory signal to written text. A text telephone displays a script of the incoming telephone speech signal on a screen for the hard-of-hearing or deaf individual to read. This device is known as a Telecommunication Device for the Deaf (TDD). These devices use standard phone lines and services. The text telephone enables individuals to call each other and type their conversations to one another. In the US, if a TDD user needs to call someone who does not have a text telephone, he or she uses the National Telephone Relay service. This service provides the transcription of spoken speech to written text delivered to the text telephone device and vice versa. This service requires a third party, called a relay operator. The relay operator listens to the speech of the person talking on the nontext telephone and transcribes the words into text that is viewed on the screen of the TDD user. Most hard-of-hearing people have speech that is clearly intelligible on the phone. For these telephone users, a variation on the TDD allows for their spoken telephone responses to bypass the relay operator and be heard directly by the listener.

Small, portable TDD keyboards can be coupled to a cell phone to allow for text transmission of conversation. However, with the proliferation of cell phones, people with hearing loss and those with normal hearing have turned to the text-messaging feature to conduct nonauditory conversations. Similarly, the use of e-mail and instant messaging, or handheld text message pagers, lends itself well to the communication needs of hard-of-hearing and deaf persons. More on this topic is covered in the chapter on telecommunications.

15.10 VOCATIONAL, DAILY LIVING, AND COMMUNICATION AIDS

In addition to the routine listening activities described throughout this chapter, there are a number of specialized vocational, recreational, and daily living activities that are acutely challenging for people with hearing loss. The same principle of enhancement or substitution is applied to these circumstances regardless of the setting or task.

15.10.1 Healthcare

Many of the tasks involved in the health professions depend on the reception of auditory signals. For example, listening for heart rate and breath sounds or measuring blood pressure are activities that depend on auditory input. Conventional stethoscope signals are largely inaccessible to those with hearing loss; however, there are a number of strategies to modify conventional stethoscope use. Modifications may include, but are not limited to, specialized ear tips that couple with custom-made hearing aids, specialty ear molds that accommodate stethoscope tips, direct audio input (DAI) or telecoil coupling from stethoscope to hearing aids or CIs; stethoscope headphones, and amplified stethoscopes. Additionally, the auditory stethoscope signal can be converted to a visual or tactile signal (Rennert et al., 2004).

15.10.2 Business Settings

Employees working in the business environment often find themselves in meetings or conferences, where multiple talkers engage in conversation simultaneously. This type of setting is very difficult for someone with a hearing loss as speech intelligibility is reduced by multiple speakers and distance between the target speaker and the listener. In this circumstance, a personal amplification device using FM transmission can be configured to achieve communication access. This scenario might include a neck-worn directional microphone/transmitter targeting the voice of the main speaker or group leader. An omnidirectional table top microphone can also be plugged into an auxiliary input jack on the transmitter. It is possible to daisy-chain multiple omnidirectional microphones along the length of a conference table so that comments from speakers seated around the room can be amplified. Regardless of the number of microphones used, the signal is transmitted to a receiver that is coupled to the listener's hearing aid. To achieve this, the listener may be wearing an inductive neck loop that converts the receiver signal to an inductive signal emanating out from the loop (Figure 15.2). The hearing aid is set to the telecoil mode and the telecoil then receives the inductive signal from the neck loop. At this point the signal travels through the hearing aid components and the amplified sound is delivered to the ear canal. One limitation of this system is the threat of electromagnetic interference (EMI). If the room has fluorescent lighting or computer equipment is in use, EMI can be picked up by the induction system and perceived by the hearing aid user as a variety of distracting and disturbing noises.

15.10.3 Large-Group Listening

Attendance at houses of worship can be especially frustrating for hard-of-hearing participants. Wide-area listening systems are a relatively easy solution, enabling attendees to hear both the choir and the clergy regardless of seating position in the congregation. This technology can be coupled to an existing public address (PA) system or it can be used alone. When used with an existing PA system, a line out from the PA mixer is fed to an infrared or FM transmitter. Input from all microphones in use at the pulpit can be delivered to the listeners' ears. The transmitter sends the signals to multiple receivers worn by the listeners in the audience. Listeners couple the receiver sound to the ear either via headphones or directly through hearing aids or CIs. An option available on some hearing aids allows for the simultaneous operation of the hearing aid microphone and telecoil or DAI input. This feature allows the hearing aid user to hear the input from the remote microphone (e.g., the pulpit) as well as the voice of his or her spouse whispering words at his or her side. Wide-area systems such as this are also used in classroom settings for children and adults with hearing loss.

15.10.4 Sports and Recreation

Athletes who are hard of hearing or deaf encounter a number of logistical challenges when they pursue sports at school, in competition, or for recreation. For any given sport, there are a variety of both high- and low-technology strategies that can be implemented to accommodate for hearing loss. Generally, the HAT solutions to sports and recreational activities address the individual listening requirements of the participants and the rules of the sport undertaken. For example, due to the amount of physical contact and the fact that helmets will cause acoustic feedback, hearing aids cannot be worn in the traditional manner while the listener is playing football. For a hard-of-hearing student athlete, the communication demands inherent in the game (exchanges between coach and players, excessive crowd noise, etc.) create barriers to participation. For a player who wishes to use a hearing aid with the helmet, modifications to the helmet and hearing aid can be applied as follows. A hole is drilled in the top of the helmet and a behind-the-ear hearing aid is secured in the top of the helmet with the microphone situated in the hole. The output of the hearing aid is fed acoustically through plastic tubing connected from the ear hook of the hearing aid to a custom ear mold in the ear. The athlete will then hear whatever sounds are picked up by the microphone at the top of the helmet.

Substitution signals can be applied in competitive sports situations in which warning and starting signals for racing are required. This would be the case in events such as swimming or track meets. For those participants who are unable to hear such signals (horn, gunshot, etc.), a visible signal is needed. For timing accuracy in competitions, the visual signal should be triggered by the audible signal via a signal detection device. The visual output could include a red light to signify "on your marks," a yellow light to signify "set," and a green light to signify "go" (Palmer et al., 1996).

15.10.5 Industry and Service Delivery

Hearing aid and HAT distribution patterns vary from country to country around the world. There are approximately 15 to 20 major manufacturers of standard hearing aids and hearing aid dispensing offices are located in at least 60 different countries in the world (2005 Hearing Health Industry World Directory, 2004). These hearing aid dispensing offices are predominantly located in the developed areas of the world. The conditions under which hearing aids are delivered to recipients vary with each country's health policy and the service delivery model under use. Many countries such as Canada and Australia have publicly funded programs that either subsidize or completely cover the cost of hearing aids for eligible individuals. Similarly, many of the European countries such as Norway, Sweden, the Netherlands, Germany, and France have publicly supported hearing healthcare programs covering all or part of the cost of hearing aids and related equipment. Eligibility and distribution models vary across these countries such that some rely on private business to supply the instruments while others are issued through government run clinics only. Spain and Portugal, as well as the United States, are countries that do not generally fund hearing aid provision, except in the case of some special populations. In all of the countries mentioned here, hearing aids for children are either subsidized or completely paid for by government-sponsored programs (Receiving Hearing Aids in Different Countries, 2005).

There is very little information available on the use and distribution of assistive listening and alerting devices around the world. In the United States, these devices can be dispensed by audiologists, in retail electronics outlets, and in a variety of online or printed catalogues. Issues of compatibility arise when consumers purchase components and are not aware of the coupling requirements. For example, a behind-the-ear hearing aid user might purchase a remote microphone system with headphones. If he or she then tries to use the headphones over the hearing aids, acoustic feedback will result. Situations such as this illustrate the need for technical expertise in the integration of compatible components.

15.11 CHALLENGES FOR THE FUTURE

When comparing rehabilitation resources available to individuals around the world, those addressing HAT fall far short of those addressing physical disabilities. Often referred to as the "invisible disability," hearing loss has not garnered the level of attention given to more apparent disabilities such as quadriplegia or blindness. One challenge that faces the hearing rehabilitation field worldwide is to make the public aware of the disabling effects of hearing loss. When this is accomplished, more widespread use of HAT will occur.

There are few international standards pertaining to hearing aids and assistive listening and alerting technology. One ISO standard describes the procedures for the measurement of acoustic properties of hearing aids. Standards for FM or infrared transmission signals have not been internationally designated at this time. Thus, users may be unable to travel from one country to another using a personal remote microphone system without encountering interference. Standardization in this realm would accelerate the further development and use of these systems. Education of all the constituencies involved in the development, distribution, and use of HAT has progressed slowly over the past few decades. These devices have been used mainly for school age children for many years, and adults with hearing loss are largely unaware of the benefits available with the use of this technology. Furthermore, funding for the fitting and use of HAT is limited in most countries. As grassroots organizations such as Self Help for Hard of Hearing People (SHHH) and the International Federation of Hard of Hearing People (IFHOH) demand more and better accommodations for their disability, rehabilitation technology for this population will steadily continue to demand engineering solutions. Areas such as miniaturization, power supply, and interference and/or integration with mainstream electronic communication technologies will continue to challenge engineers in the future.

15.12 STUDY QUESTIONS

1. Compare and contrast typical communication strategies used by a hard of hearing vs. a Deaf individual in listening situations such as one on one, telephone, and audience listening.
2. Explain some expected areas of communication difficulty typically identified in a communication needs assessment.
3. What are examples of hearing assistance technology devices that substitute for, versus enhance the reception of conversational speech?

4. Draw a simple block diagram of the major components of a hearing aid.
5. What is the impact of a remote microphone system on the signal-to-noise ratio received at the listener's ear? Why is this important in the enhancement of speech reception?
6. What advantage does a remote microphone FM system have over a conventional hearing aid for a hearing of hearing listener?
7. How would the concept of Universal Design be applied to spaces expected to be occupied by hard of hearing and/or Deaf individuals.
8. Why might a hard of hearing listening experience difficulty using a cell phone through his hearing aid telecoil?
9. Why does hearing loss receive less attention to rehabilitation than other physical disabilities?
10. What engineering challenges face those who are involved in the design of hearing assistance technology?

BIBLIOGRAPHY

Bengtsson, P., Brunved, P. When hearing aids are not enough, In: Valente, M., Hosford-Dunn, H., Roeser, R. (Eds.), *Audiology: Treatment*, New York: Thieme Medical Pub; 2000, pp. 581–599.

Bentler, R.A., Kramer, S.E. Guidelines for choosing a self-report outcome measure, *Ear Hear* 2000;21(Suppl):37–49.

Facts About Deafness. World Health Organization Web site, 2004: http://www.who.int/pbd/deafness/facts/en/

Flexer, C. The impact of classroom acoustics: listening, learning, and literacy, *Semin Hear* 2004;25:131–140.

Gagne, J., Laplante-Levesque, A., Labelle, M., Doucet, K. An audiovisual-FM system (AudiSee) designed for use in classroom settings: an evaluation of the effects of visual distractions on speechreading performance, *Semin Hear* 2002;23:43–55.

Gratton, M.A., Vazquez, A.E. Age-related hearing loss: current research, *Curr Opin Otolaryng Head Neck* 2003;11:367–371.

Healthy people 2000. Public Health Service. Washington (DC):Government Printing Office: 1990. DHHS Publication No. (PHS) 91-50213.

Hearing Health Industry World Directory (2005) *Hear J* 2004:109–173.

Heppner, C. Status of captioned movies in theaters, *Hear Loss* 2003;24:13–16.

Karchmer, M., Allen, T. The functional assessment of deaf and hard of hearing students, *Am Ann Deaf* 1999;144:68–77.

Keller, B.K., Morton, J.L., Thomas, V.S., Potter, J.F. The effect of visual and hearing impairments on functional status, *J Am Geriatr Soc* 1999;47:1319–1325.

McDermott, A.L., Dutt, S.N., Tziambazis, E., Reid, A.P. Proops DW. Disability, handicap and benefit analysis with the bone-anchored hearing aid: the Glasgow hearing aid benefit and difference profiles, *J Laryngol* 2002;28(Suppl):29–36.

Nelson, J.A., LaRue, C.B., Barr-O'Rourke M. Personal FM systems offer consumers more than before, *Hearing J* 2004;57:36–40.

Palmer, C., Mormer, E. Goals of the hearing aid fitting, *Trends Amp* 1999;4: 61–71.

Palmer, C.V., Butts, S.L., Lindley, G.A., Snyder, S.E. Time Out! I Didn't Hear You. Pittsburgh (PA): Sports Support Syndicate; 1996.

Preves, D. Hearing aids and digital wireless telephones, *Semin Hear* 2003;24:43–62.

Receiving hearing aids in different countries. Hear It Web site, 2005: http://www.hear-it.org/page.dsp?forside=yes&area=633 03;11:367–371.

Rennert, N., Morris, B., Barrere, C.C. How to cope with scopes: Stethoscope selection and use with hearing aids and CIs, *Hear Rev* 2004;Feb:34–75.

Rosenberg, G.G. Hearing and children. Access to acoustic information, *Hear J* 2003;56;51.

Thibodeau, L.M. Plotting beyond the audiogram to the TELEGRAM, a new assessment tool, *Hearing J* 2004;57:46–51.

Tye-Murray, N. Listening devices and related technology, in: *Foundations of Aural Rehabilitation 2004*, Thompson Delmar Learning; pp. 231–286.

Van den Brink R.H.S. Attitude and illness behaviour in hearing impaired elderly (Thesis). The Netherlands: University of Groningen; 1995.

Wolff, A.B., Harkins, J.E. Multihandicapped students, in: Schildroth, A.N., Karchmer, M.A. (Eds.), *Deaf Children in America*; San Diego: College Hill Pr;1986, pp. 55–82.

World Health Organization International Classification of Functioning, Disability, and Health, 2001: www.who.int

Yanz, J.L., Preves, D. Telecoils: principles, pitfall, fixes, and the future, *Semin Hear* 2003;24:29–42.

Yueh, B., Shapiro, N., MacLean, C.H., Shekelle, P.G. Screening and management of adult hearing loss in primary care, *JAMA* 2003;289:1976–1985.

16 Telecommunications, Computers, and Web Accessibility

Katherine D. Seelman, Stephanie Hackett, Bambang Parmanto, Richard Simpson, and Sailesh Panchang

CONTENTS

16.1 LEARNING OBJECTIVES OF THIS CHAPTER

Upon completion of this chapter, the reader will be able to:

Learn the basic principles of alternative telecommunications
Learn about the basic methods used in assessing a person for computer access
Understand the basic principles of Web accessibility
Gain knowledge about the applications of telecommunications, computers, and Web accessibility

16.2 INTRODUCTION

People with disabilities have engaged rehabilitation engineers in problem solving related to the accessibility and usability of telecommunications, computers, and the World Wide Web (WWW, the Web). This chapter will focus on the problems that people with disabilities have in using and accessing the Web. These problems or their solutions may include computers and telecommunication systems. On the client side, rehabilitation engineering has made important contributions to the development of assistive technology (AT). While Web-related AT is described in this chapter, other important technology are described in other chapters, such as those on the technology needs of people who have vision or hearing loss. Rehabilitation engineering has also made contributions on the server side, such as those involving universal design. Rehabilitation engineers have also been involved in transcoding and remediation that can entail rewriting the HTML or redesigning a Web site.

16.3 BACKGROUND AND HISTORY

The World Wide Web is the "set of all information accessible using computers and networking," with each unit of information being identified by a uniform resource identifier (W3C, 1999). Tim Berners-Lee created the Web, while employed as a researcher at CERN (now called European Organization for Nuclear Research), in an effort to enable collaboration between physicists and other researchers. In his original proposal in 1989, Berners-Lee incorporated three new technologies: HTML (HyperText Markup Language) to write Web documents, HTTP (HyperText Transfer Protocol) to transmit the pages, and a Web browser client software program to receive, interpret, and display the results across computer platforms (Gribble, 2005). In May 1991, all three technologies became fully operational for the first time at the CERN laboratories in Switzerland. The CERN laboratories housed the first Web server; by 2001, there were over 24 million Web servers around the globe. The Web is a part of the larger Internet, which is "the global network of networks through which computers communicate by sending packets of information" (W3C, 1999). Berners-Lee now serves as the director of the World Wide Web Consortium (W3C). The W3C is an international consortium whose mission is to "lead the World Wide Web to its full potential by developing protocols and guidelines that ensure long-term growth for the Web" (W3C, N.D.-a). These protocols and guidelines will allow for Web technologies to be compatible and allow for interoperability.

In 1997, following a White House conference, the US National Institute on Disability and Rehabilitation Research (NIDRR) and the National Science Foundation (NSF) announced support for the Web Accessibility Initiative (WAI) as a program domain within the W3C. The WAI approach to Web accessibility has revolved around three interrelated initiatives. The first is content accessibility of Web sites, so that persons with disabilities can perceive, understand, and use the sites. The second is making Web browsers and media players usable for persons with disabilities by making them operable through ATs. The third component requires Web-authoring tools and technologies to support production of accessible Web content and sites, so that persons with disabilities can use them effectively. Social, technical, financial and legal factors may impact the cost-benefit analysis that an organization must perform in designing its Web accessibility policy. Guidance for creating a business case for Web accessibility may be obtained from a W3C-WAI resource at http://www.w3.org/WAI/bcase.

Web accessibility efforts have focused on three approaches: the server side, the client side, and a transcoding intermediary process. Efforts on the server side include guidelines for writing accessible Web sites (WAI, 1999b). Developing accessible Web sites eliminates the costs of redesigning the sites later. Richards and Hanson (2004) note that the redesign of existing Web sites may be costly. Efforts on the client side include development of adaptive browsers or screen readers (Stephanidis, 2000; Zajicek & Hall, 2000). The transcoding intermediary process uses a server to transcode Web pages according to transformation rules. In considering the size and general inaccessibility of the Web, this approach has great potential in supplementing the other two approaches.

In October 2004, NIDRR again indicated the importance of Web accessibility by awarding a second 5-year engineering grant. A Rehabilitation Engineering Research Center (RERC) grant was awarded to the Trace Research and Development Center at the University of Wisconsin–Madison, working in partnership with the Technology Access Program at Gallaudet University in Washington, D.C. (Trace Center). The Trace Center has had a history of telecommunication research and development activity and enjoyed previous NIDRR support. The Telecommunications Access RERC is advancing a research and development program on accessibility and usability in existing and emerging telecommunications products, such as development of Voice over Internet Protocol (VoIP) technologies and techniques to enhance telephone usage by individuals with severe cognitive disabilities.

16.4 CONCEPTUAL MODEL: TELECOMMUNICATIONS, COMPUTERS, AND THE INTERNET

This conceptual model illustrates each of the components of this chapter. The telecommunications system is illustrated in Figure 16.1 to show transmission, emission, or reception of signs, signals, writing, images, and sounds or intelligence of any nature by wire, radio, optical, or other electromagnetic systems. The system may facilitate accessibility, for example, by incorporating a communication assistant who receives auditory messages, transforms them into visual messages, and relays them to the user who is deaf. AT is shown as a teletypewriter (TTY) for deaf people, and transcoding

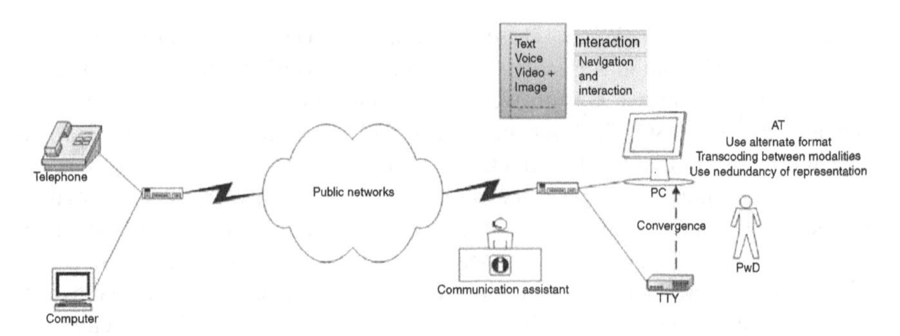

FIGURE 16.1 Conceptual model.

is shown between modalities. AT is also illustrated with text, voice, vision, and image applications for computers.

16.4.1 Need and Target Populations

The World Health Organization estimates the world's disability population to be between 7 and 10%. This means that over 500 million people in the world are disabled. It is estimated that 80% of these live in developing countries. In developing countries, difficulties accessing the Internet are already apparent. Many of these problems are related to low-bandwidth issues. These issues are closely related to those that many people with disabilities are facing accessing the Internet in general and the World Wide Web in particular (http://www.isoc.org/inet2000/cdproceedings/5c/5c_1.htm).

Increasingly, access to technology is associated with human rights as reflected in the Americans with Disabilities Act of 1990 (http://www.usdoj.gov/crt/ada/adahom1.htm), the proposed United Nations Convention on the Rights of People with Disabilities (http://www.un.org/esa/socdev/enable/rights/adhoccom.htm), and the World Summit on the Information Society (WSIS, 2005; http://www.isiswomen.org/onsite/wsis/prep3-end2.html).

Although Web access is widely acknowledged to be important in all areas of life, including health, education, and employment, it is not distributed evenly among regions of the world or among subpopulations. According to a report entitled "Computer and Internet Use Among People with Disabilities" (U.S. Department of Education NIDRR, 2000), almost three times as many people without disabilities have connected to the Internet at home than those with disabilities—31.1 vs. 11.4%. As the size of the aging population around the world has dramatically increased, the affordability, accessibility, and usability problems of older people have also become the target of study (http://www.aarp.org/olderwiserwired/oww-features/Articles/a2004-03-03-comparison-studies.html).

The causes and demographics of some subpopulations of individuals with disabilities, such as those with blindness and deafness, have been described in other chapters. The overall socioeconomic, educational, and employment status of people with disabilities is universally and significantly lower than those without disabilities. Modern communications technology is creating accessibility problems while solving others. These accessibility and usability problems challenge people with disabilities who function differently in the use of sight, dexterity, cognition, and hearing. For example, while the telephone was inaccessible to deaf users for a long period, it was mostly accessible to blind users. With the increase of visual information, access to the Web has become a challenge to people with visual disabilities. With the convergence of many applications within the cell phone, it has become more like a computer. However, cell phone use may challenge many people with disabilities who have limited or no sight, hearing, dexterity, and those who process information differently. For example, cell phone displays are often low contrast, and the buttons difficult to manipulate with complex sequencing to execute an application.

16.5 TECHNOLOGY SOLUTIONS

Rehabilitation engineering and other disciplines have contributed to technology solutions for increased access and usability of the Internet. These contributions, often involving Rehabilitation Engineering Society of North America (RESNA) and sister organizations around the world, have included universal design and related work with standards organizations within countries, such as the American National Standards Institute, and internationally, such as the International Organization for Standardization. Research and development efforts have involved the development of accessibility evaluation tools, accessibility measurements, transcoding, remediation, and AT.

16.5.1 Web Accessibility Evaluation Tools

There are various tools available to evaluate Web sites for accessibility. Several are listed in the following table:

Name of Tool	Web site
A-Prompt	http://www.aprompt.ca
Ramp Ascend 6.0	http://www.deque.com/products/ramp
LIFT Text Transcoder	http://www.usablenet.com/products_services/text_transcoder/text_transcoder.html
WAVE 3.0	http://wave.webaim.org/index.jsp
BOBBY	http://bobby.watchfire.com

The W3C maintain a list of Web accessibility evaluation tools at http://www.w3.org/WAI/ER/tools/complete.

16.5.2 Accessibility Measurements

Currently, most methods of evaluating accessibility provide absolute ratings: either the Web site complies with all guidelines or it is considered inaccessible. Most do not take into consideration the size or complexity of the Web site. Parmanto and Zeng (2005) identified characteristics of quality accessibility metrics that include:

Quantitative score providing a continuous range of values from perfectly accessible to completely inaccessible. This allows for assessment of change in Web accessibility over time as well as comparison between Web sites or between groups of Web sites.

Large discriminating power of the metric and range of values beyond simply accessible and inaccessible. This allows assessment of the rate of change of Web accessibility.

Fair, by taking into account and adjusting to the size and complexity of the Web sites.

Scalable, to conduct large-scale Web accessibility studies.

Normative, meaning that it should be derived from standard guidelines of Web accessibility such as the WCAG or Section 508.

16.5.3 Transcoding and Remediation

If Web accessibility is not taken into consideration from the planning phase of Web site development, the Web site may have be changed later on, albeit at a cost. This cost is related to resources for remediation and retrofitting. Remediation occurs after a Web site has already been developed and needs to be retrofitted to comply with accessibility standards and guidelines. The process of remediation can entail rewriting the HTML or redesigning the site.

Transcoding is the process of adapting document contents so that they may be viewed on diverse devices or accessed by users with disabilities. The adaptation can range from simple Web clippings, where images are discarded from the Web page, to advanced content summarization. Transcoding intermediary systems reformat materials that would otherwise have to be developed separately for display on different devices. Transcoding occurs at the point between the user and the browser to transform the contents of the Web page into a form that is accessible to the viewer. Transcoding tools allow a Web site to be accessible to the end user even when the original Web site has not addressed any accessibility guidelines.

16.5.4 Computer Access

Client-side ATs aid users with disabilities to access the computer. This section details the functions that users perform on the computer and the AT solutions pertaining to these functions.

It is impossible to imagine a modern computer without a mouse. The operating systems and applications of today coevolved with the mouse, and some software literally cannot be used without a mouse. The mouse is used for pointing, clicking, double-clicking, and clicking and dragging. The operations that are most troublesome for individuals with physical disabilities are often those that involve button presses. Fortunately, the mouse is but one example of a much larger class of pointing devices, all of which can perform the operations listed here. Each pointing device requires a different set of skills, which the clinician can match to a client's abilities.

As shown in Figure 16.2, pointing devices come in a variety of shapes and sizes. The most familiar pointing devices are the mouse and the trackpad (most often seen on laptop computers). Other frequently seen pointing devices include the trackball and the trackpoint. Pointing devices that are more commonly associated with

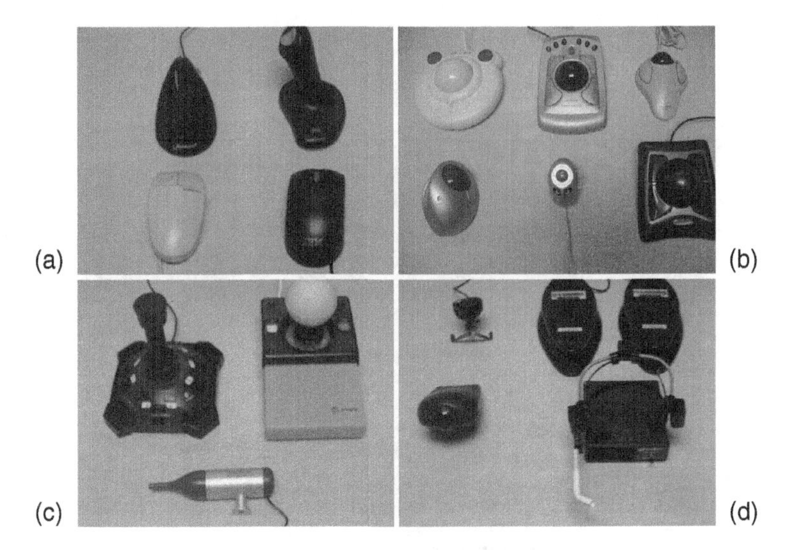

FIGURE 16.2 Pointing devices.

individuals with disabilities include touch screens, head-mounted mouse emulators, and mouse keys.

People can have problems generating mouse clicks for several reasons. Some people have trouble activating a switch. Others use a pointing device that does not have a mouse button. For example, people who use a head-mounted mouse emulator may not want to be tethered to the computer by a cord attached to a mouse button. An alternative to a physical mouse button (or buttons) is a mouse click emulation software.

16.5.5 Keyboard-Only Control

Despite the emphasis that modern applications and operating systems place on the mouse, it is possible to operate a computer (almost) entirely from the keyboard. This is a particularly useful option for individuals who have mobility limitations that make using of the mouse difficult or for individuals who are totally blind.

16.5.5.1 Text Entry

The traditional text entry device is the keyboard, with independently moving keys arranged in what is known as the QWERTY layout. As shown in Figure 16.3, text entry devices can come in a variety of shapes and sizes and, in some cases, do not even have physical keys. Virtual keyboards (also called *on-screen keyboards*) and switch-based text entry methods are covered later in this section.

16.5.5.2 Physical Keyboards

There has been much interest in the last several years in alternative "ergonomic" keyboard designs that are promoted for their ability to reduce the incidence of repetitive

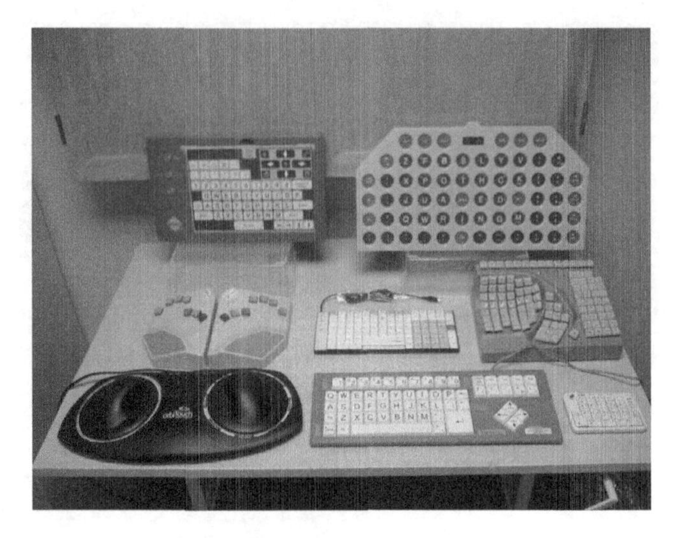

FIGURE 16.3 Text entry devices.

stress injuries (RSI) such as carpal tunnel syndrome. These keyboards often feature non-QWERTY layouts (e.g., Dvorak and Chubon) and a variety of shapes that promote a more "natural" position of the hands while typing. To date, there is no evidence that has unambiguously established a link between a specific ergonomic keyboard and either (1) reduced incidence of RSI or (2) increased productivity. This is not to say, however, that ergonomic keyboards are not useful. Most people with an RSI can find an alternative keyboard that causes less pain and increases productivity. Unfortunately, no single keyboard causes less pain and increases productivity for most people.

16.5.5.3 Virtual Keyboards

Virtual keyboards consist of text entry methods that are not operated by directly selecting from a collection of buttons. Virtual keyboards include (1) software-based on-screen keyboards that can be operated by a pointing device or switch and (2) Morse code systems. In this section, we will address on-screen keyboards operated by the mouse, and will consider on-screen keyboards operated by switch and Morse code devices later on. The advantage of on-screen keyboards is their extreme flexibility. Their size and layout can be changed almost instantaneously. The primary disadvantage of an on-screen keyboard is that it consumes screen space that could otherwise be used by an application.

16.5.6 Voice Recognition

A quality microphone is critical for good performance with voice recognition. The preferred type of microphone is the headset, which positions the microphone within an inch of the user's mouth. A disadvantage of a headset is the need to put it on and take it off, which some individuals are unable to do independently. Boom microphones

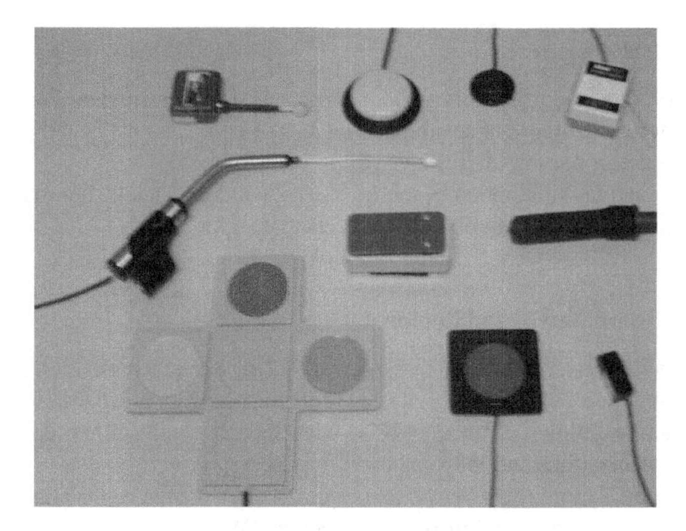

FIGURE 16.4 Switches.

and array microphones eliminate the need to wear anything, but equivalent sound quality from a boom or array microphone is more costly than for a headset.

Voice recognition users are often disappointed to learn that speaking to a computer is not at all like speaking to a person. Voice recognition software is based almost entirely on signal-processing algorithms, and does not consider context, syntax, or semantics when interpreting what a user says. Voice recognition software relies on a voice model that must be trained for each individual user, and the quality of the voice model determines the accuracy of the voice recognition software.

16.5.7 Switch Access

There are a variety of switches available, distinguished by their size, sensitivity, and the type of motion to which they respond. A number of switches are shown in Figure 16.4. The type of switch used and its location depends on what motions the client can perform both consistently and reliably.

16.5.8 Scanning

Scanning systems allow a user to select from a set of items (e.g., letters, numbers, or words) using one or two switches. The simplest form of scanning is one-switch linear scanning, in which each item is highlighted in turn, and the user presses the switch when the desired item is highlighted. A more common form of one-switch scanning is row–column scanning. A common implementation of row–column scanning with one switch requires three switch hits to make one selection from a two-dimensional matrix of items. The first switch hit initiates a scan through the rows of the matrix. Each row of the matrix, beginning with the first, is highlighted in turn until the second switch hit is made to select a row. Each column of the row is then highlighted in turn until the target is highlighted, when the third switch hit is made to select the target.

16.5.9 Morse Code

The Morse code is not used very frequently, which is unfortunate because it can be much more efficient than scanning. The Morse code also has an advantage over scanning in that it does not require visual or auditory feedback. Perhaps the greatest obstacles to increased adoption of the Morse code are the lack of familiarity among clinicians and the limited number of products that support it.

16.5.9.1 International and Regional

At every level of government and in the private sector, conventions, laws, regulations, and guidelines are being developed on Web accessibility. What follows is a brief overview and examples of these efforts, with special emphasis on identifying the names of organizations and Web sites you may access.

The United Nations (http://www.un.org/esa/socdev/enable/rights/adhoccom.htm) and the World Summit on Information Society (WSIS) (http://mailman.greennet.org.uk/public/ct/2003-September/000680.htm) (Seelman, 2003) provide examples of a human rights approach to accessible and useable information and computer technology and solutions for related problems. The WSIS disability effort has been supported by Japan's National Rehabilitation Center for Persons with Disabilities Research Institute working closely with the Daisy Consortium, a group of talking-book libraries that has been a leader in the creation and evolution of the DAISY Specification for digital talking books. The WSIS disability effort has emphasized AT, universal design, and participation by people with disabilities in the process of developing guidelines for the information society. The European Union and the Commonwealth countries have also made significant efforts to address Web accessibility, as have its member countries (http://www.w3.org/WAI/Policy/) (Freeway, 2002).

Japan published the "guidelines concerning creation method of accessible Web content over the Internet" in 1999 (MPHPT, 2001). These guidelines, based on the W3C Web Content Accessibility Guidelines, stipulate methods for Web site creators and authoring tool developers to create Web content that is easily understood by the elderly and by persons with disabilities. Japan's Ministry of Public Management, Home Affairs, Posts and Telecommunications (MPHPT) has also devised a system to evaluate and correct problems on Web sites and provide a support system to government organizations, as well as in other places, to promote the diffusion of Web site accessibility for the elderly, persons with disabilities, and for every citizen.

The U.S. has adopted laws and regulations related to the Web that are widely admitted, but much more has to be done to stimulate the private sector to develop accessible and useable products. Section 255 of the Communications Act (47U.S.C.§255, 1996), issued by the U.S. Federal Communications Commission (FCC, 2002), requires telecommunication manufacturers and service providers to make their products and services accessible to people with disabilities, if readily achievable. If such accessibility is not possible, the products and services must be compatible with peripheral devices and specialized equipment commonly used by persons with disabilities, if that compatibility is readily achievable (meaning that is can be accomplished with minimal difficulty or expense). Section 255 covers hardware and software telephone network equipment and customer premises equipment (CPE). CPE is telecommunications equipment used in the home or office for purposes of telecommunications and includes telephones, fax machines, answering machines, and pagers.

Section 504 (29U.S.C.§794, 1973) of the Rehabilitation Act mandates nondiscrimination under Federal grants and programs. No otherwise qualified individual can be excluded from the participation in, the benefits of, or be subjected to discrimination under any program or activity receiving Federal financial assistance solely by reason of his or her disability.

Americans with Disabilities Act (ADA) (42U.S.C.§§12101etseq., 1990) prohibits discrimination based on disability. Since the passage of the ADA came before mainstream use of the Internet, there is no mention of the Internet in this legislation. This makes the law subject to judicial interpretation as to whether or not Title II and Title III of the ADA apply to this medium (Kretchmer and Carveth, 2003). Title II of the ADA (USDOJ, 2004) states that "no qualified individual with a disability shall, by reason of such disability, be excluded from participation in or be denied the benefits of the services, programs, or activities of a public entity." A public entity is a state or local government and instrumentalities thereof. Title III of the ADA mandates that persons with disabilities be able to participate in "the full and equal enjoyment of the goods, services, facilities, privileges, advantages, or accommodations of any place of public accommodation by any person who owns, leases (or leases to), or operates a place of public accommodation."

Section 508 of the Rehabilitation Act of 1973 was first added to the Rehabilitation Act in 1986. Section 508 of the Rehabilitation Act (29 U.S.C. 794d), as amended by the Workforce Investment Act of 1998 (P.L. 105-220), August 7, 1998, requires all electronic technology developed or purchased by federal agencies to be accessible to persons with disabilities, as of June 2001, unless this would pose an undue burden. In the event of an undue burden, the agency must provide alternative access to the information. Section 508 does not require manufacturers to develop accessible technologies. It requires that federal agencies and departments follow accessibility regulations when procuring, developing, using, or maintaining electronic and information technology.

16.6 RESEARCH AND FUTURE DEVELOPMENT

Research is ongoing internationally in various areas of Web accessibility; including voice-driven Web browsers that enable a user to browse the Internet via voice-driven commands, transcoding gateways that transform a Web site into one that is accessible, and studies of Web accessibility and AT abandonment. However, the key to reducing the cost of accessibility is pushing accessibility into the mainstream. This is the domain of universal design, which is the discipline of designing products to maximize the number of people who can use a given product. When accessibility becomes a standard feature, rather than an add-on, costs drop dramatically. In addition to technical standards, national and international policies, such as Section 508 of the U.S. Rehabilitation Act, are important strategies. Technologies of various sorts provide an obvious approach to cost reduction. New technology includes semiautomatic Web site repair tools and Web adaptation facilities that transform existing Web content "on the fly." These tools and facilities can be deployed as part of the server, as an intermediary, or as part of the client.

Persons with disabilities have varying needs when it comes to ATs. AT solutions can be employed on the client side to aid users' in making the electronic information accessible. Text-to-speech software, or screen-reading software, translates text into synthesized voice output. This software can be used with various applications and with Web browsers. For the Web to be accessible, the users' AT must be compatible with the Web technology and the Web content must be understandable. Perspectives in design approaches, technological advances in computer access methods, and cutting-edge research have played important roles in making computers and the Web more accessible.

In August 2005, the U.S. National Science Foundation (NSF) proposed a next-generation Internet, indicating that researchers need to start thinking beyond the current Internet and consider radical new ideas for continuing challenges such as Internet security and ease of use. The NSF announced the GENI Project, which would go beyond current efforts to incrementally improve the Internet. The U.S. Department of Defense has argued for the adoption of IPv6 (Internet Protocol version 6) to replace the widely used IPv4. The GENI project would "explore new networking capabilities that will advance science and stimulate innovation and economic growth," according to its own Web page. "The GENI Initiative responds to an urgent and important challenge of the 21st century to advance significantly the capabilities provided by networking and distributed system architectures."

16.7 STUDY QUESTIONS

1. What are the business advantages of having an accessible Web site?
2. How does a communication assistant assist communication via TTY?
3. What does the most recently amended Section 508 require of government agencies?
4. How are universal design, accessibility and usability related?
5. Describe speech recognition software. Who might benefit from its use?
6. W3C and the Web Accessibility Initiative?
7. What processes can a Web site go through if accessibility is not taken into account in the initial design phases?
8. List the characteristics of a quality accessibility metric.
9. Why is Internet accessibility more profound for blind users?
10. What is the relationship between AT and new Internet technologies?

BIBLIOGRAPHY

Speech recognition software:
IBM's ViaVoice at http://www-3.ibm.com/software/speech/
CSLU Toolkit at http://cslr.colorado.edu/toolkit/main.html
Nuance at http://www.nuance.com/

Text-to-speech software:
IBM Home Page Reader at http://www-3.ibm.com/able/hpr.html
JAWS for Windows at http://www.freedomscientific.com/fs_products/software_jaws.asp
Windows Eyes at http://www.gwmicro.com
Text browser:
Lynx: http://lynx.browser.org/

29U.S.C.§792. (1973). Section 502 of the Rehabilitation Act of 1973.

29U.S.C.§794. (1973). Section 504 of the Rehabilitation Act of 1973, from http://www.section508.gov/index.cfm?FuseAction=Content&ID=15.

42U.S.C.§§12101etseq. (1990). The Americans with Disabilities Act of 1990, from http://www.usdoj.gov/crt/ada/adahom1.htm.

47U.S.C.§255. (1996). Section 255 Telecommunications Access for People with Disabilities, from http://www.fcc.gov/cgb/consumerfacts/section255.html.

Alvestrand, V., Samuels, M. (2003). A question of accessibility online. *Information World Review* (190), 10.

Astbrink, G. (1996). Web page design — Something for everyone. *Link-Up*, December 1996, 7–10.

Burgstahler, S.E. (1992). Disabled students gain independence through adaptive technology services. *EDUCOM Review*, 27, 45–46.

Chisolm, W., Vanderheiden, G., Jacobs, I. (2001). Web content accessibility guidelines. *Interactions*, 8, 34.

Coombs, N. (1991). Window of equal opportunity — online services and the disabled computer user. *Research and Education Networking*, 2.

DDA (N.D.). Disability Discrimination Act 1995, from http://www.disability.gov.uk/dda/.

e-Japan (2001). e-Japan Priority Policy Program, from http://www.kantei.go.jp/foreign/it/network/priority-all/index.html.

FCC (2002). Section 255 Telecommunications Access for People with Disabilities, from http://www.fcc.gov/cgb/consumerfacts/section255.html.

FED-STD-1037 (1996). Glossary of Telecommunication Terms, from http://www.its.bldrdoc.gov/fs-1037/.

Freeway, A. (2002). European Union Places Full Internet Access High on Its Agenda, from http://www.disabilityworld.org/09-10_02/access/internetaccess.shtml.

Godwin-Jones, B. (2001). Emerging technologies — accessibility and web design. Why does it matter? *Language Learning and Technology*, 5, 11–19.

Gribble, C. (2005). History of the Web Beginning at CERN. Retrieved April 20, 2005, from http://www.hitmill.com/internet/web_history.html.

Harrysson, B., Svensk, A., Johansson, G. I. (2004). How people with developmental disabilities navigate the Internet. *British Journal of Special Education*, 31, 138–142.

Hirsch, T., Forlizzi, J., Hyder, E., Goetz, J., Stroback, J., Kurtz, C., Scholtz, J., Thomas, J. (2000). The ELDer project: social, emotional, and environmental factors in the design of eldercare technologies. Paper presented at the Conference on Universal Usability, New York, NY.

Hochheiser, H., Shneiderman, B. (2001). Universal usability statements: marking the trail for all users. *Interactions*, 8, 16–18.

ISO (2005). International Organization for Standardization, from http://www.iso.org/iso/en/aboutiso/introduction/index.html.

ISO (N.D.). User Centered Design Process: ISO 13407. Retrieved 2/22/05, from http://www.usability.serco.com/trump/resouces/13407stds.htm.

Kaye, H. S. (2000). *Computer and Internet Use among People with Disabilities*. Disability Statistics Report (13). Washington, D.C.: U.S. Department of Education, National Institute on Disability and Rehabilitation Research.

Kretchmer, S. B., Carveth, R. (2003). Analyzing recent americans with disabilities-act-based accessible information technology court challenges. *Information Technology and Disabilities*, 9(2).

Mace, R. (N.D.). What Is Universal Design? Retrieved 02/22/2005, from http://www.design.ncsu.edu/cud/univ_design/ud.htm.

MPHPT. (2001). Toward Diffusion of Websites with Excellent Accessibility, from http://www.soumu.go.jp/joho_tsusin/eng/Releases/NewsLetter/Vol12/Vol12_11.html#2.

Mulvihill, M. L. (1995). *Human Diseases: A Systemic Approach* (4th Ed.). Norwalk, Connecticut: Appton and Lange.

NOD (2005). The National Organization on Disability. Retrieved 2/26/05, from http://www.nod.org/index.cfm?fuseaction=Page.viewPage&pageId=15.

NTIA (2002a). *A Nation Online: How Americans Are Expanding Their Use of Internet*. Washington, D.C.: National Telecommunication and Information Administration.

NTIA (2002b). *A Nation Online: How Americans Are Expanding Their Use of the Internet*, Chapter 7: Computer and Internet Use Among People with Disabilities. Washington, D.C.: National Telecommunication and Information Administration.

NY State Attorney General (2004). Spitzer Agreement to Make Web Sites Accessible to the Blind and Visually Impaired. Retrieved 10/06/2006 from http://www.oag.state.ny.us/press/2004/aug/aug19a_04.html.

Paciello, M. (N.D.). Making the Web Accessible for the Deaf, Hearing and Mobility Impaired, from http://www.samizdat.com/pac2.html.

Paciello, M.G. (2000). *Web Accessibility for People with Disabilities*. Lawrence, KS and Berkeley, CA: CMP Books.

Parmanto, B., Ferrydiansyah, R., Zeng, X., Saptono, A., Sugiantara, I. (2005). Accessibility transformation gateway. Paper presented at the HICSS: *Proceedings of the 38th Annual Hawaii International Conference on System Sciences*, Big Island, Hawaii.

Parmanto, B., Saptono, A., Ferrydiansyah, R., Sugiantara, I. W. (2005). Transcoding bio-medical information resources for mobile handhelds. Paper presented at the Hawaii International Conference on System Sciences, Island of Hawaii.

Parmanto, B., Zeng, X. (2005). Metric for web accessibility evaluation. *Journal of Society for Information Science and Technology*, 56(13), 1394–1404.

Richards, J.T., Hanson, V.L. (May 17–22, 2004). Web accessibility: A broader view. Paper presented at the WWW 2004, New York, USA.

Seelman, K.D. The information age: Participation challenges and policy strategies to include people with disabilities. *Proceedings of the World Health Organization Collaborating Centre Seminar: Creation of an Inclusive Society in an Advanced Information and Communications Society*, November 6, 2003, Tokorozawa, Saitama, Japan.

Shneiderman, B., Hochheiser, H. (2001). Universal usability as a stimulus to advanced interface design. *Behaviour and Information Technology*, 20, 367–376.

Sloan, M. (2001). Web Accessibility and the DDA. *Journal of Information Law and Technology*, 2.

Stephanidis, C. (2000). Universal access through unified user interface. Paper presented at *the CSUN 2000: Technology and Persons with Disabilities Conference*, California State University Northridge.

Takagi, H., Asakawa, C. (November 13–15, 2000). Transcoding proxy for nonvisual web access. Paper presented at the ASSETS'00. *Proceedings of the Fourth International ACM Conference on Assistive Technologies*, Arlington, VA.

Takagi, H., Asakawa, C., Fukuda, K., Maeda, J. (2002). Site-wide annotation: reconstructing existing pages to be accessible. Paper presented at the ASSETS 2002. *Proceedings of the Fifth International ACM SIGCAPH Conference on Assistive Technologies*. July 8–10, 2002. Edinburgh, UK.

Thompson, T. (2004). Universal design and web accessibility: Unexpected beneficiaries. Paper presented at the CSUN Conference on Disabilities, Los Angeles, CA.

Trace Center (N.D.). RERC on telecommunications access. Retrieved 4/26/05, from http://trace.wisc.edu/telrerc/.

US Access Board (2004). Accessible Telecommunications Product Design Technical Assistance, from http://www.access-board.gov/sec508/telecomm-course.htm

US Access Board (N.D.). About the U.S. Access Board, from http://www.access-board.gov/about.htm.

US Department of Education. National Institute on Disability and Rehabilitation Research. Disability Statistics Report 13. Computer and Internet Use among People with Disabilities. March 2000.

US Dept HHS (N.D.). Usability Basics. Retrieved February 22, 2005, from www.usabiliity,gov/basics/index.html.

US DOC (N.D.). Technology Assessment of the U.S. Assistive Technology Industry: AT Industry Composition. Retrieved April 22, 2005, from http://www.bis.doc.gov.

US DOJ (2004). ADA Home Page, from http://www.usdoj.gov/crt/ada/adahom1.htm.

W3C (1999). W3C Glossary, from http://www.w3.org/.

W3C (N.D.-a). About W3C, from http://www.w3.org/Consortium/.

W3C (N.D.-b). Web Accessibility Initiative, from http://www.w3.org/WAI/.

Waddell, C.D. (1999). The Growing Digital Divide in Access for People with Disabilities: Overcoming Barriers to Participation in the Digital Economy, from http://www.icdri.org/CynthiaW/the_digital_divide.htm.

WAI (1999). Web Content Accessibility Guidelines 1.0. Retrieved September 1, 2003, from http://www.w3.org/TR/WCAG10/.

WebAIM (N.D.). Types of Cognitive Disabilities, Retrieved April 8, 2005, from http://www.webaim.org/techniques/cognitive/.

World Summit on the Information Society: WSIS, 2005; http://222.isiswomen.org/.

Zajicek, M., Hall, S. (2000). Solutions for elderly visually impaired people using the Internet. Paper presented at the *Proceedings of People and Computers XIV Usability or Else*. September 2000. Sunderland, UK.

Zeng, X. (2004). Evaluation and enhancement of web content accessibility for persons with disabilities. Unpublished Dissertation, University of Pittsburgh, Pittsburgh.

17 Augmentative and Alternative Communication Technology

Katya J. Hill, Bruce Baker, and Barry A. Romich

CONTENTS

17.1 LEARNING OBJECTIVES OF THIS CHAPTER

Upon completion of this chapter, the reader will:

Learn the basic principles of alternative and augmentative communication (AAC)
Learn about the basic methods used in assessing a person for AAC
Understand the role of rehabilitation engineering in AAC
Gain knowledge about the applications of AAC

17.2 INTRODUCTION

Alternative and augmentative communication (AAC) is an endeavor with a goal to optimize the communication of individuals with significant communication disorders (ASHA, 2004). For individuals who have complex communication needs, the process of achieving communication success may be perceived as an insurmountable challenge by their families and those providing clinical services. However, students of rehabilitation engineering and professional rehabilitation engineers on AAC teams can contribute to the success of such endeavors. Individuals who rely on AAC are attending school, graduating from college, and participating in the workforce, because of rehabilitation technology and services.

The basic elements of a comprehensive AAC assessment and the role of rehabilitation engineers in making decisions about AAC technology are critical to achieving success. The significance of language issues and AAC language representation methods must be understood prior to evaluating solutions, emphasizing the need for AAC technology to support the spontaneous generation of language in order to optimize communication function and participation. Only by understanding language issues can rehabilitation engineering professionals appreciate the technology, device features, and human factors issues associated with AAC interventions.

17.2.1 What Is AAC?

Alternative and augmentative communication (AAC) refers to any communication approach that supplements or replaces natural speech and/or writing that may be impaired. AAC services and interventions can be considered to be interactions between components of the International Classification of Functioning (ICF) model (WHO, 2001) for improved function and participation in activities. Communication may occur within all components of the ICF model, requiring application of the principles of rehabilitation engineering to maximize outcomes. Effective communication is desired for participation at work and school or for leisure and entertainment. Although a range of AAC interventions are possible to engineer solutions for improved function and participation, the demands of communication should be analyzed in terms of the language requirements for the best outcomes to be achieved.

Alternative and augmentative communication (AAC) interventions can be classified by the methods used to transmit messages. Methods are classified as unaided or aided (Lloyd et al., 1997). Unaided symbols do not require an external device or apparatus. Nothing other than an individual's body parts are needed to transmit a message, such as using one's hands to gesture. Aided symbols, on the other hand, require some kind of an external device. Aided AAC technology can be further classified into low-, light-, and high-performance technologies. High-performance technology solutions can then be identified as nondedicated or dedicated AAC systems. Nondedicated technology generally refers to computers that are running an AAC software solution, but the primary application of the technology is computer-based. Conversely, dedicated AAC devices have been designed and evaluated specifically for communication, but frequently have secondary features that provide computer or environmental control functions. The range of aided technology increases as availability of power, voice output, electronics, and computer chips become part of the system. Table 17.1 shows the basic AAC classification taxonomy.

17.2.2 Who Uses AAC?

Individuals who use or need AAC make up a diverse group of all ages and socioeconomic, ethnic, and racial backgrounds. In addition, a variety of acquired and congenital disabilities exist, which can contribute to a person's inability to speak and maintain functional, independent communication throughout their lifetime. Acquired neurological disorders that may contribute to the need for AAC include amyotrophic lateral sclerosis (ALS), Parkinson's disease, aphasia, or traumatic brain injury. Congenital disorders that may contribute to the need for AAC include cerebral palsy, Down's syndrome, autism, Angelman syndrome, and other developmental disabilities.

Approximately 2 million, or 8 to 12 individuals per 1000 (0.8 to 01.2%), in the general population of the US can benefit from AAC (Beukelman and Ansel, 1995). Blackstone (1990) summarized results from several studies, which suggested that, at the end of the last century, 0.2 to 0.6% of the total school-age population worldwide had severe communication disorders. Demographic data from Canada and outside North America show similar prevalence estimates of 0.2 to 1.4% of populations as

TABLE 17.1
AAC Unaided and Aided Ranges of Technology Used as AAC Intervention

	Unaided	Aided		
	No tech	**Low tech (no power)**	**Light tech**	**High tech**
Description of Interventions	Vocalization Gestures Signs Eye blinks	Picture exchange Symbol/alphabet board Communication notebook Headstick	Light pointer Voice output switches Simple powered displays	Dedicated and nondedicated Electronic, computer-based Voice output systems

having severe communication disorders significant enough to warrant AAC (Enderby and Philipp, 1986; Bloomberg and Johnson, 1990).

17.3 COMPREHENSIVE AAC ASSESSMENT

Alternative and augmentative communication (AAC) assessment may be the single most important event in the life of an individual with a severe communication disorder limiting functional use of natural speech. For the beginning communicator, the AAC assessment process should establish AAC interventions to build communication competence to maximum potential across a life span. However, for an adult with a degenerative neurological disorder, the AAC assessment process needs to consider changing needs to maintain optimal communication throughout the course of the disease.

Alternative and augmentative communication (AAC) assessment is consistent with the ICF framework and definition of rehabilitation engineering. AAC assessment is a systematic application of science and technology to improve communicative functioning and participation in various activities and environments. Most authors confirm that an assessment is a client-centered, multidisciplinary team process (Hill, Glennen, and Lytton, 1998; Parette, Huer, and Brotherson, 2001; Rose and Alant, 2001). Assessment models proposed for other assistive technologies apply to AAC, such as the Human Activity Assistive Technology (HAAT) model (Cook and Hussey, 1995) and the Matching Persons with Technology Model (Scherer, 1998). These assistive technology models tend to focus on the human–technology interface and the interaction between the person, technology, activity, and context. However, for people who use AAC, the application of these feature-match approaches must follow appropriate consideration of the language issues.

Various AAC assessment models have been proposed that contain feature-match components, including the predictive assessment model (Yorkston and Karlan, 1986; Glennen and DeCoste, 1998) and the participation model (Beukelman and Mirenda, 1992). An important element of these AAC assessment models is that data are collected and information gathered to make intervention and management decisions (Lloyd, Fuller, and Arvidson, 1997). The purposes of an AAC assessment can be capsulated

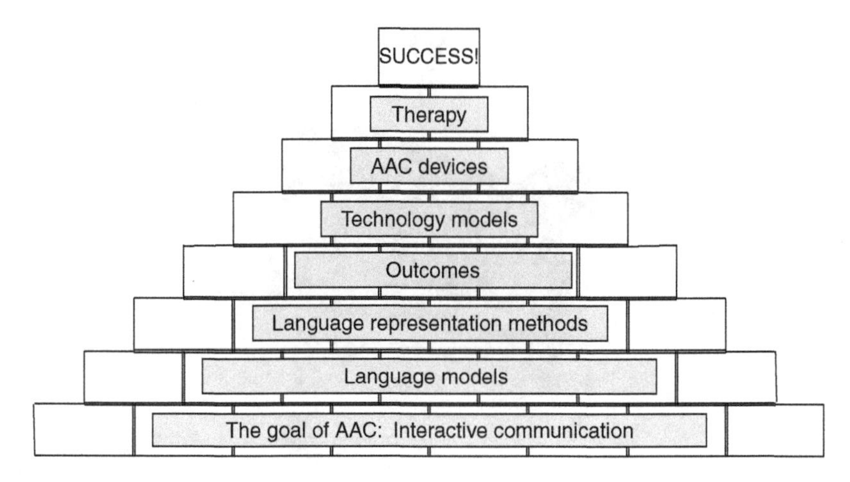

FIGURE 17.1 AAC language-based assessment and intervention model.

into three primary objectives: (1) determination of functional communication needs, (2) identification (matching) of interventions to increase or maintain interactive communication, and (3) monitor or measure the effectiveness of intervention (Wasson, Arvidson, and Lloyd, 1997; Swengel and Varga, 1993; Beukelman and Mirenda, 1992). The application of the principles of evidence-based practice should be reflected in any comprehensive AAC assessment (ASHA, 2004; Hill, 2004).

The AAC language-based assessment model conceptualizes an alternative to the technology- and feature-focused assessment and intervention process (Hill, 2001). The model serves as a metaphor for building success on a strong foundation. Starting at the bottom, each level of the process should be completed and supported with evidence before moving up to the next level (Figure 17.1). A multidisciplinary team approach is recommended with each level of decision being led by an individual with the appropriate professional knowledge and experience. Quantitative performance and outcomes measurement is essential. Success may only be reached through the application of a structured and scientific approach to assessment and intervention.

Rehabilitation engineers have a unique set of skills to bring to the AAC assessment process. These skills include not only good mathematical and data analysis abilities, analytical thinking, and problem solving but also skills in design, evaluation, and fabrication processes. However, rehabilitation engineers depend on other professionals to identify the language and communication requirements for AAC interventions. Better and/or faster technology does not necessarily fix communication function and participation when language has not been adequately considered. Following a language-based model that integrates external evidence with clinical expertise and client values produces the best outcomes.

As AAC team members, rehabilitation engineers enter the assessment process at the bottom step by committing to optimize communication. Rehabilitation engineers can contribute to the discussion on how different ranges of aided AAC technologies can contribute to optimal communication. Based on information gathered about language functioning, functional communication needs, and other physical abilities of

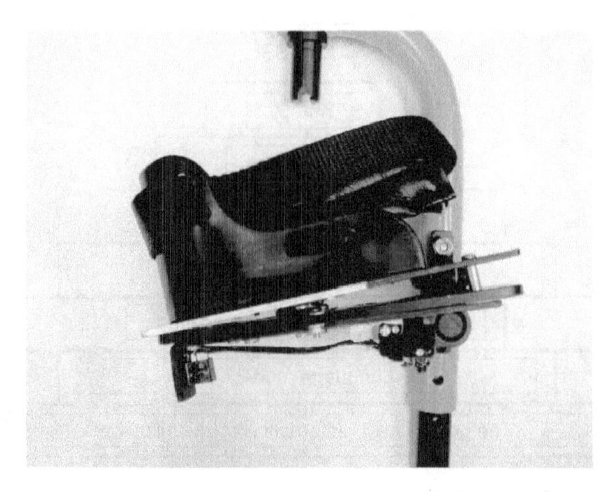

FIGURE 17.2 Custom footrest switch designed and fabricated by an engineer for row–column scanning on an AAC system (Courtesy of AAC Institute).

an individual, the rehabilitation engineer can support the evaluation of how language is represented and generated in various AAC technologies, as well as what technology features and options best support the desired outcomes. Specifically, rehabilitation engineers should be able to collect and analyze the best external evidence that answer questions about the effectiveness and efficiency of various technologies. Frequently, rehabilitation engineers will be asked to design or fabricate items that will improve positioning and access to an AAC system, such as trays, switches, or headsticks (Figure 17.2). Measurement becomes critical to maximize performance or to choose between systems.

17.4 LANGUAGE ISSUES

17.4.1 What Is "Language?"

The word "language" in English has several meanings. In a broad sense, it refers to any medium that conveys, manipulates, or records information. However, within the category of all things that can be called languages, there is a special set called natural languages (NLs). English, Chinese, Hindi, Spanish, and Portuguese are NLs. NLs are generative, recursive, ambiguous, use polysemous symbols, are typically learned in early childhood from one's immediate family, and arise naturally in human societies. For the purposes of this chapter, the term "language" will refer to an NL.

Some questions about language and AAC become easier to understand when the term *natural language* is used. The various symbol systems or methods used in AAC are all media used to convey, manipulate, and record information. However, AAC symbol systems are not NLs, because they have not arisen independently in a community through natural means. They have little or no phonetic components, and they cannot be used without external paraphernalia. The design of AAC technology should ensure that the selected symbol system can perform the functions of an NL.

There is a notable difference between communication and language. Human communication takes place through gestures, facial expressions, visual pointing, signing, body language, and silence or stillness. In a movie, during a love scene, when faces are immediately juxtaposed, a glance by one partner to the other partner's lips may communicate the intention or desire to kiss. Such body-based communication is very powerful, but it is very limited when not accompanied by NL.

NL and some of its properties and operations are central issues in augmentative communication. Two topics of importance, both in foundational knowledge of augmentative communication and current discussion, are the various methods of utterance generation and vocabulary categories. AAC technology can be engineered to address these issues, but results in wide variability in function and performance.

17.4.2 Methods of Utterance Generation

AAC technology can be engineered to support two methods of utterance generation, preprogrammed utterances and spontaneous word-by-word generation. Each method is a contrasting approach to improving functional communication and participation.

17.4.2.1 Preprogrammed Utterances

Many theorists, system designers, and clinicians (Baker, 1986; Todman, 2000) have recommended the use of preprogrammed utterances in speech-generating devices (SGD). The reasons for storing preprogrammed utterances are various. Time delays in communication interactions are a difficult barrier for people using SGDs. Much AAC practice goes on in schools, and mainstream students need to answer questions in the classroom in a timely manner. Prestored utterances seem the most effective way to actuate and communicate using an SGD. Therefore, one way of thinking about overall communication design of an SGD is to develop useful sentences for different environments (Higginbotham, 1992). These environments can include, but are not limited to, fast food restaurants, transportation, home, activities of daily life (ADL), doctors' offices, hospitals, sporting events, etc.

The communication environments just mentioned can be paired with various pragmatics categories: requesting, refusing, protesting, information giving, greetings/partings, commenting, judging, and politeness. A person with disabilities traveling on an access vehicle can have a preprogrammed transportation vocabulary with a range of utterances stating likely requests, refusals, etc., for that environment.

This same point of view holds that vocabulary stored as complete utterances "hooks" early communicators to the power of communication. The idea frequently expressed is that children and others with little communicative experience may not appreciate the environmental and interpersonal power that communication provides. Prestored utterances are thought to give "more bang for the buck" for single actuation, and thus provide an experience more persuasive of the power latent in communication than the generation of a single word. Assembling one's own utterances, no matter how short or primitive, is a crucial stage of language acquisition that cannot be ignored (Pinker, 1994).

17.4.2.2 Word-by-Word Generation

Word-by-word generation, when used on AAC technology, provides for spontaneous, novel utterance generation (SNUG). Word-by-word is more likely to "hook" early learners to the power of communication than communicating by prestored utterances. The power of communication becomes real to the communicator only when early learners choose and organize their own words. This is not only true in primary language acquisition but also in second language acquisition (Klein, 1992, 1994).

SNUG is the hallmark of human communication and should not be denied to augmented communicators at the earliest levels or later as language has developed. Generativity is one of the central characteristics of an NL, and word-by-word generation becomes easier as acquisition proceeds. Prestored, whole utterances are not generative. The language acquisition device (LAD), a term Chomsky (2000) used to describe the hypothetical neuronal complex involved with mastering one's native language, exhibits a different kind of intelligence than that measured in standardized testing. Many people with moderate to severe mental retardation who talk use the typical language structures of their nondisabled peers (Romski, Sevcik, and Adamson, 1999).

Further, word-by-word generation is more flexible than communicating through prestored utterances (Anderson and Baker, 2003). The set of frequently used simple words that make up a high percentage of our total words is the same across populations, environments, and activities. These core words allow the construction of appropriate utterances from a base of relatively few words.

17.4.3 Applying Methods of Utterance Generation

The following eight sentences were designed by an experienced clinician for an inexpensive SGD. They were designed to accompany an adolescent with cognitive impairment throughout the day. They are as follows:

I need help.	I want to watch TV.
I want to go.	Go to the bathroom.
What's that?	I'm finished.
I don't like that.	I don't want that.

These sentences will be compared in flexibility and utility with eight words:

I	go	good	more
not	eat	drink	stop

Imagine the same adolescent having a conversation while getting dressed. He needs to indicate his preferences to his personal care professional. As the care professional shows him various shirts, he could say "not good" or "good." He could modulate his preferences through saying "more good." When his hair is getting brushed, he could say "stop, not good."

The range of flexibility and simplicity of use with eight words is noticeable when compared with eight preformed sentences. How does the adolescent indicate that he likes something but he likes something else better? How does the adolescent indicate that his leg brace is too tight? Using any of the prestored sentences, "I don't want that," "I don't like that," and "I need help," would not indicate any more than the rejection of his leg brace and his desire to have it loosened.

However, with eight words, he would have a variety of ways to express his particular problem. "Not more," "Not good," "I not good," and "Stop," all could be used as springboards to a full discussion of his situation with his personal care professional. Whole utterances such as "I don't like that" are, in fact, invitations to dispute. "I need help" comes as close as any other utterance to expressing the client's need here. Yet, how much more difficult would it be for him to choose the appropriate sentence from a collection of sentences than it would be for him to assemble his own short utterances, particularly after his personal care professional engages him in further questions. The cognitive stress of weighing the nuances of the different prestored utterances may well be beyond his linguistic, reading, and reasoning skills.

With prestored sentences, there is one set of contents, each tied to a particular context. After the client has made one utterance, there is no second utterance available to him or her. Interactive communication is rendered difficult, if not impossible, by prestored whole-utterance-based vocabulary, unless the collection of prestored whole utterances is very large. A collection of 50 or more prestored utterances used interactively would be beyond most typical communicators.

A small collection of well-chosen basic words is easier and more flexible than a similar or even greater collection of prestored whole utterances, no matter how well these sentences are designed. Further, the experience of assembling one's own two-word utterances is a language acquisition experience and contributes to the development of language abilities across environments and pragmatics. Language acquisition depends on the experience and stimulation of manipulating the language code.

17.5 VOCABULARY CATEGORIES

17.5.1 CORE VOCABULARY

It is typical to see a language sample in which the 100 most frequently occurring words account for 60% of the total words used, and 80 to 85% of the words used are identified as core vocabulary. Similar results in word use were reported for toddlers (Banajee, Dicarlo, and Stricklin, 2003), preschool children (Beukelman, Jones, and Rowan, 1989), adults (Balandin and Iacono, 1999), adults with cognitive impairments (Mein and O'Connor, 1960), people who are elderly (Stuart, Beukelman, and King, 1997), and for proficient augmented communicators (Hill, 2001). Examples of core word types include pronouns, prepositions, conjunctions, or words such as "mine," "do," "under," and "this." AAC vocabulary studies show that core words are the same across subjects, environments, topics, and activities.

Core words typically include the early words that children use. The experience and stimulation of using core words is an early and important step in language acquisition.

A child or an adult acquiring language for the first time can use core words to express a wide variety of emotions, comments, ideas, etc.

Core words have such flexibility that they can replace fringe words in almost all circumstances. A sentence such as "Alcoholism is a major problem for a significant percentage of the population" can be expressed by saying "Many people drink too much." Just as with the sentence vs. prestored utterance opposition in ease and flexibility, core words also provide great flexibility and ease of use.

17.5.2 Fringe Vocabulary

Fringe vocabulary refers to those words in a language corpus that occur infrequently — about 15 to 20% of the words collected in a language sample. Fringe vocabulary consists of the thousands of words used in a language and is specific to subjects, environments, topics, and activities. Fringe vocabularies are generally very large in comparison to core vocabulary, which rarely exceeds 350 words in any sampling of speech.

Context-specific words for the different environments of daily life are typically fringe words. Many people are in a morning environment when they use the word "toaster"; they rarely say the word outside that environment. Actual language samples show that words such as "toaster" are typically not said even in the presence of a toaster. Balandin and Iacono (1999) found that even in the "foods" topic only 2.2% of the words were unique to that topic.

In AAC, there has been a tendency to focus on nouns, because they are easiest to teach and measure. Specific "food" nouns, for example, "chocolate cake," are considered highly motivating. Such nouns appear throughout systems designed for children, cognitively impaired adults, and even in systems for adults with typical cognition. However, in actual language samples from children and cognitively impaired adults, food words are infrequent in the core (Banajee, Dicarlo, and Stricklin, 2003; Beukelman, Jones, and Rowan, 1989; Mein and O'Connor, 1960).

17.6 AAC LANGUAGE REPRESENTATION METHODS

Three basic methods are used to represent language in AAC systems and are termed AAC language representation methods (LRMs). Evaluating the effectiveness of each method starts by considering the nature of a language. To be effective, AAC LRMs used with technology need to have the characteristics of an NL, such as being generative, recursive, and polysemous, in order to achieve the maximum performance. Rehabilitation engineers apply systematic observation to evaluate the performance of various methods against the characteristics of an NL.

The three common LRMs are based on (1) single-meaning or univocal (intended to have only one meaning) pictures, (2) semantic compaction™ or polysemous (intended to have more than one meaning) pictures, and (3) traditional orthography — the alphabets. Many studies have shown that univocal, polysemous, and alphabetic LRMs have differential effects on performance (Table 17.2) (Hill and Romich, in press).

TABLE 17.2
Definitions for AAC LRMs with Examples of Technology Options

LRM	Definition	Examples of AAC language application programs	Types of technology interfaces
Single-meaning pictures	The use of graphic or line-drawn symbols to represent single-word vocabulary or messages (phrases, sentences, or paragraphs). A variety of AAC symbol sets are available. A large symbol set is required to represent the typical spoken vocabulary	VocabPC Gateway Velocity	Switch Dynamic-grid, schematic Static Hybrid
Alphabet-based methods	The use of traditional orthography and acceleration techniques that require spelling and reading skills. The user selects from a standard keyboard, alphabet, or word overlay to spell or select vocabulary. Spelling letter-by-letter is a very slow process	Adult Quick Learning System (AQLS) Word Power Word Core	Dynamic-grid Static (keyboard) Hybrid
Semantic compaction™	The use of a small set of multiple-meaning icons (symbols) to select vocabulary and linguistic structures. Minspeak® is the only patented LRM (Baker, 1992). The number of symbols is reduced to represent the typical spoken vocabulary as well as the number of required keystrokes relative to spelling	Unity 128 Unity Enhanced	Dynamic-grid Static Hybrid

17.6.1 Single-Meaning or Univocal Pictures

Single-meaning pictures may be the most commonly used LRM, because of the apparent ease of use at first encounter. Traditionally, many (hundreds of) single-meaning pictures are assembled on various boards, pages, or levels. Each picture has a caption, but is intended to represent a word, concept, or idea in such a way that it is obvious or transparent to anyone using the system.

The nature of single-meaning picture programs is that the symbol sequences are short; just one picture (one selection) is needed from an array. However, the symbol set is huge for any significant vocabulary. Considering that a normally developing three-year-old child has an expressive vocabulary of around 1100 words, a symbol set of 1100 pictures is needed. An AAC system with 50 keys, for example, would require at least 22 pages to represent this small vocabulary. Furthermore, almost all words of core vocabulary are not easily represented by pictures. They are not concrete or picture-producers. Various studies have reported symbol iconicity, or the visual relationship between a symbol and its referent (Lloyd, Fuller, and Arvidson, 1997). When the referent of a symbol must be guessed, the meaning must be taught. Figure 17.3 shows how commonly used core words are not easily distinguishable or "guessable" when represented as picture symbols. To get an idea of how difficult single-meaning picture programs are to use, remove the words and ask anyone to identify the pictures. Would you be able to label each univocal symbol in Figure 17.3?

The advantage of univocal symbols is the ease of using this LRM on aided, low-, and light-technology boards for introducing AAC or for backup systems when access to a high-technology system is inconvenient or otherwise not available. Many professionals believe that the transition to aided high technology becomes easier by duplicating low-technology boards onto the arrays of high-end systems. However, many times this approach fails to appreciate the features and human factor issues involved with high-performance AAC systems. Rehabilitation engineers make a contribution to a team by discussing issues about page navigation, automaticity, and ease of use at first encounter vs. effective long-term use (refer to the section on human factors, Section 17.8, in this chapter). Nevertheless, single-meaning pictures are popular and have reported successful outcomes with various populations. A variety of commercially available univocal or single-meaning picture symbol sets and systems are used in all ranges of aided AAC technology that attempt to overcome innate difficulties in symbol iconicity, vocabulary selection, and vocabulary organization (Figure 17.3 and Figure 17.4).

17.6.2 Alphabet-Based Methods

Traditional orthography has been used in augmentative communication from the very beginning. Spelling gives the opportunity to an augmented communicator to express himself or herself by using the complete surface structure of the English language — he or she does not have to say things such as "me go."

Difficulties with alphabet methods center on the number of letters required to express a complete utterance. For instance, the statement "I think there is something wrong with my breathing" has 50 letters and spaces. Fifty key actuations for a single-finger direct selector requires a lot of energy and nearly 2 min if the communicator is able to select one key every 2 sec. For someone who is scanning, two actuations are required for each letter and the time between letters is usually more than 2 sec.

Because of effort and time, four techniques have been developed within the alphabetic LRM. The first technique involves the configuration of the interface or array configuration. Various alphabet arrays have been designed to improve the overall performance of the QWERTY keyboard. The DVORAK keyboard is one example. Other arrays attempt to: (1) position keys based on ease of visual recognition, such as alphabetic order or starting each row of an alphabet array with a vowel, and (2) position keys based on the frequency of letter occurrence by giving easier access to keys that are used more frequently. Various studies can be used as evidence that report results

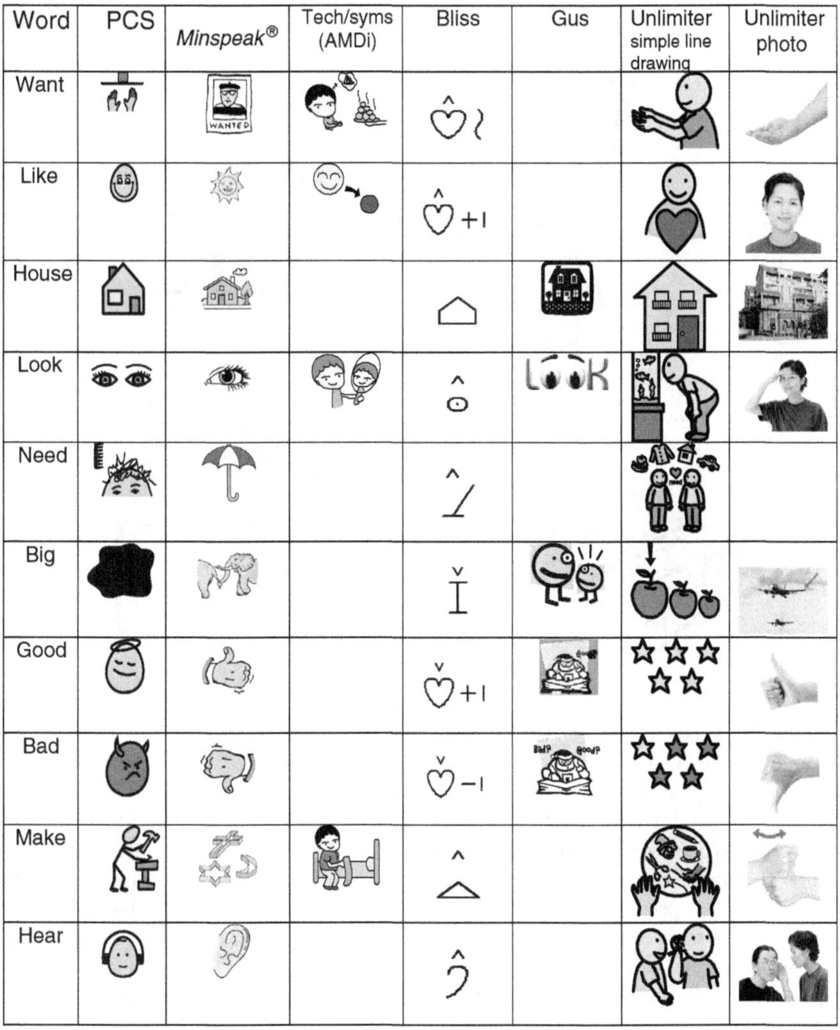

Word	PCS	*Minspeak®*	Tech/syms (AMDi)	Bliss	Gus	Unlimiter simple line drawing	Unlimiter photo
Want							
Like							
House							
Look							
Need							
Big							
Good							
Bad							
Make							
Hear							

FIGURE 17.3 Examples of language representation method symbols (single-meaning pictures from a variety of commercially available symbol sets).

on performance for different array configurations (Higginbotham, 1992; Lesher et al., 2002).

Word and character prediction functions by taking the initial and subsequent letters and predicting words or characters based on frequency, recency, grammar, and topic. Although this method may somewhat reduce the number of selections required, it typically increases latency, or time between selections (Koester and Levine, 1994). Actual measures of language generation using word prediction have never shown an overall reduction in time. Word and character prediction have not been found to be rate enhancers, but the approach does save keystrokes.

FIGURE 17.4 Examples of AAC display technology: (a) single-meaning pictures on a dynamic display with synthesized speech output (Courtesy of Saltillo Corporation), (b) single-meaning pictures on a static display with digitized speech output (Courtesy of Saltillo Corporation), (c) single-meaning pictures and alphabet-based methods on grid-type dynamic display with synthesized speech output (Courtesy of Saltillo Corporation), (d) semantic compaction™, single-meaning pictures and alphabet-based methods on grid-type dynamic display with synthesized speech output (Courtesy of Prentke Romich Company), (e) alphabet-based methods on a static display with synthesized speech output (Courtesy of Namco), and (f) semantic compaction™, single-meaning pictures and alphabet-based methods on hybrid display with synthesized speech output (Courtesy of Prentke Romich Company).

Abbreviation expansion systems were once thought to be appropriate for broad-based communication (Vanderheiden and Kelso, 1987; Light et al., 1990). Recent history has indicated that few people were able to remember sufficient abbreviations to sustain broad-based communication through abbreviation expansion. However, SGDs with full spelling capabilities often support abbreviation expansion as a feature for possible use.

Orthographic word selection is the use of whole words on an array in conjunction with the letters of the alphabet, numbers, and punctuation marks. Usually, words are chosen for the array based on high frequency of occurrence or core words. These words may be arranged on the array in word classes such as pronouns, verbs, conjunctions, and prepositions. Depending on the AAC language application program, some arrays may dedicate keys only to spelling, thus using 27 keys for one purpose. Other programs embedded in the alphabet among core word access thus increase the number of available words to select to avoid spelling. This approach requires switching between modes. The type of AAC display technology can significantly impact the performance of orthographic word selection programs in terms of automaticity and fluency. Very little quantitative data are available as external evidence that documents the performance of this technique.

17.6.3 Semantic Compaction™ or Multimeaning Pictures

With Semantic Compaction™ or Minspeak®, language is represented by a relatively small set of multimeaning symbols or icons. Semantic compaction™ makes use of meaningful relationships between the icon and the referent or the information it represents. Because of its nature, semantic compaction™ does not appear to be intuitive similar to single-meaning pictures or alphabet-based methods, and must be understood to be appreciated. This unique method holds the only patent as an AAC LRM, and is available only through limited license for use on AAC systems.

In contemporary life, people expect a public graphic to have one and only one meaning or interpretation. Its meaning is univocal (single meaning). Hieroglyphic systems, developed in antiquity, used pictures to represent concepts that were later refined by subsequent pictures. The Maya hieroglyph of the shark means "shark" in some contexts, "green" in others, and "salt water" in others. One only knows the meaning of the shark glyph when one views it in context (Baker, 1986).

The Semantic Compaction™ language representation system employs polysemous pictures to simultaneously lower the number of keystrokes needed to achieve full expression and the size of the selection set. Thus, Semantic Compaction™ systems lower both the total number of pictures on a keyboard and the number of pictures in a symbol string. Consequently, this LRM avoids many of the pitfalls of single-meaning pictures and alphabet-based methods. The symbol set is small as opposed to univocal symbols, and the sequence length is short as opposed to spelling. Figure 17.3 contrasts how personal pronouns are represented using single-meaning pictures and semantic compaction™. Note that once the motor plan for the sequence has been habituated, automaticity in accessing a pronoun is possible.

As with single-meaning pictures, polysemous picture applications have various levels for vocabulary and language. Individuals can start with a smaller set of icons and access vocabulary without sequencing and later transition to a larger set of icons and a vocabulary size requiring sequencing. The transition between levels of the language application program has been programmed to avoid relearning and to promote motor patterning. Various studies reporting performance results using semantic compaction™ report that this LRM can be significantly faster than other methods (Gardner-Bonneua and Schwartz, 1989; Hill, Holko, and Romich, 2001).

17.6.4 Multiple Methods

Using multiple methods is the choice of many individuals who rely on AAC. Observations of the most effective communicators attest to competence using the three language representation methods available with their technology. Logged data of their communication indicate that individuals who have AAC systems supporting all three of the LRMs use semantic compaction™ for 90 to 95% of everything they say, while the remaining 5 to 10% is split between spelling and word prediction. The experience of individuals who rely on AAC appears to show that LRMs that lead to faster communication are naturally selected over LRMs that are not as efficient. Available evidence suggests that semantic compaction™ is routinely three to seven times as fast as spelling.

17.6.5 Summary

More than any other technology factor, LRMs influence communication rate. Individuals who rely on AAC value being able to say exactly what they want to say, when they want to say it, and saying it as fast as they can. Communication rate is valued above other considerations in evaluating performance. Consequently, rehabilitation engineers first identify the language capabilities of an AAC system and then focus on how technology features will enhance the available LRMs and overall communication performance.

17.7 TECHNOLOGY AND DEVICE FEATURES

AAC system technology and features are selected first to support optimal communication and second to address other needs. Refer to Table 17.1 to recall the range of AAC interventions. AAC technology or SGDs are classified and recommended around display technology, selection methods, outputs, and other feature considerations that support the chosen AAC language representation methods (Romich, Vanderheiden, and Hill, 2000). AAC hardware employs an interface to select items that produce a specific output. The selected items are the available methods of utterance generation (single words or prestored messages) and LRMs (i.e., icons, alphabets, etc.) on the interface. The hardware includes the various outputs and features to enhance or record performance. The various techniques for making item selections from an AAC interface are based on direct selection, scanning, or coding. Table 17.3 summarizes the various components of AAC technology or SGDs.

17.7.1 AAC Display Technology

The interface refers to the type of AAC selection and display technology that supports the available LRMs. Today's technology-based AAC system displays are classified as static, dynamic, or hybrid (Figure 17.4). An AAC system has a static selection area where the targets on an overlay do not change when an item is selected. For example, the keys on a computer keyboard are static. Although the keys can be reconfigured from a QWERTY array to alphabetic order, once the keyboard is designed, the configuration does not change during use. This can be a significant advantage in terms of developing automatic use of the system.

Dynamic display technology is based on touch screens. The system recognizes when the display is being touched and takes appropriate action. Dynamic displays offer much flexibility because they can be reconfigured by software. Display designs available on today's technology have been classified according to how aided

TABLE 17.3
Components of Dedicated AAC Technology or an SGD

Language requirements	Hardware of dedicated SGD	Selection methods	Accessories	Additional considerations
Single-meaning pictures	**Control interface**	**Direct selection**	Mounting systems	Technical support
Alphabet-based methods	Dynamic	Keyboard	Carrying case	Training
Semantic compaction™	Static	Headpointing	Peripherals (switches, headsticks, joystick)	Repair services
Multiple methods	Hybrid	Joystick		Warranties
	Outputs	**Scanning**		Portability
	Speech/auditory	1-switch		
	Visual	2-switch		
	Electronic	**Morse code**		
	Data logging			
	Memory capacity			

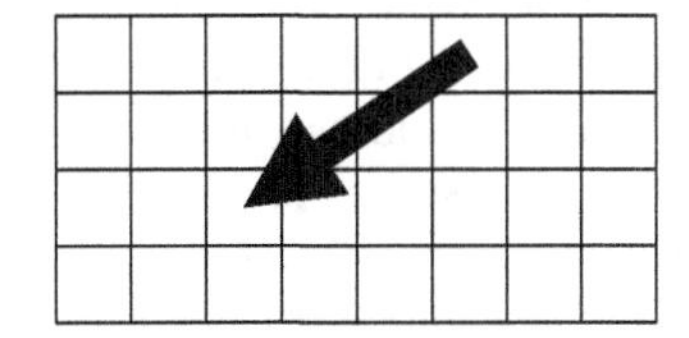

FIGURE 17.5 Direct selection.

symbols are organized. In grid designs, symbols are arranged at regular intervals (rows–columns) of an array. In schematic designs, symbols are embedded within everyday-occurring contextual scenes (Drager et al., 2003). The existence of these two alternatives, along with page navigation, makes it important for AAC teams to consider the advantages and disadvantages (Wilkinson and Jagaroo, 2004). This flexibility may not translate into improved communication performance; when the selection area of a display changes during use, the potential for becoming automatic in its use can be greatly diminished.

Hybrid systems have both a static and a dynamic selection area. The static area is typically used for access to core vocabulary, which has high frequency of use and needs to be accessed automatically. The dynamic selection area is used to access fringe or extended vocabulary using single-meaning pictures, word prediction, etc.

17.7.2 AAC Selection Methods

Direct selection requires a single action to access an item from a set of choices on the AAC system interface (Figure 17.5). Using a computer keyboard to type is a common example. Typically, each key is directly selected from the keyboard array to spell by using a finger; however, a variety of methods are available to support direct selection. Direct selection is possible using alternative parts of the body to hit a target, such as a fist or toe. Without modifying the key size or number of keys on an array, pointing can be achieved using other assistive technology, such as sticks held in the mouth

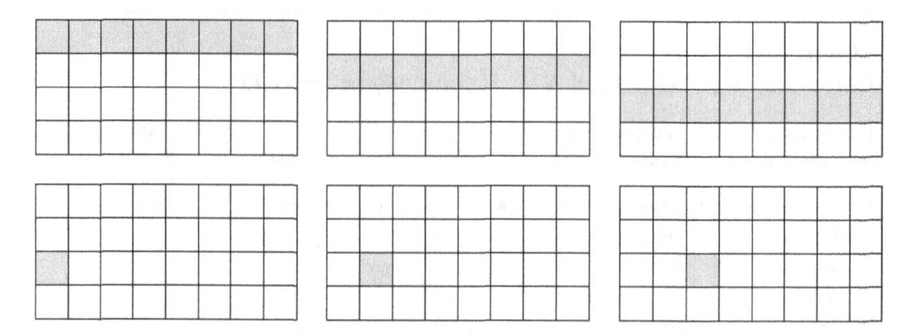

FIGURE 17.6 Row–column scanning.

or mounted on a headband or helmet. Light pointers can also be used as a method to directly select a target. Indirect pointing techniques include mouse and joystick control. Finally, enlarging the size of the interface or key or decreasing the number of available targets on an array can make direct selection possible for an individual.

With scanning, the possible selections are presented to the individual in a time-ordered sequence (Figure 17.6). The individual indicates the desired selection as it is presented, usually by activating a single momentary switch. When the size of the selection set is larger, the selections are organized into a two-dimensional matrix. In a technique termed row–column scanning, rows are scanned to select the row that contains the target, and then each item in the row is scanned to select the desired target. AAC systems that support scanning provide a variety of configurations (column–row and circular) and adjustments (scanning rate and predicative selection) to customize the approach to the individual. Since scanning is a time-dependent process, those selection items that are most frequently used should be nearest to the starting point of the scanning process in order to achieve the fastest communication.

Coding is occasionally used in AAC systems. Morse code is a common example. Morse code can be implemented using one, two, or three switches, depending on the skills of the individual.

17.7.3 AAC Device Outputs

Common outputs used in technology-based AAC systems are speech, visual, and electronic. Outputs provide specific functions and enhance interaction between the person and the AAC device, between the person and communication partners, and between the person and other technologies. AAC device outputs can serve a number of purposes. They can provide user feedback and facilitate the use of the AAC device in other ways. Outputs can be directed to a communication partner. Outputs can be used to interface with and provide control over other items. This section explores the various outputs available in AAC devices.

17.7.3.1 Speech/Auditory Output

The primary output for most AAC devices is speech. Speech output serves the purpose of replacing or augmenting the speech of the individual using the device by conveying information to a communication partner. Speech output technology can be divided into two categories: synthetic and digitized. Some AAC devices use either one or the other, while some include both.

Synthetic speech technology accepts text and converts it into speech. This is known as text-to-speech (TTS). An algorithm processes the text and generates the spoken output. TTS is commercially available in many forms and in many languages. The advantage of TTS is that the flexibility allows speaking any possible utterance. The disadvantages are that TTS is generally less natural sounding than natural speech, thus influencing intelligibility (Venkatagiri, 2003). Sometimes TTS will not speak a word properly. This may be a result of two different pronunciations for the same word, such as "read" and "read." Another possibility is that the word may not be in compliance with the rules of the algorithm. In such cases, an exception dictionary is generally part of the TTS system.

Digitized speech is essentially a digital recording. This means that anything that is to be spoken using a digitized speech system must be recorded in advance. The advantage of the digitized speech technology is that it can accurately replicate the characteristics of the original source. Some AAC devices come with a library of prerecorded items. Otherwise, the programming of the device must include a recording process. Since the timing of recordings (loudness, speed, pitch, space before and after, etc.) can be variable, the stringing together of individual words may create an utterance that sounds quite unnatural. Digitized speech is generally less expensive than synthetic speech and is, therefore, more commonly used in low-end AAC devices.

Speech can also be used to cue the individual using the AAC device, as with auditory scanning for individuals with vision impairment. In this case, the device can speak to the user through an earphone or private speaker to provide information otherwise presented visually. Related audio outputs are beeps and other audio feedback used to enhance operation of the device.

17.7.3.2 Visual Outputs

Indicators, displays, and touch screens are usually intended to provide the individual with the information needed to use the AAC device. Some systems with static selection areas (keyboards) have a small lamp associated with each key. In the case of touch screen AAC devices, the selections are presented on a display and may be highlighted. The indicators or highlighted areas of the display or touch screen may provide information to the individual using the system. This information could be guidance for the selection technique, such as pointing with a mouse or row–column scanning. Visual feedback of selection or key activation can be provided. Also, some systems can provide visual information and tutorials to help the individual learn how to use the system better.

The display can also be used by the communication partner to read or otherwise receive information generated by the device user. This may be valuable in loud environments, when speech cannot be heard, or when speech may not be used, either for privacy or to compensate for a noisy environment.

One issue relative to displays is daylight viewing. Sun shields may be fabricated to reduce glare caused by the sun. Some devices may have two displays or an add-on display, with one facing the communication partner.

17.7.3.3 Electronic Outputs

AAC devices often have electronic outputs. These can take the form of a physical connection, such as a serial, USB, or other ports. The output can also be wireless, such as infrared (IR) or radio frequency (RF). Electronic outputs can be used to connect to other items. A common application of AAC devices is to use them for computer keyboard and mouse emulation. This will be covered in greater detail in Section 17.10. IR outputs can be used for the control of entertainment devices and other related items in the environment. Many AAC devices either have IR codes included in the device and/or provide for the AAC device to learn the codes from normal remote controls. Thus, the person can use the AAC device to turn on the television, change channels, change the volume, etc. Another application is the use of the IR output with toys and games.

17.7.4 Other Technology Considerations

Other features and considerations that influence communication performance using technology-based AAC systems include rate enhancements (beyond those included in the previous discussion of LRMs), learning enhancements, data logging, and device mounting. One example of rate enhancements is a feature called predictive selection, which can make scanning faster. In predictive selection, the scanning jumps over the matrix locations that do not contain valid selections.

Data logging is a feature that records the time and content of language events generated using the AAC system (Lesher et al., 2000). This information can be invaluable in understanding how the system is being used and in providing the foundation for measuring communication performance. The following example shows the events that led to the utterance "It's faster than spelling everything out which is what I used to do":

16:26:05 SEM "It's"
16:26:08 SEM "faster"
16:26:14 SEM "than"
16:26:41 SPE "sp"
16:26:42 SPE "e"
16:26:45 SPE "l"
16:26:45 SPE "l"
16:26:46 SPE "i"
16:26:47 SPE "n"16:26:48 SPE "g"
16:26:49 SPE " "
16:26:58 SEM "everything"
16:27:02 SEM "out"
16:27:05 SEM "which"

16:27:08 SEM "is"
16:27:11 SEM "what"
16:27:14 SEM "I"
16:27:19 SEM "used"
16:27:22 SEM "to do"

The three-letter mnemonic code indicates the method used to generate the event (SEM = semantic compaction™; SPE = spelling). Data logging supports evidence-based practice, which is the expected approach used for professional service delivery. However, data logging is also used by individuals who use AAC and their family members. Logged data can be analyzed to provide a summary measure report of communication performance.

When the individual who uses AAC also uses a wheelchair or walker, mounting systems are used to position the AAC device so that it can be used most effectively. The mounting system generally has a clamp or other attachment mechanism for connecting to the wheelchair. An arm, usually detachable, is connected to the clamp, and the AAC device is attached to the arm using some type of adapter that allows the AAC device to be positioned appropriately. Powered mechanisms can be used to allow the individual to adjust the position of the AAC device or to move it out of the way.

17.8 HUMAN FACTORS AND AAC

Human factors refer to issues regarding the interaction between a person and a machine or an object, such as a tool, appliance, vehicle, computer, etc. When a person uses an AAC system, various human factor issues come into play. Appropriate consideration of these issues is required for achievement of the most effective communication possible.

17.8.1 Ease of Use at First Encounter

Ease of use at first encounter may be contrary to most effective long-term use (Norman, 1980; King, 1999). Therefore, following a short evaluation, if one chooses the AAC system that appears easiest, one may also be choosing the system that will be least effective in the long term.

17.8.2 Learning Time

Proficiency in performing a task develops over time. Even for able-bodied adults with no cognitive disability, hundreds or thousands of iterations of a task may be necessary to reach the achievable level of skill (Jagacinski and Monk, 1985). Therefore, performance following a short trial on a selection method (keyboard, headpointer, etc.) or LRM may offer little or no indication of what can be achieved.

17.8.3 Automaticity

When arrays of selections change, such as with word prediction lists or pages of single-meaning pictures, it constitutes a discontinuous cognitive process (Romich, 1994). A discontinuous process prevents the development of automaticity, a basic requirement of fluency. Therefore, core vocabulary may be most effectively accessed using a static (nondynamic or nonchanging) set of symbols. When the size of the vocabulary exceeds the number of selections directly available, then icon sequencing may be required to maintain automaticity.

17.8.4 Speed Factors

Selection rate, measured in bits per second (bps), is the rate at which a person can enter information into a system. The upper limit of selection rate, using physical techniques, is thought to be about 100 bps (Lucky, 1991). Selection rate directly impacts communication rate.

The time required to make a selection is proportional to the logarithm of the distance to the target and inversely proportional to the logarithm of the size of the target (Fitts, 1954). Therefore, the relative locations and sizes of elements to be selected can have a profound impact on the communication rate.

17.8.5 Design Trade-Offs

Design trade-offs are inherent in nearly everything (Petroski, 2003). Rarely does a product design address all possible interests. When a product is designed to meet the needs of a funding agent or reimbursement policy, for example, features that can result in more effective communication may need to be omitted.

17.9 PERFORMANCE MEASUREMENT

AAC clinical decision making is moving toward increased application of evidence-based practice (EBP). The American Speech-Language-Hearing Association (ASHA) has recognized the shift toward scientific methods in its Scope of Practice (ASHA, 2001), which places the expectation on practitioners to use instrumentation to collect data in accordance with the principles of EBP. EBP requires the conscientious, explicit, and judicious use of current best evidence in making decisions about the care of individuals (Sackett et al., 1997).

Rehabilitation practitioners are expected to weigh the external (field) evidence with that collected at the personal level before making recommendations for any assistive technology solutions. The ICF model provides a framework for structuring performance and outcomes measurement (Mandich et al., 2002). AAC team members can target measures based on the ICF model components of function, structures, activities, and/or participation. Various assessment approaches using qualitative and quantitative data can be used to monitor the change and rate of change of an AAC intervention.

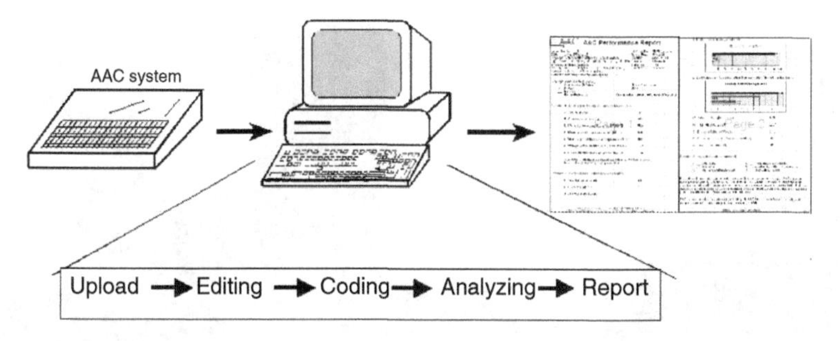

FIGURE 17.7 Diagram representing language activity monitoring (LAM) procedure (Courtesy of AAC Institute).

17.9.1 LANGUAGE ACTIVITY MONITORING

Language activity monitoring (LAM) provides an automated approach to measure how AAC technology solutions are used. Early efforts on data logging were based on the development of research tools (Miller, Demasco, and Elkins, 1990; Koester and Levine, 1994; Higginbotham, Lesher, and Moulton, 1998). Early work on LAM was initiated at the School of Health and Rehabilitation Sciences at the University of Pittsburgh. LAM was developed under a grant from the United States National Institute for Deafness and Other Communication Disorders of the National Institutes of Health (NIH) (Romich and Hill, 2000a). The basic procedures for LAM are illustrated in Figure 17.7. LAM log file data can be collected in two ways: (1) using a built-in feature in an AAC device, and (2) using a software application that allows a PC to function as an LAM, such as the Universal LAM (Hill and Romich, 2003a) (Figure 17.8). Log files can be analyzed for a variety of performance measures. Specific analysis software has been developed to generate an AAC performance report (Hill and Romich, 2003b). Rehabilitation engineers find value in using log files to measure the effectiveness of specific design features and options on language performance using a high-technology AAC system.

17.9.2 COMMUNICATION RATE

Measuring communication rate provides information on several performance parameters. Communication rate is measured in words per minute, and is calculated for each utterance, using the first event in the utterance as the start time and the last event as the end time (Romich and Hill, 2000b). Only communication generated through spontaneous use of individual words and commonly used phrases is analyzed for reporting. The average communication rate is the average of all utterances in the language sample, weighted according to the number of words in each utterance. The peak rate is the highest rate for an utterance longer than the mean length of all utterances in the sample. Summary measures can be reported for average and peak communication rates in words per minute, selection rate in bits per second, and rate index in words per bit. The elements of communication rate include selection rate, LRMs, and errors (in the ability to use the system).

FIGURE 17.8 Photo showing collecting of LAM data at home (Photo courtesy of AAC Institute).

17.9.3 Selection Rate

Selection rate refers to the speed of the human–machine interface and is measured in bits per second (Romich, Hill, and Spaeth, 2001). For example, if a person were able to make one choice per second from a keyboard with 128 keys, the selection rate would be 7 bps. Many factors influence selection rate, such as the severity of a physical disability. Optimizing selection rate is the realm of the occupational therapist (OT), physical therapist (PT), and sometimes, the rehabilitation engineer.

17.9.4 Rate Index

The rate index provides a measure that compensates for selection rate differences, providing a measure for the comparison of communication rates (Hill and Romich, 2002). The rate index (RI) is the average communication rate (CR) in words per minute, divided by the selection rate (SR) in bits per second, divided by 60 (seconds per minute):

$$\mathrm{RI} = (\mathrm{CR}/\mathrm{SR})/60$$

Thus, the unit of measure for the rate index is words per bit. For example, an average communication rate of 21 words per minute generated at a selection rate of 7 bps (one keystroke per second from an 8×16 matrix) would result in a rate

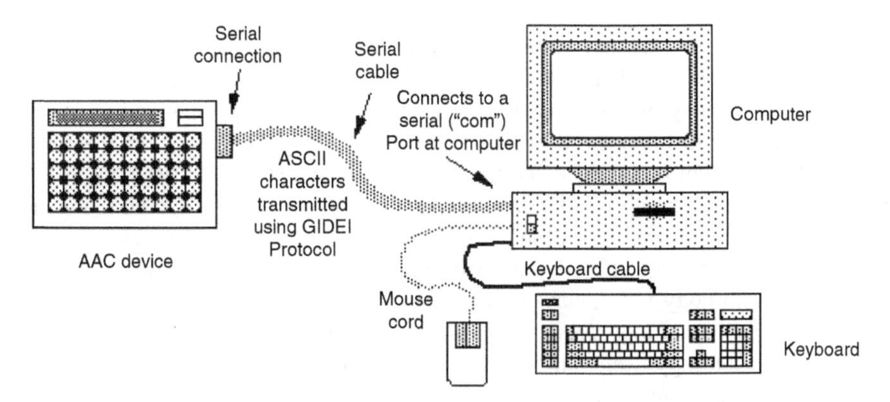

FIGURE 17.9 Schematic of AAC device being used to emulate a keyboard and a mouse (Graphic courtesy: Trace Center, University of Wisconsin–Madison).

index of 0.05 words per bit. The rate index offers a valid means of comparison of communication rate performance among individuals with different selection rates and can be used to facilitate subject selection for research.

17.10 COMPUTER ACCESS

Many people who use AAC also use computers. Many of these individuals have difficulty using a traditional keyboard or mouse. While alternative methods of accessing the computer may be available, many people use their AAC systems as their interface to the computer. There are a number of advantages to doing this. First, they may find that the physical access method through which they make selections on the AAC system is more effective than using the keyboard or mouse. Second, the traditional text entry process on a keyboard, i.e., spelling one letter at a time, is far slower than other methods. The LRMs used to make communication faster may also make computer text entry faster.

AAC systems that have electronic outputs (serial, USB, IR, Bluetooth, etc.), wired or wireless, can provide the information needed to emulate the computer keyboard and mouse. With appropriate software, such as SerialKeys or AAC Keys, characters and commands entered into the computer can be treated as though they are keyboard and mouse inputs to the computer (Vanderheiden, 1980, 1981). The General Input Device Emulating Interface (GIDEI) protocol defines the translations of incoming character strings to computer commands. Figure 17.9 shows typical connections. The original keyboard and mouse generally remain functional.

17.11 SUMMARY

Rehabilitation engineers can play a significant role in the lives of individuals who rely on AAC. Rehabilitation engineers are responsible for applying engineering principles to the design, modification, customization, fabrication, and/or support of AAC assistive technology. The Code of Ethics of the Rehabilitation Engineering and Assistive Technology Society of North America (RESNA) states that credentialed rehabilitation engineering technicians (RET) hold the welfare of persons served professionally to be of paramount importance and maintain the highest standards (RESNA, 2005). These values are upheld by appreciating the importance of language and language representation methods in the design, assessment, and intervention processes, and by applying the principles of EBP and performance measurement in making decisions.

New technology will always be on the horizon for individuals benefiting from AAC assistive technology. However, dedicated professionals are not seduced simply by newer technology. Rehabilitation engineering professionals know how to apply systematic and scientific approaches to measuring the performance of technology to ensure maximum benefits, and can contribute to a transdisciplinary team in this regard. Rehabilitation engineers know the systematic application of these principles to design and evaluate AAC technology that addresses the human desire for the most effective communication possible.

17.12 STUDY QUESTIONS

1. What is meant by augmentative and alternative communication (AAC)? What types of disabilities benefit from AAC assistive technology?
2. How does AAC assistive technology fit into the ICF model? What is the importance of language and communication for optimizing an individual's ability to achieve his or her potential?
3. Identify the two major categories for AAC interventions. How would you describe these types and ranges of technology?
4. List the components of an AAC assessment. What skills does a rehabilitation engineer contribute to the AAC assessment team? How would you explain to an AAC team your role as a rehabilitation engineer?
5. What are the three AAC language representation methods? Compare their characteristics. What conclusions can you draw about each method's ability to represent language based on these descriptions?
6. What are the basic components of dedicated AAC technology? What are the parts or features that would lead to selecting one AAC system over another?
7. How would you explain the differences among the three AAC display technologies? What advantages and disadvantages would one type of display have over another?
8. Identify two selection techniques available to access vocabulary using an AAC system. How would a rehabilitation engineer contribute to selecting a method for an individual with a disability?
9. What are the relative advantages and disadvantages of synthetic and digitized speech outputs?
10. Give two different examples of how an infrared (IR) output on an AAC system might be used.

11. Can an AAC system be used to provide input to a computer? If so, what are the requirements of the AAC system? What advantages might this offer to the individual?
12. Discuss the importance of human factor consideration in AAC assistive technology. Which human factor concept seems particularly critical for rehabilitation engineering?
13. What is the importance of performance measurement? Calculate the communication rate for the first utterance in the LAM log file example in this chapter. How would you use measures related to communication rate and selection rate to recommend an AAC solution?
14. How would you take advantage of the language representation methods and accessibility features on an AAC system to perform other tasks to improve function and participation according to the ICF model?
15. What are the basic principles of evidence-based practice? How would you apply these principles to providing AAC services as a rehabilitation engineer?

BIBLIOGRAPHY

American Speech-Language-Hearing Association (ASHA), 2001, *Scope of Practice*, Rockville, MD: Author.

American Speech and Hearing Association (ASHA), 2004, Evidence-Based Practice in Communication Disorders: An Introduction [Technical report], Rockville, MD: Author.

American Speech Language Hearing Association, 2004, *Preferred Practice Patterns for the Profession of Speech-Language Pathology*, Rockville, MD: Author.

Anderson, A., Baker, B., 2003, *Navigating AAC Devices for Early Communicators*, Presented at Assistive Technology Industry Association (ATIA) Annual Conference, Orlando, FL, ATIA.

Baker, B., 1986, Semantic compaction™ lets speech-impaired people quickly and effectively communicate in a variety of environments, *Byte*, 11, 160–168.

Baker, B., 1986, Using images to generate speech, *Byte*, 11, 160–168.

Balandin, S., Iacono, T., 1999, Adolescent and young adult vocabulary usage, *Augmentative and Alternative Communication*, 14.

Balandin, S., Iacono, T., 1999, Crews, wusses, and whoppas: Core and fringe vocabularies of Australian meal-break conversations in the workplace, *Augmentative and Alternative Communication*, 15, 95–109.

Banajee, M., Dicarlo, C., Stricklin, S.B., 2003, Core vocabulary determination for toddlers, *Augmentative and Alternative Communication*, 19, 67–73.

Beukelman, D., Ansel, B., 1995, Research priorities in augmentative and alternative communication, *Augmentative and Alternative Communication*, 11, 131–134.

Beukleman, D.R., Jones, R., Rowan, M., 1989, Frequency of word usage by nondisabled peers in integrated preschool classrooms, *Augmentative and Alternative Communication*, 5, 243–248.

Beukelman, D.R., Mirenda, P., 1992, *Augmentative and Alternative Communication: Management of Severe Communication Disorders in Children and Adults*. Baltimore: Paul H. Brookes Publishing.

Blackstone, S., 1990, In pursuit of opportunities and interdependence, *Augmentative Communication News*, 1–4.

Bloomberg, K., Johnson, H., 1990, A statewide demographic survey of people with severe communication impairments, *Augmentative and Alternative Communication*, 6, 50–60.

Chomsky, N., 2000, *New Horizons in the Study of Language and Mind*, Cambridge, UK: Cambridge University Press.

Cook, A.M., Hussey, S.M., 1995, *Assistive Technologies: Principles and Practice*, St. Louis: Mosby–Year Book, Inc.

Drager, K.D.R., Light, J.C., Spetz, J.C., Fallon, K.A., Jeffies, L.Z., 2003, The performance of typically developing 2 1/2 year olds on dynamic display AAC technologies with different system layouts and language organizations, *Journal of Speech, Language, Hearing Research*, 46, 298–312.

Enderby, P., Philipp, R., 1986, Speech and language handicap: Toward knowing the size of the problem, *British Journal of Disorders of Communication*, 27, 159–173.

Fitts, P.M., 1954, The information capacity of the human motor system in controlling the amplitude of movement, *Journal of Experimental Psychology*, 47, 381–391.

Gardner-Bonneau, D.J., Schwartz, P.J., 1989, A Comparison of Words Strategy and Traditional Orthography, in *Proceedings of the 12th Annual RESNA Conference*, pp. 286–287.

Glennen, S.L., DeCoste, D.C., 1998, *Handbook of Augmentative and Alternative Communication*, San Diego, CA: Singular Publishing Group.

Higginbotham, D.J., 1992, Evaluation of keystroke savings across five assistive communication technologies, *AAC Augmentative and Alternative Communication*, 8, 258–272.

Higginbotham, D.J., Lesher, G., Moulton, B., 1998, R4-Evaluating and enhancing communication rate, efficiency and effectiveness, in *Communication enhancement AAC-RERC: Engineering advances for communication enhancement in the new millennium*, National Institute on Disability and Rehabilitation Research, CFDA: 84.133E.

Hill, K., 2001, The development of a model for automated performance measurement and the establishment of performance indices for augmented communicators under two sampling conditions, *Dissertation Abstracts International*, 62(05), 2293 (UMI No. 3013368).

Hill, K., 2004, Augmentative and alternative communication and language: Evidence-based practice and language activity monitoring, *Topics in Language Disorders*, 24, 18–30.

Hill, K., Glennen, S., Lytton, R., 1998, The role of the manufacturers' consultants in delivering AAC services, Poster presented at the 1998 International Society for Augmentative and Alternative Communication (ISAAC) Conference, Dublin, Ireland: ISAAC.

Hill, K.J., Holko, R., Romich, B.A., November, 2001, AAC performance: The elements of communication rate, Presented at the 2001 ASHA Annual Convention, New Orleans, LA.

Hill, K., Romich, B., 2002, The AAC rate index in clinical practice, in *Proceedings of RESNA 2002 Conference*, pp. 81–83, Minneapolis, MN: RESNA Press.

Hill, K.J., Romich, B.A., 2003a, U-LAM (Universal Language Activity Monitor): A computer program for collecting AAC language samples [Computer software], Edinboro, PA: AAC Institute.

Hill, K.J., Romich, B.A., 2003b, *PeRT (Performance Report Tool): A Computer Program for Generating the AAC Performance Report* [Computer software], Edinboro, PA: AAC Institute.

Hill, K., Romich, B., *AAC Handbook for Evidence-Based Practice*, Pittsburgh, PA: AACI Press, in press.

Jagacinski, R.J., Monk, D.L., 1985, Fitts' law in two dimensions with hand and head movements, *Journal of Motor Behavior*, 17, 77–95.

King, T., 1999, *Assistive Technology: Essential Human Factors*, Boston: Allyn and Bacon.

Klein, W., 1994, *Second Language Acquisition*, Cambridge, UK: Cambridge University Press.

Koester, H.H., Levine, S.P., 1994, Modeling the speed of text entry with a word prediction Interface, *IEEE Transactions on Rehabilitation Engineering*, 2, 177–187.

Lesher, G.W., Moulton, B.J., Higginbotham, D.J., Alsoform, B., 2002, Acquisition of scanning skills: The use of and adaptive scanning delay algorith across four scanning displays, in Proceedings of RESNA 2002 Conference, pp. 75–77. Minneapolis, MN: RESNA Press.

Lesher, G., Moulton, B.J., Rinkus, G., Higginbotham, D.J., 2000, A universal logging format for augmentative communication, in Proceedings of the 2000 CSUN Conference. Los Angeles, CA: CSUN, retrieved February 14, 2003, from http://www.csun.edu/cod/conf/2000/proceedings/0088Lesher.htm.

Light, J., Lindsay, P., Siegel, L., Parnes, P., 1990, The effects of message encoding techniques on recall by literate adults using AAC systems, *Augmentative and Alternative Communication*, 6, 184–201.

Lloyd, L.L., Fuller, D.R., Arvidson, H.H., 1997, *Augmentative and Alternative Communication: A Handbook of Principles and Practices*, Boston: Allyn and Bacon.

Lucky, R., 1991, *Silicon Dreams: Information, Man, and Machine*, New York: St. Martin's Press.

Marvin, C.A., Beukelman, D.R., Bilyeu D., 1994, Vocabulary-use patterns in preschool children: Effects of context and time sampling, *AAC Augmentative and Alternative Communication*, 10, 224–236.

Mein, R., O'Connor, N., 1960, A study of the oral vocabularies of severely subnormal patients, *Journal of Mental Deficiency Research*, 4, 130–143.

Norman, D., 1980, *The psychology of everyday things*, New York: Basic Books, Inc.

Mandich, A., Miller, L., Law, M., 2002, Outcomes in evidence-based practice, in M. Law (Ed.), *Evidence-Based Rehabilitation: A Guide to Practice*, pp. 49–70, Thorofare, NJ: SLACK Incorporated.

Miller, L.J., Demasco, P.W., Elkins, R.A., 1990, Automatic data collection and analysis in an augmentative communication system, in *Proceedings of the Thirteenth Annual ASHA Conference*, pp. 99–100, Washington D.C.: ASHA.

Parette, H.P., Huer, M.B., Brotherson, M.J., 2001, Related service personnel perceptions of team AAC decision-making across cultures, *Education and Training in Mental Retardation and Developmental Disabilities*, 36, 69–82.

Petroski, H., 2003, *Small Things Considered*, New York: Alfred A. Knopf.

Pinker, S., 1994, *The Language Instinct*, Harper-Collins Publishers, New York.

Rehabilitation Engineering and Assistive Technology Society of North America (RESNA), *Code of Ethics*. (2005), Arlington, VA: Author.

Romich, B., 1994, Knowledge in the world vs. knowledge in the head: the psychology of AAC systems, *Communication Outlook*, 16, 19–21.

Romich, B.A., Hill, K.J., 2000a, *Language activity monitor feasibility study*, National Institute for Deafness and Other Communication Disorders of the National Institutes of Health (NIH Grant No. 1 R43 DC 4246-01).

Romich, B.A., Hill, K.J., 2000b, AAC communication rate measurement: Tools and methods for clinical use, in *Proceedings of the RESNA 1999 Annual Conference*, Arlington, VA: RESNA Press.

Romich, B.A., Hill, K.J., Spaeth, D.M., June 2001, AAC: a selection rate measurement: a method for clinical use based on spelling, in *Proceedings of the RESNA 2001 Annual Conference*, pp. 52–54, Arlington, VA: RESNA Press.

Romich, B., Vanderheiden, G., Hill, K., 2000, Augmentative communication, in J.D. Bronzine (Ed.), *Biomedical Engineering Handbook, 2nd ed.*, pp. 101–122, Boca Raton, FL: CRC Press.

Romski, M.A., Sevcik, R.A., Adamson, L.B., 1999, Communication patterns of youth with mental retardation with and without their speech-output communication devices, *American Journal on Mental Retardation*, 104, 249–259.

Rose, J. Alant, E., 2001, Augmentative and alternative communication: Relevance for physiotherapists, *South African Journal of Physiotherapy*, 57, 18–20.

Sackett, D.L., Richardson, W.S., Rosenberg, W.M., Haynes, R.B., 1997, *Evidence-Based Medicine: How to Practice and Teach Evidence-Based Medicine*, New York: Churchill Livingstone.

Scherer, M.J., 1998, *The Matching Person & Technology (MPT) Model Manual, 3rd ed.*, Webster, NY: The Institute for Matching Person & Technology, Inc.

Stuart, S., Beukelman, D.R., King, J., 1997, Vocabulary use during extended conversations by two cohorts of older adults, *AAC Augmentative and Alternative Communication*, 13, 40–47.

Swengel, K., Varga, T., 1993, Assistive Technology assessment: the feature match process, Paper presented at Closing the Gap Conference, Minneapolis, MN.

Todman, J., 2000, Rate and quality of conversations using a text-storage AAC system: a training study, *Augmentative and Alternative Communication*, 16, 164–179.

Vanderheiden, G., January 1981, Practical applications of microcomputers to aid the handicapped, *Computer*, IEEE Computer Society.

Vanderheiden, G., April 2–3, 1980, Microcomputer aids for individuals with severe or multiple handicaps: barriers and approaches, in *Proceedings of the IEEE Computer Society of the Application of Personal Computing to Aid the Handicapped*, Johns Hopkins University.

Vanderheiden, G.C., Kelso, D.P., 1987, Comparative analysis of fixed-vocabulary communication acceleration techniques, *AAC Augmentative and Alternative Communication*, 3, 196–206.

Venkatagiri, H.S., 2003, Segmental intelligibility of four currently used text-to-speech synthesis methods, *Journal of the Acoustical Society of America*, 113, 2095–2104.

Wasson, C.A., Arvidson, H.H., Lloyd, L.L., 1997, AAC assessment process, in L.L. Lloyd, D.R. Fuller, H.H. Arvidson (Eds.), *Augmentative and Alternative Communication: A Handbook of Principles and Practices*, pp. 169–198, Needham Heights, MA: Allyn and Bacon.

Wilkinson, K.M., Jagaroo, V., 2004, Contributions of principles of visual cognitive science to AAC system display design, *Augmentative and Alternative Communication*, 20, 123–136.

World Health Organization, 2001, *ICIDH-2: International Classification of Functioning & Disability* (Pre-final draft), Geneva: Author.

Yorkston, K., Karlan, G., 1986, Assessment procedures, in S. Blackstone (Ed.), *Augmentative Communication: An Introduction*, pp. 163–196, Rockville, MD: American Speech Language Hearing Association.

18 Adaptive Sports and Recreation Technology

Mary E. Buning, Ian M. Rice, Rory A. Cooper, and Shirley G. Fitzgerald

CONTENTS

18.1 LEARNING OBJECTIVES OF THIS CHAPTER

Upon completion of this chapter, the reader will be able to:

Understand the value of adaptive sports and recreation in intrinsically motivating and preparing individuals to return to productive human occupation and participation in the presence of physical, sensory, or cognitive impairment

Consider the essential role of the physical context (temperature, moisture, surface characteristics, etc.) in which activities typically occur in conceptualizing, designing, and engineering technology for sports and recreation

Recognize the interplay between adaptive equipment, rule modification, and physical training by athletes that leads to true competition among athletes with disabilities

Utilize economical approaches that account for variances in human skills and abilities in the design of sports and recreation technology in order to maintain affordability

18.2 INTRODUCTION

Play and leisure activities are much more interest driven than other human occupations. They are chosen for the excitement, challenge, enjoyment, or feeling of competence that they provide. The more our life is filled with things that interest us, the higher will be our satisfaction with the quality of life (Kielhofner, 1985).

The human need for re-creation or recreation is especially important in the presence of physical, sensory, or cognitive impairment that affects the ability to function in everyday activities and environments. Primary rehabilitation interventions place emphasis on self-maintenance, education, and employment activities. Yet, the theory of human occupation cites the importance of the "volitional elements of the personality" as critical factors in a client's motivation for this habilitation or rehabilitation process (Kielhofner, 1985). This volitional component is the process by which a person experiences, interprets, anticipates, and chooses occupational behaviors. It supports an individual's sense of personal ability, effectiveness, and judgment of what is important and meaningful. It is reflected in what one is attracted to (interests), what is important within one's worldview (values), and the belief that success is possible. Because these volitional factors influence choice about action and behavior, they have a key role in enabling an individual to adapt to disability and re-engage in life (Kielhofner, 1985).

18.3 THE CONTRIBUTION OF ADAPTIVE SPORTS AND RECREATION

Recreation allows humans to challenge their physical limits in a setting that accepts a new participant as a novice, provides instruction and emotional support for learning, and by its nature implies social participation and the promise of fun and escape from the realities of everyday life. Recreation puts people in touch with their imagined selves — a person who is adventurous or strong or graceful — and does not need to be justified in the same way as work and self-maintenance activities — they are chosen just because they appeal (Kielhofner, 1985).

For persons with acquired disability, recreation can represent a return to a valued form of daily life activity. Persons with congenital disability see it as a way to experiment and develop new and socially valued skills and abilities. For both groups, recreation is a valuable strategy for inclusion in activities that are culturally valued (Buning, 1996). Healthy competition with self and others through sports and recreation creates an arena for continuing the gains of medical rehabilitation, challenging personally held ideas learned from the culture about disability and handicap, and testing out a new self-concept that hopefully includes acceptance of disability (Schlein et al., 1997).

There is another reason to support adaptive sports and recreation — personal fitness. Today's medical advances have increased survival rates; yet, lack of opportunity and information about adaptive fitness make it more likely that these same individuals will fall prey to the negative health consequences of inactivity, repetitive strain injury, and obesity (Taylor et al., 1998; Rimmer et al., 1996; Heath and Fentem, 1997). Involvement in recreation helps maintain good health.

18.3.1 The History of Adaptive Sports and Recreation

Over the past 50 years, there has been movement from Europe to the United States to Japan to recruit the benefits of active sport and recreational activity for people with disabilities (Dummer, 2002). Warm Springs Hospital in Georgia became renowned in the days of the polio epidemic for using water therapies to restore function (Pelka, 1997). In Europe and the US, veterans of World War II returning home with amputations and spinal cord injuries looked for ways to resume involvement in skiing and court sports such as basketball (Schweikert, 1954). The U.S. Veterans Administration hospitals and organizations such as Paralyzed Veterans of America and Disabled American Veterans were aware of this and began to include adaptive sports in rehabilitation and post-discharge programs. The Vietnam War created another surge of interest and recipients for this "recreation therapy" approach to rehabilitation. These veterans, motivated by their personal recovery experiences through adaptive sport and recreation, went on to create nonprofit organizations such as Disabled Sports USA. Today, a network of local chapters reaches individuals of all ages. Original programs expanded to include individuals who experience disability due to accident, disease, and congenital disability.

Persons with innate talent and natural drive for competition have taken adaptive sports to highly competitive levels that range from regional to international events. The Paralympics, affiliated to the Olympic Games, offer the ultimate adaptive sports competition. The Paralympics movement has had a positive impact on society's perception of people with disabilities (Steadward and Peterson, 1999). Through the achievements of Paralympics athletes, society is able to see disability in a new way. Adaptive sports demonstrate that disability does not diminish the human spirit.

With this background of the importance of adaptive sports and recreation in mind, we arrive at the focus of this chapter: an overview of assistive technology (AT) devices that enable participation in team and solo sports and recreational activities with family and friends. When people lose the skills and abilities that enable them to engage in recreation and sport, modifications and adaptations are necessary to enable continued participation in these activities. The basis for sport or recreational activity remains the same, but the rules, equipment, and adaptations that enable participation vary greatly. In general, most adaptive approaches to recreational and sports equipment and participation in recreational activities are products of the motivation and creativity of persons with disabilities. An overview of common adaptive recreation and sports devices and approaches follows. Categories of indoor, summer and winter sports, and activities will organize the discussion.

18.4 INDOOR RECREATION AND SPORTS

18.4.1 BASKETBALL

Basketball, one of the first sports adapted to disability, started in the 1940s with the interest of veterans returning from World War II (Strohkendl, 1996). Today, players have various diagnoses such as paraplegia, cerebral palsy, amputations, postpolio syndrome, etc., and may or may not use a wheelchair on a daily basis. This sport is attractive to many individuals who use wheelchairs because they are already familiar with manual wheeled mobility and simply use it for a new activity. A basketball court is a wheelchair-accessible environment, and success in this sport is attributed to upper body strength, skill in ball handling, and having access to a wheelchair frame that fits and supports high-level play.

A basketball wheelchair is similar to a typical manual wheelchair, but it incorporates features to enhance stability and maneuverability. They are lightweight to allow for speed, acceleration, and quick braking. Although basketball is not a contact sport, some incidental contact is inevitable, so spoke guards cover the rear wheel spokes to prevent wheel damage. Made of high-impact plastic, spoke guards provide several additional benefits. They allow players to pick up the ball from the floor by pushing it against the spoke guard and rolling it onto their lap. Spoke guards protect hands and fingers from aggressive play when reaching for the ball, and provide space to place team and sponsor names. Stability comes from the camber in the wheels, which creates a broader wheelbase. Additionally, camber brings the top of the wheel closer

FIGURE 18.1 Wheelchair Basketball demands fitness and skill and is promoted by organizations such the United Spinal Association at http://www.unitedspinal.org/.

to the body, makes the wheelchair more responsive to turns, and protects a player's hands during collision, as hands are located away from the plane of contact.

Regulations now describe basketball wheelchair features (International Wheelchair Basketball Federation [IWBF], 2004). A chair must have four wheels — two large rear wheels for power and two front casters for pivoting and turning. The casters are 5 cm in diameter and usually made from extremely hard plastic, similar to inline skate wheels, with precision roller bearings. The rear wheels use high-pressure tires (827.4 to 1379 kPa [120 to 200 psi]) with minimal tread, have pushrims, and are less than 66 cm in diameter. The footrest must be no higher than 11 cm when the front wheels are in their forward movement position and must also be designed to avoid damage to the playing surface.

To gain an advantage for shooting baskets, forwards usually try to make wheelchair seats as high as possible, though 53 cm or less from the floor. Seats typically angle toward the rear, creating "squeeze" or seat bucketing of 0.085 to 0.255 rad (5° to 15°) to increase pelvic stability for the player. Guards prefer lower seat heights, which, combined with increased seat angle, makes chairs faster and more maneuverable for ball handling. Cushions may be used if made of flexible material, that is the same size as the seat of the wheelchair and no higher than 10-cm thick. If a player is classified as a 3.5-, 4.0-, or 4.5-point player (classification follows), then the cushion cannot exceed 5 cm. Black tires, gears, brakes, or steering devices are not allowed on the chair. Referees check each player's wheelchair before each game to ensure it meets requirements.

Wheelchair basketball incorporates many of the rules of typical play, for example, two halves of 20 min, five players per team, and the same court dimensions and hoop height (IWBF, 2004). To accommodate varying levels of impairment, a player classification system is used to configure teams. Each player's trunk movement is assessed as he or she performs basketball skills such as pushing the wheelchair, dribbling, passing, receiving, shooting, and rebounding, and is given a point value or classification. Point classes range from the lowest functional ability of 1.0 to 4.5 (the highest) in 0.5 increments. A 1.0 player might be a person with a thoracic spinal cord injury (SCI), whereas a 4.5 player may have a sacral SCI or single-leg amputation. The point values of the five participating players yield a team total. For IWBF World Championships, Paralympics, and zonal championships and qualifiers, team total may not exceed 14 points at any time during the game, even with substitutions. Each player is issued a player classification card that also describes any modifications to sitting position, use of straps, and orthotic and prosthetic devices.

Game rules do vary slightly to accommodate propelling a wheelchair and dribbling, passing, and shooting. To cover "traveling," the player in possession of the ball may not push more than twice in succession with one or both hands in either direction without tapping the ball to the floor. A player may, however, wheel the chair and dribble the ball simultaneously just as an able-bodied player runs and dribbles the ball, but there is no double-dribble violation in wheelchair basketball. A player may only remain within the opponents' restricted area (key) for 3 sec or less, otherwise they incur a "three-second violation." This restriction does not apply while the ball is in the air during a shot, during a rebound, or while the ball is dead (Vines, 2004). Because of the varying degrees of disability among players, a basic rule of remaining firmly seated in the wheelchair at all times and not using a functional leg or leg stump for physical advantage over an opponent, is strictly enforced. An infraction of this rule (rebound, jump ball, etc.) constitutes a physical advantage foul. It is also recorded in the official score book. Three such fouls disqualify a player from the game. Two free throws are awarded, and the ball is given to the opposing team out-of-bounds.

This rather intensive look at wheelchair basketball is used to demonstrate how the spirit and the goals of the game are preserved through modifications to both the wheelchair (equipment) and the rules of the game. The performance requirements and subsequent features of the wheelchair derive from the demands of the game, player safety, and compensation for the performance limitations of the players. Since equipment can be used to create competitive advantage, it is carefully regulated. Because equipment cannot completely equalize player performance, modified rules and classification work to create functional equivalence between teams. All of these parameters work together to create a sport that is equal in challenge and excitement to any basketball competition.

18.4.2 Wheelchair Rugby

Wheelchair rugby, also known as Quad rugby, was originally called "murder ball" because of the aggressive nature of the game. In 1981, Brad Mikkelsen teamed up with University of North Dakota's Disabled Student Services to create the first team, and ultimately introduced Rugby to the United States. The United States Quad Rugby Association (USQRA) was formed in 1988 to regulate and promote the sport on both a national and an international level. Rugby was introduced into the Paralympics in 1996, first as a demonstration event and then in the Sydney Paralympic Games 2000 as a full medal sport.

This sport differs from basketball in that the players must have both an upper and lower extremity impairment to be eligible to participate. Most players have sustained cervical SCI and some degree of tetraplegia. Similar to basketball, players are given a classification number, based on seven classifications ranging from 0.5 (greatest impairment) to 3.5 (least impairment, e.g., injury level of spinal cord injury), and a team may only have a maximum of 8 points on a floor at one time. Four-player teams may be coed, although, with the classification system, gender advantage is minimal (U.S. Quad Rugby Association, 2004).

FIGURE 18.2 Quad rugby players using wheelchair especially designed for aggressive play. A volleyball is used to play the game.

The object of wheelchair rugby is to cross the opponent's goal line with two wheels while in possession of the ball. While the offensive team tries to advance the ball, the defense works to halt their progress with turnovers. When a player has possession of the ball, he or she must bounce the ball every 10 sec while carrying the ball across the midline in 15 sec or less. The court "key area" extends 1.75 m from the goal line, and within it, certain restrictions apply. Only three defensive players are allowed in the key, and if a fourth enters, a penalty is assessed or a goal awarded. Offensive players can only stay in the key for 10 sec or a turnover is awarded (Lapolla, 2000).

In terms of equipment, players tend to use extreme wheelchair configurations, elastic binders, foam arm protectors, and special gloves with a tacky surface to improve performance. Similar to basketball, players become extremely proficient in adapting their equipment to promote balance and speed; so much so that they often appear to possess more functional capacity then their diagnosis would suggest. Wheelchair styles are strictly regulated to ensure fairness but vary considerably depending on a player's preferences, functional level, and role in the team.

Players with the most extreme upper body deficits tend to take on defensive roles (blocking and picking) and use chairs with additional length and hardware that enable them to grab other player's chairs. Players with more function usually take on the roles of ball handlers and use offensive chairs that are built for scoring and are fast and maneuverable. The chair design allows players to deflect or slide off other chairs to minimize getting caught or stopped. Regardless of functional abilities and classification, all rugby chairs have extreme amounts of camber (0.27 to 0.34 rad, or 16° to 20°), significant bucketing, and anti-tip bars. Again, camber provides lateral stability, hand protection, and ease in turning. The bucketing helps with trunk balance and protection of the ball.

18.4.3 Tennis

Men and women, in either singles or doubles matches, play wheelchair tennis on a typical tennis court — another accessible environment for wheelchairs. The rules of tennis vary in two ways. The ball may bounce twice before it is returned and rules that apply to the athlete now apply to the wheelchair as an extension of his or her body. So, the wheelchair may not be stabilized with brakes, and the athlete must keep one buttock in contact with the seat while hitting the ball.

Serious tennis players use a wheelchair especially designed for tennis that is equipped with three or four wheels — two large ones for propulsion and one or two 5-cm casters under the feet that allow for quick turning. The large wheels are typically 61 to 66 cm (24 to 26 in.) in diameter and use high-pressure tires (827.4 to 1379 kPa; 120 to 200 psi) to reduce rolling resistance on the court. Wheels are set with extreme camber to maximize mobility and stability on the court, especially when making shots. Players with high-level SCI play with a power wheelchair and with longer rackets to compensate for the length taken up by strapping the racket to the hand.

Tennis players, as with rugby and basketball players, use a steep seat angle to gain greater pelvic control and, hence, balance and stability. Knees are flexed, and the feet are tucked behind the player's knees. With the body in this relatively compact

FIGURE 18.3 David Hall, a world-class Australian wheelchair tennis player. His wheelchair has an anti-tip wheel in the rear.

position, the combined inertia of rider and wheelchair is reduced (similar to figure skaters bringing their arms in to spin faster) and the chair becomes more maneuverable (Cooper, 1990; International Tennis Federation, 2004). Many players use chairs with handles incorporated in the front of the seat to assist them in leaning for a shot. The handles also help keep the players knees in place during quick directional changes. Players can also use straps around their waist, knees, and ankles to increase balance.

Wheelchair skills; getting to top speed quickly, and quick stopping and changes in direction to deal with difficult bounces are the keys to success in tennis, and players must focus and work out strategies to develop these skills. Tennis is another sport that demonstrates the need for specialized equipment, modified rules, custom adaptations based on the functional ability of the player, and attention to the player's physical conditioning, strength, agility, and competitive spirit.

18.4.4 Archery and Marksmanship

Archery has a long history as an Olympic sport and as a sport for persons with lower extremity disabilities. The first international wheelchair archery competition occurred in 1948 in Stoke Mandeville, UK, through the efforts of Sir Ludwig Guttman, a neurosurgeon. He had developed a sports program for returning World War II veterans with SCI, and this led to the development of the Stoke Mandeville Wheelchair Games. The games eventually became the first Paralympic Games in Rome in 1960 (Scruton, 1998).

Archery competition is coed, undertaken as singles or teams, and offers participants the option of either sitting in a wheelchair or standing. The rules for the competition describe the types of adaptive equipment that may be used, based on the competitor's level of impairment (Harris, 2005). For example, archers may use a body support to stand, an authorized strapping system, a mechanical release aid, an approved compound bow, an elbow or wrist splint, or an assistant to load arrows into the bow if unable to do it independently. The scoring procedures are identical to those used by athletes competing in the Olympic Games (Harris, 2005). Approved adaptive archery equipment can be found through various organizations, and local and national tournaments exist for both recreational and competitive athletes (Rice et al., 2005).

Marksmanship is also used in other shooting sports such as rifle and pistol shooting. As in archery, competitors may use assistive devices to create equivalence among athletes with differing functional abilities. Shooting events with firearms are comprised of free and supported rifle and free pistol, both air and 0.22 caliber, shooting at distances from 10 to 50 m. Competition occurs in men's, women's, mixed, and team events, although team events are not held at the Paralympic Games (International Shooting Committee for the Disabled [ISCD], 2004).

18.4.5 Swimming

Swimming is a sport and recreational activity that is compatible with many types of impairment and persons of all ages and ability. With flotation and/or personal assistance, even nonswimmers or persons with severe impairment can enjoy movement in water as a result of the effects of apparent reduction in gravity. Swim aids used to train competitive swimmers, such as kickboards, flippers, pull-buoys, and hand paddles, or flotation used in water aerobics classes, can be used to compensate for specific impairments.

Despite the origins of Swimming for persons with disability originated in physiotherapy and rehabilitation. However, swimming competition is the largest and most popular competitive event in the Paralympic Games. Swimming events are open to athletes with all types of impairments. Athletes are classified and compete against others based on how they move in the water. The methods used for starts and turns and the strokes possible for the swimmer are dictated by whether impairment affects hands, arms, trunk, and/or legs. No prostheses or assistive devices may be worn; however, stroke adaptations are permitted and specified within each stroke category. For example, in the breaststroke competition, a swimmer may either start in the water or off the block. After the signal to start is given, one asymmetrical stroke is permitted so as to allow the swimmer to attain the breast position. Swimmers in certain classes may roll over on their back to breathe but may not attempt forward propulsion during that period. The swimming events at the Paralympic level include freestyle (50 to 1500 m), backstroke (50 to 200 m), breaststroke (50 to 200 m), butterfly (50 to 200 m), individual medley (150 and 400 m), and relay (4 × 50 m freestyle, 4 × 200 m freestyle, 4 × 50 m medley, and 4 × 100 m medley) (International Paralympic Committee, 2004).

18.5 SUMMER RECREATION AND SPORTS

It is arbitrary to divide sports by summer and winter because this concept is based on climate and hemisphere and culturally based notions of season. Regardless, it is used here to differentiate sports and recreation played indoors and on snow and ice.

18.5.1 Athletics

The track and field sports, a long standing Olympic tradition, have been adapted to disability since the Stoke Mandeville Games and include throwing and racing events (Scruton, 1998). Instead of wheelchairs, the throwing events allow competitors to use specially designed throwing chairs attached to a holding device that offer greater stability and support. Throwing chairs are taller and eliminate the large wheels that could potentially interfere with the dynamic upper body movements required in throwing events. When a wheelchair is used, the seat and cushion must not exceed 75 cm in height (Ewing, 2005). The chair design is important because it can significantly enhance performance, depending on how well it matches the thrower's body and functional abilities. Consequently, athletes select chairs of different configurations to optimize their throwing performance. Competitors in wheelchair athletics are classified based on the neurological level of injury and their ability to effectively carry out movements necessary to complete a throwing task (Ewing, 2005). These functional or "F" categories range from 1 (severe tetraplegia) to 8 (minially disabled). These functional classes also regulate who may compete in an event. For example: the javelin throw is not held for the F1 class due to the absence of sufficient hand function to control a javelin safely, whereas athletes with higher F ratings are allowed to throw from a standing position and use a heavier javelin. Athletes with diminished hand function tend to use "resin" or adhesive-like substances to augment grip and throw implements such as the club rather then the javelin (Ewing, 2005).

18.5.2 Wheelchair Racing

Wheelchair racers and others with physical impairments compete in short distance sprints through to the marathon. Wheelchairs used in racing are customized and designed to fit the body of each user. This equipment used in combination with proper propulsion biomechanics results in an extremely efficient means of movement (Cooper, 1989a, 1989b, 1992). Others are able to run with specially designed prosthetic limbs or with assistive devices such as orthotics or crutches (Davidson, 2005; Cheskin, 2004).

The design of a racing wheelchair optimizes the abilities of each user, incorporating features such as three-wheeled design,high-pressure tubular tires, lightweight rims, precision hubs, carbon disc/spokes wheels, compensator steering, small pushrings, ridgid frame construction, and 0.034 to 0.084 rad (2° to 15°) of camber. The fit of the racing chair to one's body and motor abilities is critical to overall performance. Racing-chair manufacturers require many body measurements when a chair is ordered because the frame and seat cage are made to fit each individual (Cooper, 1990). Body measurements include: hip width, chest width, thigh length, arm length, trunk length, height, and weight. Wheelchair racers position themselves to promote biomechanical efficiency and aerodynamics. Legs without control are brought close to controllable portions of the body (chest, shoulders, or arms) to increase stability for wheelchair control and propulsion (Cooper, 1990).

Athletes with paraplegia, tetraplegia, or amputated limbs have differing position preferences. Both disability and racer experience determine the location of rear axles

FIGURE 18.4 Ian Rice in racing chair.

with respect to the seat cage. Whereas experienced athletes with paraplegia prefer 15 to 25 cm from the seat back to the rear axles inserts, those with tetraplegia prefer 5 to 20 cm. Novices choose more stable configurations. The seat cage upholstery and rear axle positions are adjusted to allow optimal access to the front edge of the pushrims and allow both arms to complete a full stroke that reaches to the bottom of the pushrims.

In contrast to wheelchairs for community mobility, racers use one of three body positions: the kneeling bucket, kneeling cage, and upright cage. The kneeling bucket is the lightest, most aerodynamic setup. Over the years, it has allowed athletes with SCI to make tremendous performance gains. Racers with cervical SCI pull their knees up against the chest to further enhance stability, balance, and breathing. Athletes inexperienced with the kneeling bucket position may use a kneeling cage, which allows either sitting upright or kneeling and so permits flexibility of body position. Upright cage seats work well for athletes with lower limb amputations, and for athletes with low levels of paraplegia; that is, those who have sufficient trunk control to adjust body position while racing (Cooper, 1990).

A racing chair is steered by applying pressure to small handlebars that turn the front wheel. Minor steering adjustments are accomplished when racers swing their upper body and or hips within a properly fitted racing chair. Racers call this maneuver "hipping" the chair. When racing on an oval track, athletes use a steering component called a compensator that is engaged with one hand while in the curve of a track and then disengaged on the straight section.

Wheelchair racers can maintain chair control at high speeds for long distances because they use a highly specialized propulsion stoke and special gloves. These gloves are constructed using combinations of leather, foam, rubber, and self-molding plastics. Racing gloves provide solid contact with the pushrim, reduce errors when grasping and releasing the wheel at high speeds and allow the delivery of more force per stroke. The propulsion stroke can best be described as a punching motion. It can take years to master, and is designed to deliver maximum force (Cooper, 1989).

18.5.3 Running Prosthesis

Today, amputee athletes around the world are setting near-Olympic records with Flex-Foot® legs designed for sprinting and other sports. In 1982, Van Phillips, teamed up with Dale Abildskov, an aerospace composite engineer at the University of Utah. He was seeking to design a prosthesis that would allow him to run again. Abildskov's knowledge of materials led to the development of a flexible carbon-graphite shaped in a long C-curve that allowed it to flex and return energy (Davidson, 2005). When force is applied via body weight landing on the heel, the graphite arc converts it into energy that simulates the spring action of the normal foot, creating rebound and allowing the wearer to run and jump. Today, more than 90% of amputee athletes worldwide use some model of the basic Flex-Foot® (Cheskin, 2004). Speeds of runners wearing these prostheses are very close to those of able-bodied runners, and because of the materials used, the Flex-Foot® adapts to sports such as triathlons, which combine running, swimming, and bicycling (Brown, 2001).

18.5.4 Off-Road Wheelchairs

Wheelchairs have also been redesigned for use in backcountry and wilderness. Athletic individuals with a desire to hike, camp, and go where conventional wheelchairs cannot, have designed adaptations to make rough terrain navigable. They use 66-cm (26-in.) knobby tires such as those on mountain bikes to gain traction on soft, wet, or difficult terrain. Front casters are significantly larger to decrease the chance of getting stuck on obstacles. Larger casters required frame redesign, and so an off-road wheelchair looks more like a four-wheel buggy than a typical wheelchair.

Today, there are several classes of off-road wheelchairs. Features can include: knobby tires, independent four-wheel suspension, hand brakes, and ultralightweight frames. Off-road fans, recognizing their affinity with mountain bikers, often compete in dual slalom and downhill mountain bike events and reach speeds well over 80.5 km/h (50 mph).

18.5.5 Cycling

Bicycling has been a popular form of leisure for many years. Adaptive cycling through adaptations to typical bicycles, use of tandem cycles, or use of a handcycle opens this recreational and competitive sport to individuals with many types of impairments.

FIGURE 18.5 Keith Moore at the College of Engineering, University of Cambridge, designed for himself a lightweight cycling prosthesis.

Using toe clips, altering the size of the arc of the pedals, or modifying the handlebars, handgrips, or placement of the gears and brake levers are the only modifications needed by some (Buning, 1996). An additional benefit of cycling is its contribution of aerobic activity to overall fitness (National Center for Chronic Disease Prevention and Health Promotion, 1996).

For those with lower extremity impairment such as spinal cord injury, multiple sclerosis, hemiplegia, and amputation, a handcycle is used. Handcycles use the upper extremities for propulsion and provide the stability of three wheels and placement closer to the ground. For persons with reduced grip strength, cuffs can be mounted to the arm crank handles. Elastic abdominal binders can stabilize the trunk. A handcycle allows the user to propel, steer, break, and change gears, all with the upper extremities and trunk.

Two types of handcycle designs are typically used: upright and recumbent. In an upright handcycle, the rider is in a vertical position as in a wheelchair. Upright handcycles use a pivot steer in which only the front wheel turns while the rear wheels of the cycle follow. It is easier to transfer into and balance on an upright cycle. In the recumbent handcycle, the rider's torso is semireclined and the legs are positioned in knee extension. Steering occurs in one of two ways. In one, as the rider leans toward the desired turn, force is transferred through a linkage bar, causing the frame to pivot and turn. This would be challenging for riders without trunk stability. The other style — a pivot steering recumbent handcycle — uses arms and shoulders to execute turns like a typical bike. Recumbent handcycles are ideal for racing as they are lighter and faster (Rice et al., 2005).

18.5.6 Golf

Golf has become a popular sport for many individuals with upper and lower limb amputation and spinal cord injury. Many assistive devices are already in use, including

FIGURE 18.6 Handcycle competitors at the 23rd National Veterans Wheelchair Games, 5–9 July, 2003, Long Beach, CA.

clubs designed specifically to be swung from a seated position, one-person golf carts intended to replace a wheelchair on the green and fairway, gripping aides, and practice equipment such as automated ball-teeing devices. Specialized golf pros, trained in techniques used by able-bodied golfers, are able to analyze the capacities of persons with physical and sensory limitations and adapt instruction and technique to fit. The Americans with Disabilities Act (ADA) has helped this sport grow by including golf courses in the definition of public accommodation. Several US and international golf associations exist just to support and inform golfers with disabilities and to help golf courses develop accessible facilities (National Center on Accessibility, 2003; National Alliance for Accessible Golf (NAAG), 2003).

18.5.7 Waterskiing

Waterskiing is a popular water sport for individuals with physical and sensory impairment. Those with visual impairment appreciate the towrope and their proximity to the towboat. Water-skiers with amputations or leg impairment and strong upper bodies are able to use slalom techniques to ski on one leg. New skiers often appreciate learning to ski with the help of a rigid boom that is attached to the side of an inboard motorboat. In this technique, the skier is adjacent to the boat driver and rides over smooth water in front of the boat's wake (Buning, 1996). Those with significant weakness or with spinal cord injury are able to use sit-down skis designed specifically for waterskiing.

Sit-waterskis can compensate for trunk instability and hand weakness, and so incorporate a variety of adaptive features to suit a wide range of functional levels. Furthermore, some skis adjust vertically, horizontally, and diagonally at the fin, allowing users to fine-tune their equipment to meet various skiing styles, body weights, and boat velocities. At the competitive level, skiing events include men's and women's slalom, tricks, and jumping events (Rice et al., 2005). Water skis generally provide a stiff ride as skiers skim over wakes and turbulence, and so equipment improvement is still needed.

FIGURE 18.7 Water sit-skiing adapted the concepts of stand-up skiing to the sitting position.

18.5.8 Camping

Individuals interested in camping, fishing, and hiking can make use of adaptations that are generally available. Camping requires moving away from the type of accessible, supportive environments (i.e., level, paved surfaces, availability of running water, stable surfaces for transfers and postural supports, etc.) that enable independent functioning for individuals with physical and sensory impairment. Yet, humans repeatedly seek rugged environments to "get away" and challenge their ability to adapt.

Accessible campsites that are level and located near water and toilet facilities are found in most U.S. state and national parks. "Car camping" gives proximity to an automobile and reduces the need for carrying supplies and equipment over distances. Camping in remote locations becomes possible when canoes, rafts, and kayaks are used to traverse rivers and streams. In this situation, a boat provides accessible travel into remote locations and camping occurs at beach campsites. Because riverbanks are constantly changing, this method of camping relies heavily on creativity and support from others who are able-bodied.

Depending on an individual's functional abilities, more or less assistance from family members, friends, and volunteers is needed to make "roughing it" at a campsite fun rather than an ordeal. Individuals interested in camping are usually very creative in recruiting others or developing strategies and devices to make camping possible.

18.5.9 Hunting and Fishing

Hunting and fishing are recreational activities with a strong social and symbolic component for many individuals, especially males. A disability can make it difficult to handle air weapons or firearms without some means of assistance. To overcome these difficulties, stabilization devices such as freestanding or tabletop shooting rests are used. In addition, people with limited finger dexterity can use sip-and-puff trigger activators to safely discharge a rifle. Prosthetic appliances have been developed to allow an arm amputee to hold a bow, and manual trigger activators are compatible with several types of crossbows (Buning, 1996).

Hunting is a recreational activity that lends itself well to partnering between able-bodied hunters and those with disabilities. Four-wheel-drive vehicles can get hunters with disabilities into the backcountry. Off-road wheelchairs, adaptations to weapons, blinds adapted for wheelchairs, and ingenuity make it possible for hunters with disabilities to continue enjoying their sport.

Access to fishing has been increased by building barrier-free fishing piers in parks, along streams, and in wilderness areas. Fishing adaptations range from specialized grips to devices that mount rods and reels to power and manual wheelchairs or boats and railings. Numerous companies sell adapted equipment designed to compensate for hand and upper extremity weakness, such as one-handed reels, electric reels, and knot ties (Rice et al., 2005).

18.5.10 Paddle Sports

Families and friends who enjoy kayaking, canoeing, rafting, and rowing can easily include individuals with mobility, cognitive, and sensory impairments. Paddle sports can be enjoyed on flat water in tidal basins and lakes or in tumbling streams and surf. Paddle sports provide exercise, are safe when personal flotation devices and helmets are used, and are generally affordable.

Kayaks and canoes are easily adapted to accommodate limitations in balance, stability, and hand skills. Common adaptations include modification to technique for boat exits and entry, grip modifications for paddles, and use of outriggers to add stability to boats. For individuals requiring more assistance, tandem kayaks are helpful. A paddling partner seated in the stern of tandem kayaks can help with steering and paddling when necessary. Tandem kayaks benefit beginners, individuals with limited paddling strength, and those with visual impairments (Rice et al., 2005).

18.5.11 Sailing

In coastal regions and areas with large lakes, sailing holds interest for many, including persons with disabilities. Sailors are often willing to support adaptive sailing programs whether recreational or competitive in orientation. Adaptive sailing, now a Paralympics event, uses two classes of boats: single-handed and crew. In the single-handed boat, the sailor faces forward in the boat, with all controls within an arm's reach. In crew boats, sailors generally take on a role with a certain part of the boat — tiller, sheets, or lines.

In crew sailing, the large cockpit and balanced, inboard rudder add stability to the boat, making the sport ideal for individuals with a variety of impairments. In competitive sailing, individuals on the crew are assigned a point classification based on their level of disability. The classification test rates the athlete's ability to compensate for the movement of the boat (stability), operate the control lines and tiller (hand function), move about in the boat (mobility), and see while racing (vision). When sailors present for classification, they are required to bring all the personal assistive devices, adaptations, prosthesis, and orthotics they intend to use during racing. A crew's total score must not exceed a set point value.

Seats on sailboats are designed to allow the sailors to position themselves so they can control the tiller and sheet without fear of falling. Seats can be as basic as a modified lawn chair or as complex as a translating seat that allows a sailor to switch sides of the boat. Sailors often require some sort of surface to serve as a transfer bench for switching sides when tacking or jibing. Competition rules do not allow permanently modifying a boat by drilling holes or installing fixtures, nor can the sail be modified in any way that raises it more than 20 cm above the existing seat. Safety practices such as radio for communications, the presence of support boats, and the use of personal flotation devices are also required.

18.5.12 Scuba Diving

Many individuals with disabilities participate in scuba diving and join dive clubs, which are found even in landlocked communities. Most individuals with disabilities enjoy greater freedom when under water due to the effects of apparent reduction in gravity. The amount of support a diver will require is largely a function of residual ability and diving experience (Rice et al., 2005). Many disabled divers augment their abilities via adaptive equipment such as webbed diving gloves for extra push, powered devices for moving over greater distances, and adaptive buoyancy compensators. The Handicapped Scuba Association certifies individuals with significant disabilities for open-water diving. This organization has established guidelines for indicating the amount of support a disabled diver will need in order to dive safely (Boyd, 2001).

18.6 WINTER RECREATION AND SPORTS

Frigid environments often translate into isolation and reduced community independence for people with disability. Adaptive recreation and sport on snow and ice can be an important means of making cold weather more enjoyable and providing challenges.

18.6.1 Skiing

Skiing is an activity in which the physical context of the sport supports adaptation. Chairlifts carry *all* skiers up the mountain, and skiers use techniques in combination with specialized equipment to control the effects of gravity and carve turns gracefully through the snow. Adaptive ski equipment (e.g., outriggers, mono- and bi-skis) emerged from analyzing the physics of skiing and applying mechanical concepts to compensate for a skier's movement limitations. Compensation for sensory and cognitive impairment is taught through specialized instructions and ski guides or ski buddies.

Although adaptive skiing started with Austrian veterans returning home to their villages in the 1940s, it got established in the United States only during the 1970s. Veterans with limb loss learned stand-up methods that added outriggers. Originally, outriggers were the tips of damaged skis and were mounted to the ends of forearm crutches. Austrians used outriggers with the ski tips in a continuous running position, whereas Americans developed the "flip ski." Flip skis use a spring-loaded mechanism to allow the tip to either parallel the snow in a running position or, when released, flip up on the heel of the ski tip so outriggers can function more like ski poles or crutches. A metal claw bolted to the heel of the ski tip provides traction in icy lift lines. In the

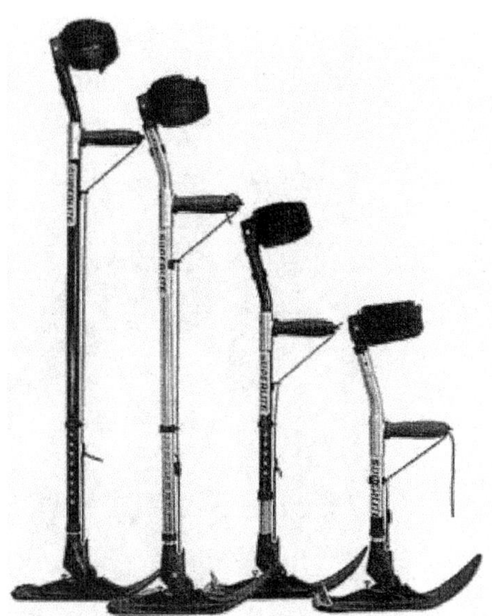

FIGURE 18.8 Three-track skiing is accomplished with the help of outriggers.

late 1980s, outriggers were further redesigned to be lighter and available in either recreational or racing models.

The sit-ski was developed to provide an equivalent adaptation for those who could not stand. The sit-ski was little more than a sled with a roll bar and a neoprene spray skirt, and yet it afforded sit-down skiers the opportunity to ride a chair lift to the top and, using short, handheld poles, link S-turns to make their way down a hill. New sit-skiers were tethered to a strong skier until they mastered the technique and could ski untethered. Sit skiers continued to need strong lifters to raise them to the level of the chairlift for each ride to the top. Later, a three-strap system was designed that allowed emergency evacuation from the chairlift if needed.

Further progress on sit-skiing and the benefits of the outrigger led to the development of the monoski. This device combined a fiberglass, form-fitting cab sitting over a suspension mechanism that attached to a single ski. A lever mechanism was added to allow the skier to raise the height of the cab up to the level of a chairlift. Once up, the monoskier could push forward in the lift line by using outriggers in the pole position and allow the chairlift and its momentum to scoop up the monoski and carry it to the top. Dismounting from the chairlift was accomplished by leaning forward just as the chairlift reached the dismount down slope, causing the momentum of the chairlift to drop.

Monoskiing requires good upper body strength and trunk control. Skiers need to be able to grasp the outriggers and lean and reach forward to create the forces needed to put the ski on its turning edge and initiate a turn. Skiers with less trunk control and hand strength were unable to ski until the bi-ski was developed in the 1990s. This device also sits the skier in a cab designed to snugly hold the pelvis, but instead of a single ski it uses two skis (placed parallel to each other) joined by a linkage

FIGURE 18.9 A monoskier using outriggers reaching into a turn.

bar to create a broader base. The linkage bar responds to the forces generated by a simple lean to one side by lowering the turning edge of the correct ski into the snow, resulting in a carved turn to the right or left. Skiers use handgrips placed at midline to brace their upper body as they lean from one side to the other to create linked turns. Optional "outriggers" can be attached to the base of the cab just in front of the skier to prevent the extremes of leaning that would lead to a fall.

18.6.2 Sledge Hockey

Sledge hockey derives its name from the Norwegian word *sledg,* meaning sled. The ice sledge or sled was invented at a rehabilitation center in Stockholm, Sweden, in the early 1960s by a group who, despite their physical impairment, wanted to continue playing hockey.

Ice hockey rules apply with some changes because of the nature of the game and its participants, but the ice surface, goal net, and pucks are all the same (United States Sled Hockey Association, 2004). The primary piece of equipment used in the game consists of the sledge (a metal-framed oval sled with two blades under the pelvis and a small runner in the front), a seat with a backrest, leg straps, and optional push handles. For push handles a typical hockey stick is shortened to 73.6 cm (29 in.) and modified with two picks (metal pieces with a minimum of three teeth measuring a maximum of 4 mm) attached to the end (United States Sled Hockey Association, 2004). Picks provide traction to move the player down the ice and give the player leverage for shooting the puck with the blade end of the stick. With a quick flip of the wrist, the players are able to propel themselves using the spikes and then play the puck with the blade end of the stick. Players generally use two sticks with blades to facilitate both propulsion and shooting with either hand.

Sticks must be made of wood or other approved material such as aluminum or plastic. The stick must not have any projections, and all edges must be beveled. The dimensions of the stick are as follows: maximum length, 100 cm measured in a straight line from the toe to the pick end; maximum width, 3 cm; maximum thickness, 2.5 cm. The shaft must be straight. The maximum blade length must be 32 cm from the heel to the toe, and the blade must have a width of between 5 and 7.5 cm at the toe (front of the blade). A goalkeeper's stick may be equipped with a larger blade that must not exceed 35 cm in length and 9 cm in height or be less than 7.62 cm (3 in.) anywhere along the blade. Players are allowed to secure their hands to the stick if their grip is impaired.

As in able-bodied ice hockey, sledge teams are limited to three forwards, two defenders, and a goalie, and games consist of three 15-min periods. Protective gear must be worn at all times. The required protective gear includes a helmet with cage or shield, shin guards, shoulder pads, gloves, elbow pads, neck guard, and hockey pants. In addition, straps are used to secure a player's feet, ankles, knees, and hips to the sledge. Repeated loss or adjustments of straps causing delay of game is penalized accordingly. The sledge may be equipped with a backrest, but it must not extend beyond the armpits of the player seated on the sledge. The backrest may be padded and should have rounded edges or corners with no hard or sharp obtrusions to the sides.

Sledge hockey is a unique adapted sport in that it does *not* have a classification system. The sport only requires that a player must have a disability that would prevent him or her from playing able-bodied hockey.

18.7 ISSUES IN THE DEVELOPMENT OF ADAPTIVE RECREATION TECHNOLOGIES

Most recreation technologies are simply tools that help compensate for a missing human performance ability or skill, e.g., impaired balance, strength, visual acuity, endurance, or loss of bipedal mobility. As stated earlier, people with disabilities using adaptive instincts and creative solutions have developed many, if not most, recreation technologies. The desire for the thrill, camaraderie, beauty, speed, and the fun of a particular recreational or sports activity is a powerful motivator. Most recreational adaptations fall into three categories:

- Changes in the equipment technology used in the activity
- Modifications that make the environment more supportive
- Modifications to instruction, technique, or rules to maintain equity in sport

Adaptations have been developed because someone analyzed skiers and ski boots, thought creatively, and came up with monoski, which is a "boot for the butt" attached to a ski with a shock absorber. They looked at stand-up waterskiing and thought of ways to accomplish it sitting down. The forces for turning derive from the forward propulsion of a powerboat in combination with opposing forces from the arms and shoulders. These forces in two planes are transmitted through the torso to the hips and onto the water sit-ski, allowing a skier to link S-turns behind the boat.

Many kinds of adaptive recreation happen in environments that actually minimize the requirement for performance skills. Swimming, water aerobics, scuba diving, snow skiing waterskiing, rafting, canoeing, and sailing all take advantage of the ability of water to support, transport, or propel. Carefully selecting the terrain for the first snow ski lesson can make the difference between success and failure for a beginner.

Adaptation to instruction technique is also an important part of recreational technology. Although a blind skier might use normal ski equipment, it is difficult to convey ski concepts when the student cannot see. One successful technique has a visually impaired skier glide down a gradual slope side-by-side an instructor while grasping opposite ends of the same pole. Suddenly the idea of linking turns is conveyed through the movement of the pole and close physical and conversational support. Kinesthetic and auditory channels are used to substitute for vision.

18.7.1 Development of Adaptive Instruction and Certification

The diverse number of sports and activities has led to program development and centers for excellence and instruction focused on particular sports or activities. With greater interest in adaptive recreation, centers have popped up in many places. New England has adaptive sailing, the Rocky Mountains have adaptive skiing, and California has adaptive scuba diving and tennis. Persons interested in acquiring skills travel to these locations or join in regional "learn to" events. Centers of excellence have done a lot to spread adaptive techniques and promote instructional competence through certifications. For example, Professional Ski Instructors of America (PSIA) has three levels of adaptive certification with specializations in blind, developmental disability, and stand-up and seated methods of skiing. Adaptive ski instruction now occurs in all PSIA regions.

18.7.2 Development of Sport Organizations

In the US, there are four major organizations that control sports competition for elite athletes. These groups include the International Paralympic Committee (IPC),

International Olympic Committee (IOC), United States Olympic Committee (USOC), and United States Paralympics.

18.7.2.1 International Paralympic Committee

This is an international nonprofit organization representing athletes of all sports and disabilities that is run by individuals from nearly 160 national Paralympic committees worldwide. The IPC conducts summer and winter Paralympic Games every 4 years, as well as world and regional championships. It creates the rules of conduct applied to each sport, including athlete classification rules and regulations, classifier coaching and officials training, and promotion of Paralympic sport worldwide (United States Paralympics, 2005).

18.7.2.2 International Olympic Committee

A relationship exists between the IPC and IOC. This is essential because the Paralympic Games occur in the same city (since 1988) and typically use the same sporting venues and facilities as the Olympics. In addition, the international sports federations that govern able-bodied sports often provide the framework for the guidelines applied to Paralympic sports. Some Paralympic sports also require athletes to be members of these sports federations (United States Paralympics, 2005).

18.7.2.3 The United States Olympic Committee

The USOC serves as the National Paralympic Committee (NPC) in the United States and is responsible for a variety of functions that include coaching, administration and classification, training, doping control, athlete development programs, and delegation representation at the Paralympics Games and other major international IPC-sanctioned competitions (United States Paralympics, 2005). The structure and function would be similar in other countries.

18.7.2.4 The United States Paralympics

In the United States, United States Paralympics, a division of the United States Olympic Committee, was created in May 2001 to focus efforts on enhancing programs, funding, and the opportunities for persons with physical disabilities. United States Paralympics has developed comprehensive and sustainable elite programs integrated into Olympic national governing bodies, and has functioned as a platform to promote athletic excellence for persons with disabilities. Some of the specific goals of United States Paralympics as stated in their official Web site include:

1. Identify Paralympic sports organizations (PSOs) for each of the Paralympic sports. This typically is the USOC-member national governing bodies and disability sport organizations, as well as other organizations that can demonstrate the capability to direct a sports program for elite athletes with disabilities.

2. Develop a funding philosophy and secure financial support for Paralympic sports. Unlike the USOC, which is essentially restricted to seeking corporate support, private donations, and grant funding, United States Paralympics also has the flexibility to seek financial support from governmental agencies.
3. Facilitate the development of performance plans for each of the Paralympic sports. Performance plans typically include performance goals and plans for athlete identification and development, coach education, sports science support, and policy development (Rice et al., 2005; United States Paralympics, 2005).

It takes talent, time, and support for any athlete to attain elite status. National Disabled Sports Organizations (DSOs) with support from the USOC help promote the development of athletes with disabilities in "grassroots" programs. In the US, there are six DSOs involved in the Paralympics movement. These organizations include the National Disability Sports alliance (NDSA), US Association of Blind Athletes (USABA), Special Olympics International (SOI), Wheelchair Sports USA (WSUSA), Disabled Sports USA (DS/USA), and Dwarf Athletic Association of America (DAAA). Although the existing DSOs differ in the services they provide, common features include athlete development camps, competition, and organization of networks of local clubs offering opportunities for athlete and official education and services (Dummer, 2002).

18.7.3 Developing New Adaptive Recreation Technologies

This review of adaptive sports and recreation activities currently used by persons with disabilities may lead a rehabilitation engineering student to think that no challenges remain. Yet, developments in materials science as well as in design and manufacturing processes have continued to change the equipment used by able-bodied athletes and sports enthusiasts — simply consider the evolution in skis, tennis racquets, bicycles, and hiking/climbing equipment over the years.

The best sources of ideas for innovation are participants in the sport or recreational activity. A focus group provides an ideal setting for an engineer to collect data for a requirements analysis. Potential users will provide information about current adaptive equipment specific to its features and limitations or discuss gaps that need filling. This same setting would be a good place to test conceptual designs and get reaction and suggestions from potential users. Conceptual designs would consider the environmental demands (temperature, moisture, etc.) as well as user needs (adjustability, stability when mounting or dismounting, etc.).

In fact, user needs is an area of particular focus since there is no "normal distribution" when designing products for persons with disabilities. Diversity in diagnosis or functional ability in any focus group would uncover difference in use patterns, needs, and potential risks. If good ideas are proposed, then preliminary design and prototyping follows. Trials with prototypes allow users to identify product strengths and weaknesses. This may redefine product requirements and lead to further modifications.

Acknowledgement of the contribution of physical activity and fitness to persons with disabilities has created greater interest in devices for adaptive sports, recreation, and fitness. Federal agencies fund this research and development process through phase 1 and phase 2 Small Business Innovative Research (SBIR) grants for development and testing of product concepts.

Of course, even good product concepts need the benefit of technology-transfer processes to make their way into manufacturing, sales, and distribution. One drawback to the development of AT for recreation and sports applications is the absence of any true funding system. The VA Healthcare System is the only "system" that currently funds recreational devices. Otherwise, the purchasers of these devices are individuals with the means to purchase them, individuals with sponsors who buy the equipment, and large and small sports organizations with the resources to purchase and then lend or rent adaptive equipment to their members.

18.7.4 Market Size and Risks

Although the size of the recreation technology market is relatively small, it has potential to grow larger and stronger. The scarcity of external funding sources means that product purchasers often come from the ranks of those who are employed and/or have discretionary income for purchasing sports gear. As the number of employed people with disabilities increases, the strength and size of this market will develop further. A clear risk in developing a recreation technology product and taking it to the level of sales and distribution is the cost of product liability insurance for high-risk sports activities. This issue, over the years, led to halting the manufacture of some recreation technology products.

A common feature of manufactured products is the use of standards to regulate manufacturing quality and assure consumers that they are getting a quality product. Currently, standards for most adaptive recreational products are nonexistent. The interest in skiing has led to the development of a draft American National Standard for Adaptive Skiing Equipment and proposals regarding ski area accessibility (Axelson, 2004).

18.8 SUMMARY AND CONCLUSION

Through this chapter, we intended to show clear examples of adaptive equipment by showing devices and approaches that compensate for limitations, considering relevant features of the environment, and maintaining safety while preserving the spirit of the sport. Although we focused on recreational and sport activities for which adaptations have already been created, there is still need for adaptive technology development, innovation, and improvement.

For instance, there is still need for seating technologies that can be used in all types of recreational settings. Those who have skied in water sit-skis often talk about the need for a shock absorption technology that is lightweight, low-cost, and effective. More needs exist, but they will best be uncovered through conversations with the sportsmen and sportswomen who dream of ability and participation. This will occur through the transfer of mainstream technologies to adaptive use and through new adaptations for new recreational activities and extreme sports such as climbing walls, ski bicycling, roller blading, kite surfing, and freestyle skiing.

18.9 STUDY QUESTIONS

1. What are the recreational activities that you engage in that are the most meaningful to you? What have you learned about yourself and your personality from participating in these activities? How does participating in these activities make you feel?
2. Using the World Wide Web as a search tool, locate disability sports organizations where you could refer a person with the following disabilities:

   Athletes with hearing loss
   Athletes who are blind or visually impaired
   Athletes with cognitive disabilities
   Athletes with spinal cord injuries, spina bifida, poliomyelitis, or other lower limb impairments
   Athletes with amputations
   Athletes with cerebral palsy, stroke, or head injury

3. Ask persons with disabilities about their background in sport and physical activity. Ask them about the sports programs in which they participate, the coaching available in those programs, and the accessibility of these facilities.
4. Think about the importance of sports and recreation for typically developing children. List five benefits of access to adaptive sports and recreation for children with disabilities.
5. What are your ideas for increasing the number and variety of sport and recreation opportunities for persons with disabilities? How would these ideas apply to the needs of children who are members of families?
6. Consider a sports program in which you participate as an athlete, coach, official, or administrator. What are the implications of the ADA for that sports program?
7. List two or three questions for future investigation regarding the psychological and physiological benefits of adaptive sports that occurred to you based uon your reading on this topic.
8. Locate a manufacturer for three competing products that currently allow an individual with a disability to:

   Play tennis
   Monoski
   Handcycle

9. What safety concerns might arise regarding the development of AT devices for recreation? Would these concerns hinder product development and create worries about manufacturer's liability?
10. Can you think of a sport or recreational activity for which you would like to design a piece of adaptive equipment? How would you start? List your product design and performance criteria. How would you test your prototype?

BIBLIOGRAPHY

Axelson, P.W., 2004, *Beneficial Designs: Improving Access for People of All Abilities*, Minden, NV, Beneficial Designs, Inc.

Boyd, J., 2001, Underwater wonders, *Sports 'n Spokes*, 27, 9–11.

Brown, S., 2001, Sarah Reinertsen: Don't give up…not once…not ever, in Brown, S. (Ed.) *Transition Times*, Robert Walters.

Buning, M.E., 1996, Recreation and Play Technology (Chap. 10), In Hammel, J. (Ed.), *Assistive Technology and Occupational Therapy: A link to function*, Bethesda, M.D., American Occupational Therapy Association.

Cheskin, M., 2004, Paralympic athletes: Equipped for success, *inMotion.*

Cooper, R.A., 1989a, Racing wheelchair crown compensation, *Journal of Rehabilitation Research and Development,* 26, 25–32.

Cooper, R.A., 1989b, Racing wheelchair rear wheel alignment, *Journal of Rehabilitation Research and Development,* 26, 47–50.

Cooper, R.A., 1990, Wheelchair racing sports science: A review, *Journal of Rehabilitation Research and Development,* 27, 295–312.

Cooper, R.A., 1992, Contributions of selected anthropometric and metabolic parameters to 10K performance: A preliminary study, *Journal of Rehabilitation Research and Development,* 29, 29–34.

Davidson, M., 2005, Artificial parts: Van Phillips, in *Innovation*, T. L. C. F. T. S. O. I. A. (Ed.) Washington, D.C., Smithsonian National Museum of American History.

Dummer, G.M., 2002, *Disability Sports*, East Lansing, Michigan State University.

Ewing, B., 2005, Athletics: Wheelchair track & field and road racing, in *Wheelchair Sports*, U. (Ed.) Earlham, IA, Wheelchair Track & Field, USA.

Harris, G., 2005, *Wheelchair Archery: USA 2005 Official Rules*, Earlham, IA, Wheelchair Sports USA.

Heath, G.W., Fentem, P.H., 1997, Physical activity among persons with disabilities — a public health perspective, *Exercise and Sport Sciences Reviews*, 25, 195–234.

International Paralympic Committee, 2004, *International Paralympic Swimming Rules*, in Green, A. (Ed.) St. James WA, Australia.

International Shooting Committee for the Disabled (ISCD), 2004, Shooting rules and functional classifications, Waasmunster, Belgium, International Paralympic Committee.

International Tennis Federation, 2004, *Wheelchair Tennis Handbook*, London, International Tennis Federation.

International Wheelchair Basketball Federation (IWBF), 2004, Official wheelchair basketball rules 2004, in Council, I.E. (Ed.) Athens, Greece, International Wheelchair Basketball Federation.

Kielhofner, G., 1985, *A Model of Human Occupation: Theory and Application*, Baltimore, William & Wilkins.

Lapolla, T., 2000, International rules for the sport of wheelchair rugby.

National Alliance for Accessible Golf (NAAG), 2003, National alliance for accessible golf webpage, Bloomington, IN, Indiana University.

National Center for Chronic Disease Prevention and Health Promotion (1996), Physical activity and health: A report of the surgeon general, US Department of Health and Human Services.

National Center on Accessibility (2003), *National Center on Accessibility: Recreation, Parks, and Tourism*, Bloomington, IN, Indiana University.

Pelka, F., 1997, *ABC-CLIO Companion to the Disability Rights Movement*, Santa Barbara, CA, ABC-CLIO, Inc.

Rice, I., Cooper, R.A., Cooper, R., Kelliher, A., Boyles, A., 2005, Sports and recreation for persons with spinal cord injuries.

Rimmer, J.H., Braddock, D., Pitetti, K.H., 1996, Research on physical activity and disability: an emerging national priority, *Medicine and Science in Sports and Exercise*, 28, 1366–72.

Schlein, S.J., Ray, M.T., Green, F.P., 1997, *Community Recreation and People with Disabilities: Strategies for Inclusion*, 2nd ed., Baltimore, Paul H. Brookes Publishing Co.

Schweikert, H.A., 1954, History of wheelchair basketball, *Paraplegia News*.

Scruton, J., 1998, *Stoke Mandeville Road to the Paralympics*, Aylesbury, The Peterhouse Press.

Steadward, R.D., Peterson, C., 1999, Paralympics: where heroes come, in Peterson, B. (Ed.), Edmonton, Alberta, CA, One Shot Holdings Publishing Division.

Strohkendl, H., 1996, *The 50th Anniversary of Wheelchair Basketball: A History*, New York, Waxman Publishing Co.

Taylor, W.C., Baranowski, T., Rohm Young, D., 1998, Physical activity interventions in low-income, ethnic minority, and populations with disability, *American Journal of Preventive Medicine*, 15, 334–343.

United States Paralympics (2005), US Paralympics Homepage, in Paralympics, U.S. (Ed.) Colorado Springs, CO.

United States Sled Hockey Association (2004), US Sled Hockey Homepage, Colorado Springs, CO, Hockey.com.

United States Quad Rugby Association (2004), Quad Rugby. What's it all about? A brief history, Lake Worth, FL, US Quad Rugby Assn.

Vines, H., 2004, National wheelchair basketball association official rules and case book 2003–04, in (NWBA), N. W. B. A. (Ed.) Indianapolis, IN, National College Athletics Association (NCAA).

Selected Terms and Definitions

Accessible design: A design approach that focuses on making products and the built environment accessible to individuals with disabilities and the elderly.

Acuity: Visual resolution, expressed as a fraction of "normal" (e.g., 20/20, 20/200).

Abduction: Moving a body part away from the midline of the body; opposite of adduction. When the arm is held out to the side, the shoulder is abducted. The scapula is abducted when it protracts or moves forward and around the ribcage.

Acquired amputation: A limb deficiency that results from disease or trauma after birth.

Adaptive recreation: Activity that ranges from passive to active that is freely chosen based on one's interests because it is creative, satisfying, amusing, relaxing, or renewing to the health, spirit, or intellect. Humans are drawn to recreation because they enjoy it and like to repeat the experience.

Adaptive sport: In the context of this book, sport refers to the organized, active competitive interaction between individuals or teams, in which the object of the activity (e.g., ball across the goal line, arrows in the bulls-eye, etc.) and the rules are clearly understood. In this book, sport refers to an amateur, unpaid activity undertaken for the love of the game or the demonstration of a skill.

Adduction: Moving a body part closer to midline; opposite of abduction.

Alerting device: A device used to alert a deaf person or one who is hard of hearing to environmental, safety, or alarm sounds via alternative sensory inputs. For example, a vibrating alarm clock or a strobe light smoke detector.

ALT tags:	HTML tag that provides alternative text when nontextual elements, typically images, cannot be displayed.
Anchor point:	Point (area) on a vehicle interior component, floor, or wall, wheelchair, or wheelchair tie-down to which an anchorage is attached.
Anchorage:	Assembly of components and fittings by which loads are transferred directly from the wheelchair tie-down to the vehicle, or from the occupant restraint to the vehicle, wheelchair, wheelchair tie-down, or vehicle interior component.
Anthropomorphic test device (ATD):	Articulated physical analog of a midsize male used to represent a wheelchair occupant in a test.
Armrest:	The component of a seating system that provides support to the upper extremities. They can be classified by their shape, length, or height adjustability.
Assistive listening devices:	Devices used by persons with hearing loss to enhance signal-to-noise ratio. Examples include personal amplifiers, FM systems, and television amplifiers.
Assistive technology:	Any item, piece of equipment, or system, whether acquired commercially, modified, or customized, that is commonly used to increase, maintain, or improve functional capabilities of individuals with disabilities.
Back support:	The component of the seating system that provides posterior and lateral postural support without inhibiting freedom of movement of the trunk and upper extremities needed for functional activities. They can be categorized by their shape, height, and stiffness.
Belt:	A length of webbing material used as part of an occupant restraint.
Blindness:	Absence of usable vision; often wrongly used to mean statutory blindness.
Blindness, statutory:	Best corrected acuity 20/200 or worse, or 20° visual field.
Body systems:	A group of interacting, interrelated, or interdependent elements of the human body forming a complex functional physiological unit.
Braille:	A tactile code for reading using the sense of touch.
CCTV:	Closed circuit TV magnification system.
Computer vision:	Use of computing power to extract information from images.
Comfort:	A condition or feeling of pleasurable ease, well-being, and contentment.

Cortical implant:	Electrodes implanted to stimulate the visual cortex.
Congenital amputation:	A limb deficiency present at birth.
Deaf:	Persons with severe hearing loss who primarily communicate via nonauditory means.
Deformity:	A change or misalignment of structure or the joints of the skeletal or other body systems.
Direct audio input:	A hearing aid coupling option allowing for the direct delivery of an audio output signal into the hearing aid, bypassing the microphone.
Disarticulation:	An amputation that occurs through a joint.
Distal:	Distant, or away from the center of the body; opposite of proximal. For example, the ankle is distal to the knee.
Docking-type securement:	Method of wheelchair securement by which portions of the wheelchair structure, or add-on components fastened to the wheelchair, align, mate, and engage with a docking device fastened to the vehicle upon maneuvering the wheelchair into position in the vehicle. *Note:* Securement of the wheelchair may occur automatically during wheelchair engagement, or may require manual intervention through operation of a mechanical lever or electrical switch. Release of the wheelchair will usually require operation of a mechanical lever or electrical switch.
Docking device; docking securement device:	Assembly of fixtures and components designed for installation in motor vehicles for the purpose of securing a wheelchair by engaging with, and locking onto, securement points on the wheelchair frame or on wheelchair securement adaptors attached to the wheelchair frame.
Dorsiflexion:	Moving the ankle joint such that the forefoot is pointing upward; opposite of plantarflexion.
Dual sensory Loss:	Both vision and hearing impaired.
Elevating leg rest:	Leg supports whereby the angle can be adjusted to provide various degrees of knee flexion or extension.
End effector:	a device or tool connected to the end of a robot arm.
Envelopment:	Describes a support surface's ability to deform around irregularities on the surface (such as creases in clothing, bedding, or seat covers and protrusions of bony prominences) without causing a substantial increase in pressure. A fluid support medium would envelop perfectly. The elastic property of its cover, however, would play an important role in envelopment. For example, a fluid-filled support

surface, such as a water bed, would not envelop as well as water alone because the membrane or cover containing the water has elastic properties. A support surface cover under high tension (e.g., overstretched or very taut bed covers) may cause locally high peak pressures.

Eversion: Turning the sole of the foot outward; opposite of inversion.

External limb prostheses: Externally applied devices consisting of a single component or an assembly of components used to replace wholly, or in part, an absent or deficient lower- or upper-limb segment. (Taken from ISO http://isotc.iso.ch/livelink/livelink.exe/fetch/2000/2122/547446/547447/547448/TC_168_Draft_Business_Plan.pdf?nodeid=3590970&vernum=0 on 11/10/04)

Extension: Moving the surfaces of two body parts away from each other; opposite of flexion. When the leg is straight, the knee is extended. The shoulder extends when the arm moves behind the body.

Extremity: The distal portion of an arm or leg, i.e., the hand or foot. This term should not be used interchangeably with "limb."

Extremities: The two upper extremities include the arms, forearms, and hands. The two lower extremities include the thighs, legs, and feet.

Flexion/extension of the wrist: Back and forth movement of the hand.

Flexion: Moving the surfaces of two body parts closer together; opposite of extension. When the arm is bent at the elbow, the elbow is flexed. The shoulder flexes when the arm moves in front of the body.

Foot support: The component of a seating system that provides support to the legs and feet. They can be classified by their angle, length, adjustability, or ability to be removed.

Four-point tie-down: A wheelchair tie-down system that attaches to the wheelchair frame at four separate securement points and also attaches to the vehicle at four separate anchor points.

Four-point strap-type tie-down: A four-point tie-down that uses four strap assemblies to secure the wheelchair in the vehicle.

Friction: A result of a force acting tangential to the interface and opposes shear force. When the skin is not sliding along the support surface (a static condition), friction and shear forces are equivalent. The maximum friction is determined by the coefficient of static friction of the support surface and the pressure. The coefficient of static friction is the ratio of the magnitudes of two forces: the maximum static frictional force (parallel to the skin's surface) and the normal force (the force perpendicular to the skin's surface). The coefficient of static friction depends on the type of material from which each surface is made, the condition of the surfaces (rough, lubricated, etc.), and other variables such as temperature. This is why support surfaces with a high coefficient of friction have the potential for high shear.

Some friction is necessary to prevent a patient from simply sliding off a support surface while sitting in bed or in a wheelchair. However, to prevent pressure ulcers, the friction necessary to prevent sliding should be applied in low-risk regions of the support surface and minimized near high-risk areas such as those surrounding bony prominences. In sitting patients, for example, the fleshy parts of the buttock and the posterior thighs would be areas of low pressure while the ischial tuberosities would be areas of high pressure.

Function: Activities or interactions that work properly, smoothly and prevent physical stress.

g: Abbreviation for acceleration due to gravity measured at sea level — 1g is equal to 9.8 m/s^2

Gimbal: A device consisting of two rings mounted on axes at right angles to each other so that an object will remain suspended in a horizontal plane between them regardless of any motion of its support.

Hall effect sensor: An electronic device that varies its output voltage in response to changes in magnetic field density. These can be used to measure proximity, speed, or current.

Haptics: The use of the sensation of touch (tactile) for control of and interaction with computer applications.

Hard of hearing: Persons with hearing loss who primarily communicate via oral or aural means.

Head support: The component of a seating system that provides posterior, lateral, and, in some instances, anterior head support especially for individuals with poor head control, individuals who use tilt and/or recline seating systems, or who require one for vehicle transportation purposes. These should support the

	head but not inhibit voluntary movement or field of vision. They can be categorized by shape and size.
Hearing assistance technology:	A variety of devices used to enhance participation and safety for deaf and hard-of-hearing persons.
Hearing aid:	A personal amplification device worn at ear level.
Hearing aid coupling:	A means by which the output of an assistive listening device is received and processed in a listener's hearing aid.
Hemiparesis:	Paralysis along one side of the body (e.g., right leg and right arm).
Hemipelvectomy prostheses:	Prosthetic devices appropriate for people with amputations at the pelvic level.
Hip disarticulation prostheses:	Prosthetic devices for individuals whose amputations occurred through the hip joint.
HTML:	Hypertext Markup Language, the primary language of the Web and of most Web documents.
Hydraulic knee:	Knee joint of a prosthesis that contains a hydraulic component that permits safer ambulation at variable speeds.
Immersion:	Immersion into a support surface increases the contact area between the person and the surface, allowing pressure concentrated beneath a small area to be spread over the surrounding tissue. Significant pressure reduction is possible when increased immersion causes additional bony prominences to come into contact with the support surface. For example, when a person is sitting on a relatively hard cushion, a disproportionately large amount of body weight is born by the tissue beneath the ischial tuberosities. On a softer surface, the protrusions of the ischial tuberosities become immersed in the cushion and weight is distributed to the area beneath the greater trochantors also. With this greater immersion, the body weight is divided among the bony prominences and pressure is decreased on the ischial tuberosities.
Impact simulator:	A device for decelerating, accelerating, or a combination of decelerating and accelerating a section of a vehicle or simulated vehicle structures, including instrumentation for measuring pertinent data.
Impact sled:	That part of an impact simulator to which components can be mounted for impact testing.
Implantable hearing devices:	A variety of personal amplification devices of which parts are implanted at or near any part of the ear.

Intercalary limbs:	A limb deficiency in which bony elements exist distal to the affected portion.
Inversion:	Turning the sole of the foot inward; opposite of eversion.
Joint contracture:	Decreased and restricted range of motion to a joint, compared to normal range, due to joint capsule stiffness or ligament or muscle shortening.
Knee disarticulation prostheses:	Prosthetic devices for individuals whose amputations occurred through the knee joint.
Kyoyo-Hin:	A design approach that redesigns existing products to be used by as many people as possible, including the elderly and those with disabilities.
Large print:	Print size of 14 point or more.
Limb:	An arm or a leg. This should not be used interchangeably with "extremity."
Low vision:	Best corrected acuity 20/70 or worse.
Mainstream products:	Simple and affordable products designed for the majority of users, mostly consisting of able-bodied people and not necessarily including the needs of people with disabilities or the elderly.
Metric map:	A map of the environment that encodes the distance between points.
Moisture control:	Moisture appears to be another key extrinsic factor in pressure ulcer development. The sources of skin moisture that may predispose skin to breakdown include perspiration, urine, feces, and fistula or wound drainage. Excessive moisture may lead to maceration. Some investigators have theorized that the risk of skin damage increases fivefold in the presence of moisture. This may result from the slight increase in friction that occurs with light sweating. Or it may be due to a decrease in the normal, slightly acidic nature of the skin that occurs when urea is broken down into ammonia by bacteria from fecal contamination. The skin's protective "acid mantle" has antimicrobial effects that are subsequently lost as the pH becomes more alkaline. Moisture vapor transmission rate is an important support-surface characteristic related to moisture control. This term refers to the rate at which moisture vapor is transferred through a specific distance and surface area of a material.
Muscle weakness:	Decreased muscle strength and/or endurance whereby muscles become weak abnormally fast.
Muscle tone:	The resistance to passive elongation or stretch provided by neural activity, viscoelastic properties of

muscle and joints, and sensory feedback to the central nervous system.

Muscular system: The bodily system that is composed of skeletal, smooth, and cardiac muscle tissue, and functions in movement of the body or of materials through the body, maintenance of posture, and heat production.

Neurological system: The system of cells, tissues, and organs that regulates the body's responses to internal and external stimuli. It consists of the brain, spinal cord, nerves, ganglia, and parts of the receptor and effector organs.

Occupant restraint: A system or device designed to restrain a motor vehicle occupant to prevent ejection and prevent or minimize contact with the vehicle interior components and other occupants during an impact and/or rollover event.

Orientation and mobility: Finding one's way safely from one place to another.

Orphan products: Products that are solely designed for use by people with disabilities, and includes under 200,000 users on the United States market.

Orthosis: A device that passively supports the joint or limb of a person but does not replace a missing body part.

Orthotic: Adjective of orthosis.

Paralympic: Refers to the summer and winter Olympic-type games for athletes with physical and sensory disability that occur every 4 years in parallel with the International Olympic Games.

Patellar-tendon-bearing socket: The most common socket used in for individuals who have an amputation between the knee and ankle. This socket encloses the residual limb and is held in place by a ridge at the knee cap.

Peak pressure measurement: Peak pressure measurement refers to the highest average pressure over a small area of interface (variable definitions, but generally less than or equal to 1 $in.^2$).

Pelvis: The base of support in a seated position that is comprised of two hip joints, the sacrum, and pelvic ring.

Personal FM System: A device comprised of a remote microphone and transmitter and a body-worn receiver. The signal from the microphone is transmitted via FM radio waves to the ear of the listener at an enhanced signal-to-noise ratio.

Phantom pain: Pain distal to a residual limb. For example, a person with a transfemoral amputation may experience a burning sensation in his amputated foot. This is to be contrasted with phantom sensation, which is a sensation that is not painful.

Plantarflexion: Moving the ankle joint such that the forefoot is pointing down; opposite of dorsiflexion.

Point class: A concept used in competitive team sports to create baseline equivalence between teams comprised of athletes with objectively different levels of ability due to diagnosis or impairment (i.e., amputation site or spinal cord injury level). Athletes are given a number rating based on their function, and the team limit is confined to a total number of points.

Polycentric knee: Knees in lower-limb prostheses that have moving centers of rotational movement to improve stability.

Posture: A position of the body or of body parts.

Preparatory socket: Temporary socket used to help shape the end of the residual limb until a permanent prosthesis is fabricated.

Pressure distribution: Pressure is the force per unit area (pounds per square inch) exerted on the body by a mattress, seat cushion, or other body support. The distribution of pressure on a seat cushion or mattress depends on the relative fit between the body and the support surface, the mechanical characteristics of the body tissues and the cushion or mattress, and the distribution of weight in the body.

For example, the ideal pressure distribution would be one in which soft tissue shape was not altered relative to its unloaded condition. This would minimize many of the effects believed to lead to the development of pressure ulcers, including capillary blood flow occlusion, impaired lymph flow, and excessive interstitial fluid flow.

Pressure gradient: Pressure gradient is the amount of change in pressure over a distance (generally over a short distance of less than 1 in.) from one section of the support surface to another. However, in general, pressure gradients represent a continuum of values over the distance measured. If the pressure across a surface were plotted, the pressure gradient would be the slope of the curve. Skin and other soft tissue at risk for breakdown consist of a mixture of a fibrous collagen network, interstitial fluids, blood vessels, lymphatic vessels, and other elements. Therefore, a pressure gradient in the support surface will result in a flow of the tissue's fluid elements from one area to the other.

	Knowledge of peak pressure and pressure gradients is needed to minimize the mechanisms that are presumed to lead to pressure ulcer development. Given the large number of pressure ulcers that occur on bony prominences, it is logical to want to know which areas of a support surface have high peak pressure and high pressure gradients. This would allow the clinician to avoid positioning a patient's bony prominences in these areas.
Pronation/supination of the forearm:	Twisting of the forearm about its long axis.
Proprioception:	The sensation that allows us to determine where our joints are in space.
Prosthesis:	A device that replaces all or part of the function of an internal body organ or external body member.
Prosthetic:	Adjective of prosthesis.
Proximal:	Toward the center of the body; opposite of distal.
Pylon:	A rigid post in the shank, positioned from hip to knee or knee to ankle in a lower-limb prosthetic device.
Reclining back:	A component of a seating system that provides a change in seat-to-back angle orientation while maintaining a constant seat angle with the ground that assists with opening the hip angle for recumbent positioning.
Reflex:	A reflected action or movement; the sum total of any particular automatic response mediated by the nervous system.
Remotely readable signage:	Signs that can be read at a distance.
Residual limb:	The portion of a leg or arm that remains following an amputation.
Retinal implant:	Means of artificially stimulating the retina.
Screen reader:	Software that informs a blind user what is on a computer screen.
Screen reader software:	Reads aloud the textual information displayed on the computer screen.
Seat surface:	The component of a seating system that provides a base of support for the pelvis and upper body. Provides a balance of pelvic stabilization, pressure distribution, and allows for some degree of movement to remain functional to the individual's needs.
Seating system:	A system designed to provide seated postural support specific to the user's medical, functional, and personal preference and needs.

Section 255 of the U.S. Telecommunications Act:	Requires telecommunication manufacturers and service providers to make their products and services accessible to people with disabilities.
Section 504 of the U.S. Rehabilitation Act:	Protects qualified individuals from discrimination based on their disability. The nondiscrimination requirements apply to the federal agencies and organizations that receive grants or contracts from any federal department or agency.
Section 508 of the U.S. Rehabilitation Act:	Requires electronic and information technology in federal agencies be accessible to people with disabilities.
Securement points:	Specific portions of the wheelchair or seat frame that are designed and designated for attachment of a wheelchair tie-down.
Sensory enhancement:	Enhancing or making better use of a sensory input.
Sensory substitution:	Replacing information from one sense with input via another.
Shear:	Shear commonly occurs when the skin surface remains stuck to a support surface while the underlying bony structure moves in a direction tangential to the surface. For example, suppose the skin over the sacrum does not slide along the surface of the bed or the bed does not absorb the resulting shear force by deforming in the horizontal direction when the head of a bed is raised or lowered. The effect would be a shearing of the soft tissue between the sacrum and the support surface.
Signal-to-Noise ratio:	The difference, in decibels, between the intensity of a desired signal and that of a competing noise.
Skeletal system:	The internal structure composed of bone and cartilage that protects and supports the soft organs and tissues.
Solid-ankle, cushioned heel (SACH):	A prosthetic foot that contains a solid-ankle joint and cushioned heel, the most commonly prosthetic foot used in the United States.
Solid-ankle, flexible endoskeleton (SAFE):	A prosthetic foot with a solid ankle joint and a flexible keel, which makes walking over irregular surfaces easier for some amputees than with the SACH foot.
Socket:	The interface between a person and his or her prosthetic device. Sockets are comprised of rigid or semirigid materials.
Spasticity:	A state of increased muscle tone, with heightened deep tendon reflexes; also referred to as *hypertonic*.
Speech recognition software:	Enables the computer to recognize spoken words, to take dictation, but not understand what is being said.

Spine: A structure of vertebrae (cylindrically shaped bones) that form the spinal column and provide structure to the trunk as well as protection to the spinal cord. The spine is divided into four sections (from top to bottom): cervical spine (neck with seven vertebrae), thoracic (rib area with twelve vertebrae), lumbar (lower back with five vertebrae), and sacral (a structure of the pelvis with five fused vertebrae).

Stability: The ability of a person to maintain equilibrium or resume his or her original, upright position after displacement.

Telecoil: An induction coil in a hearing aid that receives electromagnetic signals emanating from a telephone or electromagnetic loop system.

Telecommunications : Any transmission, emission, or reception of signs, signals, writing, images, sounds, or intelligence of any nature by wire, radio, optical, or other electromagnetic systems.

Temperature control: The role of temperature in pressure ulcer development has not been definitively investigated. It is generally believed that any increase in temperature combined with pressure makes tissue more susceptible to injury, either from ischemia or reperfusion when pressure is relieved. However, researchers have reached varying conclusions, depending on the amount and duration of the pressure and temperature applied.

Important characteristics related to temperature control include heat flux, heat transfer rate, and specific heat. *Heat flux* is simply heat flow or transfer via convection, conduction, or radiation. The time for heat to transfer may be measured for each method. *Heat transfer rate* refers to the rate at which heat flows through a specific distance and cross-sectional area. *Specific heat* refers to the capacity of a material to retain internal energy and is dependent on the nature of the material.

Terminal limbs: Limbs with deficiencies that develop normally to a level, past which no further bony elements exist.

Text to speech: Technology for conversion of ASCII code to speech.

Tilt: A component of a seating system that provides change in seat angle orientation while maintaining a constant seat-to-back angle to redistribute pressure, realign posture, improve comfort, and increase sitting tolerance.

Topological map:	A map of the environment that encodes the connections between locations but does not contain distance information.
Transfemoral amputation:	An amputation above the knee joint but below the hip joint, across the femur. This term now replaces "above knee" amputation.
Transhumeral amputation:	An amputation above the elbow but below the shoulder joint, across the humerus. This term now replaces "above elbow" amputation.
Transtibial amputation:	An amputation below the knee but above the ankle, across the tibia. This term now replaces "below knee" amputation.
Transradial amputation:	An amputation below the elbow but above the wrist, across the radius and ulna. This term now replaces "below elbow" amputation.
Transverse amputation:	An amputation that occurs across the shaft of a bone.
UDIG adaptor:	A wheelchair securement adaptor that conforms to the UDIG specifications.
Universal docking interface geometry (UDIG):	Specifications set forth in ISO 10542-3 for the size, shape, and location of wheelchair securement points, including surrounding clear zones, intended for use with a variety of docking devices installed in a wide range of vehicles.
Universal design; barrier-free design; design for all:	A design approach that includes the needs of the largest percentage of potential users in our society, including individuals with disabilities and the elderly. See the chapter on universal design.
Usability:	Is comprised of several factors pertaining to the quality of the user's experience when interacting with a product or system.
U.S. Access Board:	Independent federal agency whose primary mission is accessibility for people with disabilities.
User-centered design:	An iterative process involving the user in the design activities until objectives are satisfied.
Video description:	Method of describing the visual content of videos and movies.
Visually impaired:	Any reduction of best corrected acuity below 20/40.
Web Accessibility Initiative (WAI):	A program office within the World Wide Web Consortium (W3C) that develops strategies, guidelines, and resources to help make the Web accessible to people with disabilities.
Wheelchair adaptor; wheelchair securement adaptor:	Hardware that is attached temporarily or permanently to the wheelchair frame to accommodate wheelchair securement by a wheelchair tie-down device.

Wheelchair tie-down and occupant-restraint system (WTORS):	A complete restraint system for use by wheelchair-seated occupants, comprised of a system or device for wheelchair tie-down, as well as a belt-type system for restraining the occupant.
Wheelchair tie-down; wheelchair securement; wheelchair lockdown:	A device or system used to secure a wheelchair in place in a motor vehicle.
World Wide Web Consortium (W3C):	Develops interoperable technologies (specifications, guidelines, software, and tools) to lead the Web to its full potential.
World Wide Web:	The set of all information accessible using computers and networking.

Abbreviations

AFO	ankle foot orthosis
ARM Guide	Assisted Rehabilitation and Measurement Guide
ASHT	American Society of Hand Therapists
BAM	Brake Actuated Machine
CO	cervical orthosis
CP	cerebral palsy
CPM	continuous passive motion
CTLSO	cervical, thoracic, lumbar, sacral orthosis
CTO	cervical, thoracic orthosis
DIP	distal interphalangeal joint
DOF	Degree of Freedom
EO	elbow orthosis
ETA	Electronic Travel Aid
EWHO	elbow, wrist, hand orthosis
FES	functional electrical stimulation
FO	finger orthosis
FO	foot orthosis
HKAFO	hip, knee, ankle, foot orthosis
IMA	Intelligent Mobility Aid
ISO	International Standards Organization
KAFO	knee, ankle, foot orthosis
KO	knee orthosis
LE	lower extremity
LSO	lumbosacral orthosis
M3S	Multiple Master, Multiple Slave
MIME	Mirror-Image Motion Enabler
MP	metacarpophalangeal joint
MULOS	motorized upper limb orthotic system
OA	osteoarthritis
PIP	proximal interphalangeal joint
RGO	reciprocating gait orthosis
SCARA	Selective Compliance Assembly Robot Arm
SEWHO	shoulder, elbow, wrist, hand orthosis

SOMI	sternal occipital mandibular immobilizer
TLSO	thoracolumbo, sacral orthosis
UCBL	University of California Biomechanics Laboratory
UE	upper extremity
WHO	wrist, hand orthosis

Index

Note: Page numbers in italics refer to *figures* while that of bold refer to **tables**.